CONSTRUCTIVE LINEAR ALGEBRA

CLA CONSTRUCTIVE

LINEAR ALGEBRA

Allan Gewirtz

Brooklyn College
of the City University
of New York

Harry Sitomer

C. W. Post College
of Long Island University

Albert W. Tucker

Princeton University

PRENTICE-HALL INC., Englewood Cliffs, New Jersey

Library of Congress Cataloging in Publication Data

GEWIRTZ, ALLAN.
Constructive linear algebra.

Bibliography: p.
1. Algebra, Linear. I. Sitomer, Harry, joint author. II. Tucker, Albert William, 1905- joint author. III. Title.
QA184.G49 512'.5 73-20284
ISBN 0-13-169276-3

CONSTRUCTIVE LINEAR ALGEBRA

A. Gewirtz, H. Sitomer, A. W. Tucker

10 9 8 7 6 5 4 3 2 1

Printed in the United States of America

Prentice-Hall International, Inc., *London*
Prentice-Hall of Australia, Pty. Ltd., *Sydney*
Prentice-Hall of Canada, Ltd., *Toronto*
Prentice-Hall of India Private Limited, *New Delhi*
Prentice-Hall of Japan, Inc., *Tokyo*

Contents

Whenever you can settle a question by explicit construction, be not satisfied with purely existential arguments.

HERMANN WEYL
Princeton University Bicentennial Conferences
Series 2, Conference 2, 1946

There is a wide range of problems in the biological, medical, psychological, economic, and political sciences, which are . . . problems of organized complexity. *These new problems, and the future of the world depends on many of them, require science to make a third great advance, an advance that must be even greater than the 19th century conquest of [physical] problems of* simplicity, *or the victory in the first half of the 20th century over [statistical] problems of disorganized complexity. Science must . . . learn to deal with these problems of organized complexity.*

WARREN WEAVER
"Science and Complexity"
American Scientist (36), 1948

At a great distance from its empirical source, or after much "abstract" inbreeding, a mathematical subject is in danger of degeneration. . . . Whenever this stage is reached, the only remedy seems to me to be the rejuvenating return to the source: the reinjection of more or less directly empirical ideas.

JOHN VON NEUMANN
The Mathematician
The World of Mathematics, Vol. IV
edited by J. R. Newman
New York: Simon and Schuster, 1956

The exchange-method . . . used . . . by Steinitz for the foundation of linear algebra . . . is a shining illustration of the fact that there is only one single field of mathematics and not two separate ones—pure and applied.

EDUARD L. STIEFEL
An Introduction to Numerical Mathematics
New York: Academic Press, 1963

Preface

This book presents a new approach to the elements of linear algebra. We think you'll like it.

During recent summers the three of us have helped plan and write the linear algebra chapters of text material prepared for mathematically gifted high-school students by the Secondary School Mathematics Curriculum Improvement Study (SSMCIS). From this start we have gone on to develop this text for a first course in linear algebra, suitable for general use in the first or second year of college.

By dealing mainly with mappings from R^m to R^n, we take a middle road between numerical linear algebra and abstract linear algebra. Most of the principal theorems of elementary linear algebra appear in the text in a constructive manner. At the same time, our algorithms are translatable into efficient computer programs.

Much care has gone into arranging the sequence of topics. We begin with linear programming for two reasons. First, our teaching experience shows that students enjoy linear programming. Second, the simplex method enables students to gain experience with tableaus, which constitute our modus operandi throughout the book. These tableaus have great "pictorial" value for n-dimensional linear algebra, just as sketches of lines and planes have for solid geometry. Therefore, early and extensive experience with tabular manipulation is imperative. If necessary, the linear programming chapter may be postponed or omitted, but Section 1.4 must be mastered. We treat tableaus and mappings (Chapter 2) before matrices (Chapter 3) because our preference is to motivate matrix multiplication and inversion by composition and inversion of mappings. This organization is supported by good historical and pedagogical reasons. In Chapter 4 we use the fundamental algorithm, developed in Chapter 2, to calculate solution sets of systems of linear equations. So it is not feasible to interchange Chapters 4 and 2, but it is feasible to interchange Chapters 4 and 3.

Chapter 5 deals with R^n as a vector space and with its vector subspaces. In Chapter 5 the ideas developed in Chapters 1–4 are put to use to help construct a mathematical structure. The notions of basis and dimension are developed. Chapter 6 is a continuation of the development of Chapter 5. The notion of affine spaces and subspaces is developed, as is the relationship between vector and affine subspaces. The concepts of the sum and intersection of two subspaces are also presented. In addition, the inner product is introduced and the concepts of Euclidean spaces explored.

Chapter 7 is multipurpose. In it, linear mappings, affine mappings, and Euclidean mappings are studied, as are their respective relationships to vector, affine, and Euclidean spaces. The chapter is written in such a way that the links between the tableaus of this book and functional notation are made clear. Thus a student who studies this book will be able to read other linear algebra texts without difficulty. Chapter 8 is an attempt to resurrect determinants. To our mind, predictions of the demise of determinants, except as a historical curiosity, are premature. We present the pivot procedure for calculating determinants so that computation of determinants of order greater than 3 is feasible. Examples of recent uses of determinants are in Section 8.6.

Following are a few suggestions on how to use this book for courses with differing aims and with students of differing interests. In courses for prospective students of mathematics, the physical sciences, or engineering, we recommend that the entire book be covered with the possible exception of Section 5.4, which is more abstract than the rest of the book. Exclusion of Section 5.4 does not affect the continuity of the rest of the text. If circumstances dictate, Chapter 1 on linear programming may be postponed or deleted, provided that Section 1.4, which is basic to the rest of the text, is covered. If time is pressing, sections of Chapter 7 and Sections 8.4 and 8.6 of Chapter 8 may also be omitted.

For courses in finite mathematics (elementary linear algebra is finite mathematics in the sense that limits and continuity are not used), the first four chapters, possibly excluding Section 2.8, plus Chapter 8, are a substantial text. The text can be supported with material related to biology, economics, business administration, sociology, psychology, or other fields of interest to the students.

This book may be used by students who have completed three years of high-school mathematics. It presents material for a course that precedes or accompanies calculus. Chapters 1–4, possibly excluding Section 2.8, and Chapter 8 supplemented by parts of Chapter 5, form a text for a substantial course in matrix algebra for high-school seniors.

We wish to express our deepest appreciation to many people for the help they have given us with this book. For their professional advice and assistance we thank George B. Dantzig, Bernard Eisenberg, Howard F. Fehr, Richard A. Good, Ronald J. Graham, Kenneth M. Hoffman, Meyer Jordan, Steven B. Maurer, Torrence D. Parsons, Henry O. Pollak, Louis V. Quintas, Robert R. Singleton, Gerald P. Thompson, Bruce R. Vogeli, and Paul A. White. For their help with the planning and preparation of this book we thank Deborah Gold, Harvey Halpert, Florence Joyce, Lyn Kanterman, Margaret McNeily, Anna Preschle, Murray Rosenbaum, Bernard Rosenberg, Menachem Rosenberg, Arthur Wester, and Joel Zipper.

ALLAN GEWIRTZ

HARRY SITOMER

ALBERT W. TUCKER

CONSTRUCTIVE LINEAR ALGEBRA

CLA

CHAPTER 1

Linear Programming

Linear programming is a new field of mathematics initiated in 1947 by the American mathematician George B. Dantzig, although earlier work had touched on the field—as early as 1826 by the French mathematician J. B. J. Fourier. An extensive account of its origins can be found in *Linear Programming and Extensions* by George B. Dantzig, Chapters 1 and 2. Its initial applications were in economics, where companies (and governments) aim to maximize profits and minimize costs. Before long, many other applications of linear programming were discovered.

Today urban planners are interested in maximizing their ability to remove waste matter subject to restrictions on the amount of equipment available and available labor. Dieticians want to minimize food costs subject to meeting certain caloric and nutrient demands, and airlines wish to maximize flight time per plane subject to restrictions of maintenance, pilot availability, and so on.

We have three purposes in this chapter:

1. To learn how to solve simple linear programs.
2. To perceive the duality theory underlying these programs.
3. To get acquainted with tableaus and pivot exchanges, major tools used throughout the book.

1.1 A Maximum Linear Program

A machine company makes special types of nuts, taper pins, and bolts for use in space shuttles. The company soon discovered that these parts were suitable for other situations in which extreme temperature and vacuum conditions prevailed. Each of the parts was sold in 100-pound kegs. After calculating all costs it was found that the profit was $3 for each keg of nuts, $2 for each keg of taper pins, and $2.50 for each keg of bolts. A small lathe, a large lathe, and manual labor were required in the production process. The sales force tried to sell as much of the most profitable item as possible and neglected to push sales of the other two items. Before long the production force found that the two kinds of lathes were idle longer than an efficient schedule of work called for. Each day the factory had available 12 man-hours of manual labor, 2 hours on the small lathe and 4 hours on the large lathe. Each keg of nuts required 4 man-hours and 1 hour on the small lathe. Each keg of pins required 2 man-hours and 1 hour on the large lathe. Each keg of bolts required 2 man-hours, 1 hour on the small lathe, and 3 hours on the large lathe. The question before the company is: To obtain the greatest profit, how many kegs of each part must we manufacture daily? In brief, the company's production problem (called a *linear programming problem*) is to find a production schedule that maximizes profit subject to meeting various production constraints.

Let us imagine that a mathematician is consulted to solve this problem. He translates the problem into a mathematical format called a *maximum linear program*. Since there are many data, he organizes them in a table as follows:

	Nuts	Pins (per keg)	Bolts	Available (per day)
Labor (hours)	4	2	2	12
Small lathe (hours)	1	0	1	2
Large lathe (hours)	0	1	3	4
Profit (dollars)	3	2	2.5	

The next step is to translate the relations in the problem into mathematical sentences, selecting only those that bear on the question. To begin with, the numbers of kegs of nuts, taper pins, and bolts to be made are represented by n, t, and b, respectively. It is immediately stipulated that $n \geq 0$ (n is greater than or equal to zero), $t \geq 0$, and $b \geq 0$, for the manufacturer does not produce a negative number of kegs.

Each keg of nuts calls for 4 man-hours of manual labor; thus $4n$ represents the number of man-hours needed for the n kegs of nuts. Similarly, $2t$ represents the number of man-hours needed for t kegs of pins and $2b$ the number of man-hours needed for b kegs of bolts. The total number of man-hours spent on all parts is

$$4n + 2t + 2b$$

This number cannot exceed 12, since only 12 man-hours are available. Let S_1 be the number of unused man-hours (possibly 0). Thus

$$4n + 2t + 2b + S_1 = 12$$

It is convenient to write this equation in the equivalent form

$$4n + 2t + 2b - 12 = -S_1 \tag{1}$$

Note that S_1 can be zero but not negative. So $S_1 \geq 0$. S_1 is called a *slack variable* because it "takes up the slack" between used time and available time.

Now consider the total time on the small lathe. The n kegs of nuts need 1 hour each. The t kegs of pins need no hours and the b kegs of bolts need 1 hour each. If S_2 is the unused number of hours on the small lathe, then

$$1n + 0t + 1b + S_2 = 2$$

where $S_2 \geq 0$, or

$$n + 0t + b - 2 = -S_2 \tag{2}$$

Using similar reasoning the equation for the large lathe (slack variable $S_3 \geq 0$) is

$$0n + 1t + 3b + S_3 = 4 \quad \text{or}$$

$$0n + t + 3b - 4 = -S_3 \tag{3}$$

Equations (1), (2), and (3) are the conditions that must be satisfied for any production schedule, regardless of profit. They are called the *constraint equations*.

How about the profit? It is composed of three parts: profit from nuts, from tapers, and from bolts. Since there are n kegs of nuts, each returning a profit of $3, the profit on nuts is $3n$. Similarly, the profit on tapers is $2t$ and that on bolts $2.5b$. The total of these profits we symbolize by M. Thus

$$3n + 2t + 2.5b = M$$

This equation for M is called the *objective equation*. We try to maximize M.

We summarize our analysis in the following way:

Find nonnegative numbers $n, t, b, S_1, S_2,$ *and* S_3 which maximize

$$3n + 2t + 2.5b = M$$

subject to

$$(1) \qquad 4n + 2t + 2b - 12 = -S_1$$

$$(2) \qquad n + 0t + b - 2 = -S_2$$

$$(3) \qquad 0n + t + 3b - 4 = -S_3$$

This linear program is a mathematical model of the linear programming problem of the machine company.

Linear programs are written in a *tableau* form, which we now describe. The format of equation (1) is the conventional one. We also represent (1) as

n	t	b	-1	
4	2	2	12	$= -S_1$

In this form multiply each number in the box or tableau with the variable or -1 directly above it and set the sum of the products so obtained equal to the variable at the right. This form is convenient for a set of equations whose left members have the same variables, for then these variables need be written only once at the top of a tableau. For instance, the three constraint equations are written as

n	t	b	-1	
4	2	2	12	$= -S_1$
1	0	1	2	$= -S_2$
0	1	3	4	$= -S_3$

For each equation, multiply each number in a row of the tableau by the variable or -1 directly above it and set the sum of the three products equal to the variable at the right. (A tableau is also called a *schema.*) The next tableau takes us one step further. In addition to the constraint equations (1)–(3) of the linear program, the objective equation is written in the bottom row of the tableau.

	n	t	b	-1	
Constraint equations	4	2	2	12	$= -S_1$
	1	0	1	2	$= -S_2$
	0	1	3	4	$= -S_3$
Objective equation	3	2	2.5	0	$= M$

This tableau represents a linear program. Dashed lines are used in the tableau form of a linear program to set off the bottom row and last column. The reason for this will become clear as we proceed.

This brings us to the point at which we think about a solution. To begin with, the tableau houses a *system of four linear* equations, involving six variables in the three constraint equations, seven variables in all four equations. A *solution* to the system is a set of numbers, one for each variable, for which each equation is true. For instance, if $n = 1, t = 0, b = 2, S_1 = 4$, then the first row of the tableau becomes

1	0	2	-1	
4	2	2	12	$= -4$

This reads $4 \cdot 1 + 2 \cdot 0 + 2 \cdot 2 + 12(-1) = -4$, which is true. If, in addition, $S_2 = -1$, the first two rows are

1	0	2	−1	
4	2	2	12	= −4
1	0	1	2	= −1

The second equation is also true. Let $S_3 = 2$. The third row becomes

1	0	2	−1	
0	1	3	4	= −2

This is false for $0 \cdot 1 + 1 \cdot 0 + 3 \cdot 2 - 4 \neq -2$.

If, instead, $S_3 = -2$, then $n = 1$, $t = 0$, $b = 2$, $S_1 = 4$, $S_2 = -1$, and $S_3 = -2$ satisfy the three constraint equations, and this set of numbers is a solution of the system of constraint equations. However, $S_2 = -1$ and $S_3 = -2$ violate the requirement that $S_2 \geq 0$ and $S_3 \geq 0$. So while we have a solution to the system of linear equations, it is not acceptable from the viewpoint of linear programming because the solution does not meet the nonnegativity conditions of a linear program. We discuss this further in the next section.

Exercises

For each of the problems in Exercises 1–6, write a maximum linear program in two forms:

(a) Maximize ________
subject to ________.
(b) The tableau form.

1. A manufacturer makes two kinds of tables, model A and model B. Each is put together by a carpenter and finished by a finisher. The carpenter takes 2 hours to assemble one model A table and 4 hours for one model B table. The finisher takes 4 hours to finish one model A table and 2 hours for one model B table. Each worker may work 12 hours per day but no more. The profit on each model A is \$20 and on each model B table is \$30. Assuming that all tables made can be sold, how many of each model table should be made per day to produce the maximum profit?

2. A manufacturer makes two kinds of radios, model A and model B. Radios are assembled in shop I and carefully tested in shop II. Three man-hours are needed to assemble one model A radio and 1 man-hour to test it. One and one-half man-hours are needed to assemble one model B radio, and

2 man-hours to test it. Shop I (the assembly) is capable of doing no more than 30 man-hours of work per day, and shop II, having a smaller force, can do no more than 20 man-hours of work per day. Profits per radio are $30 for model *A* and $20 for model *B*. How many (to the nearest unit) of each model should be made per day to produce maximum profit?

3. The radio manufacturer in Exercise 2 organized a third group (shop III) that packaged each radio and promoted sales. It took 1 man-hour to process each radio of both models, and this group could produce no more than 12 man-hours per day. How many (to the nearest unit) of each model should he make now per day to produce maximum profit?

4. A manufacturer makes three chemicals *A*, *B*, and *C*, each of which has to be processed in three shops, I, II, and III. The respective number of man-hours needed to process a ton of *A*, *B*, and *C* are 5, 2, and 1 in shop I; 3, 3, and 2 in shop II; and 2, 4, and 2 in shop III. The daily available number of man-hours in the respective shops are 180, 135, and 120. The profits per ton of the respective chemicals are $200, $150, and $120. How many tons of each chemical make for the maximum profit per day?

5. A manufacturer makes three kinds of tables, *A*, *B*, and *C*. Each table is assembled by a carpentry shop and finished by a finishing shop. To make one table *A* requires 4 man-hours of carpentry and 1 man-hour of finishing; to make one model *B* table requires 9 man-hours of carpentry and 1 man-hour of finishing; to make one model *C* table requires 7 man-hours of carpentry and 3 man-hours of finishing. The carpentry shop is capable of doing 600 man-hours of work per day; the finishing shop can do no more than 400 man-hours of work per day. If profits per *A*, *B*, and *C* tables are $15, $20, and $25, respectively, how many of each table should be made each day to maximize profits?

6. A manufacturer makes chemicals *A* and *B* using three machines I, II, and III in the process. The number of hours each machine must be used for each chemical, and the profit, in dollars, on each ton of each chemical is given by the table:

	I	II	III	Profit
A	20	15	4	40
B	4	15	24	35

Machine I cannot be used more than 32 hours per week; machine II cannot be used more than 30 hours per week; machine III cannot be used more than 20 hours per week. How many tons of each chemical should be made, per week, to realize the maximum profit?

7. Compose a linear programming problem for which the tableau is

x_1	x_2	x_3	-1	
1	2	3	12	$= -S_1$
2	1	1	10	$= -S_2$
3	5	6	0	$= M$

1.2 Solving the Maximum Linear Program: First Steps

Now that we have experience in setting up linear programs, we proceed to solve one. Let us return to the nut–pin–bolt problem. Its tableau is repeated for ready reference:

n	t	b	-1	
4	2	2	12	$= -S_1$
1	0	1	2	$= -S_2$
0	1	3	4	$= -S_3$
3	2	2.5	0	$= M$

Recall that in Section 1.1 we found that $n = 1, t = 0, b = 2, S_1 = 4, S_2 = -1$, and $S_3 = -2$ comprise a solution of the constraint equations of the tableau. However, this solution is not feasible for the linear program because the linear program requires that n, t, b, S_1, S_2, and S_3 be nonnegative. Solutions to the constraint equations, which also satisfy the nonnegativity condition, are called *feasible solutions*.

We seek feasible solutions among which we hope to find a best feasible solution that maximizes M. One feasible solution is found by setting the top variables, n, t, and b, equal to zero. From the first row (equation) it follows that $S_1 = 12$, from the second $S_2 = 2$, from the third $S_3 = 4$. For these values, $M = 0$. (The solution is feasible but hopefully we will find a better one, that is, a feasible solution for which M is greater than zero.)

The tableau

0	0	0	−1		
4	2	2	12	=	−12
1	0	1	2	=	− 2
0	1	3	4	=	− 4
3	2	2.5	0	=	0

is true in all rows. $n = 0$, $t = 0$, $b = 0$, $S_1 = 12$, $S_2 = 2$, and $S_3 = 4$ is a solution of the system of constraint equations. Moreover, it is a feasible solution since none of the variables is negative. The six equalities $n = 0$, $t = 0$, $b = 0$, $S_1 = 12$, $S_2 = 2$, and $S_3 = 4$ are written as

$$(n, t, b, S_1, S_2, S_3) = (0, 0, 0, 12, 2, 4)$$

with the understanding that the first named in each parentheses are equal, the second named are equal, and so on. Such a set of numbers, in which order is specified, is called an ***ordered set of numbers***. ***Equality*** of two (or more) ordered sets means equality between numbers occupying corresponding positions, for all positions.

Here, then, we have our first feasible solution. For this solution $M = 0$. It is a start that can be used to find a second feasible solution. To increase the profit from $M = 0$, we hold two of the top variables at zero and increase the third. Which third? At this stage the decision is arbitrary. Let us agree to keep $n = 0$ and $b = 0$ and try to increase t. The tableau reflecting this state of affairs is

0	t	0	−1		
4	2	2	12	=	$-S_1$
1	0	1	2	=	$-S_2$
0	1	3	4	=	$-S_3$
3	2	2.5	0	=	M

The increase in t must not cause S_1, S_2, or S_3 to become negative, for in that case a nonfeasible solution would emerge. The best that can be done is to increase t so that one of S_1, S_2, and S_3 is zero while the other two remain nonnegative. From row 1, if t surpasses 6, then S_1 is negative. From row 2, t does not affect S_2. From row 3, if t surpasses 4, then S_3 is negative and a solution is no longer feasible. To avoid this loss of feasibility we hold t at 4 and obtain the tableau

0	4	0	−1	
4	2	2	12	$= -S_1$
1	0	1	2	$= -S_2$
0	1	3	4	$= -S_3$
3	2	2.5	0	$= M$

We see from this tableau that $S_1 = 4$, $S_2 = 2$, and $S_3 = 0$. This yields a second feasible solution:

$$(n, t, b, S_1, S_2, S_3) = (0, 4, 0, 4, 2, 0)$$

For this solution we see that $M = 8$, up from 0. We have made some progress.

Can we get a better feasible solution? A review of what has happened so far may suggest a tactic. Starting with a feasible solution with all zeros at the top, one of the top variables (t) was increased, and because of this increase one of the right-side variables (S_3) became zero. This suggests that the two variables t and S_3 exchange their roles.

To effect this exchange of roles between t and S_3 we proceed as follows. The third equation,

$$t + 3b - 4 = -S_3$$

contains the two variables that are to be exchanged. Solve for the variable t.

$$t = -S_3 - 3b + 4$$

Substitute this expression for t in each of the other three equations. The first equation,

$$4n + 2t + 2b - 12 = -S_1$$

becomes

$$4n + 2(-S_3 - 3b + 4) + 2b - 12 = -S_1$$

or, rearranging terms,

$$4n - 2S_3 - 4b - 4 = -S_1$$

The second equation is

$$n + b - 2 = -S_2$$

Since t does not appear in (2), no substitution takes place. From the third equation we have already derived $t = -S_3 - 3b + 4$, or

$$S_3 + 3b - 4 = -t$$

The fourth equation is

$$3n + 2t + 2.5b = M$$

On substituting for t and rearranging it becomes

$$3n - 2S_3 - 3.5b + 8 = M$$

This *solve–substitute–simplify* procedure is the essence of the method for exchanging the roles of the right-side and top variables.

The four rearranged equations, in tableau form, are

n	S_3	b	-1		
4	−2	−4	4	=	$-S_1$
1	0	1	2	=	$-S_2$
0	1	3	4	=	$-t$
3	−2	−3.5	−8	=	M

On setting the new set of top variables equal to zero the feasible solution for this tableau is

$$(n, t, b, S_1, S_2, S_3) = (0, 4, 0, 4, 2, 0)$$

This, of course, is the same solution obtained earlier for which $M = 8$, since all we have done is to rearrange the equations. We now try to derive a third feasible solution, hopefully increasing M. Which of the top variables should be increased? We get our answer from the new objective equation

$$3n - 2S_3 - 3.5b + 8 = M$$

Any increase in S_3 or b will cause M to diminish since S_3 and b have negative coefficients. Only an increase in n can cause an increase in M. So we keep $S_3 = 0$ and $b = 0$ as in the next tableau and try to increase n.

n	0	0	-1	
4	-2	-4	4	$= -S_1$
1	0	1	2	$= -S_2$
0	1	3	4	$= -t$
3	-2	-3.5	-8	$= M$

From the second row n cannot exceed 2 or S_2 will become negative. From the first row, if n exceeds 1, then S_1 becomes negative. Therefore, we set n equal to 1, as in the following tableau:

1	0	0	-1	
4	-2	-4	4	$= -S_1$
1	0	1	2	$= -S_2$
0	1	3	4	$= -t$
3	-2	-3.5	-8	$= M$

We see that $S_1 = 0$, $S_2 = 1$, $t = 4$, and M has increased to 11. Thus our third feasible solution $(n, t, b, S_1, S_2, S_3) = (1, 4, 0, 0, 1, 0)$ has increased our profit M to \$11. If, as previously, we exchange n with S_1 and rearrange the tableau, the last line of the new tableau is

$$-\tfrac{3}{4}S_1 - \tfrac{1}{2}S_3 - \tfrac{1}{2}b + 11 = M$$

From this equation we see that increasing b, S_1, or S_3 decreases M, which we certainly do not want to do. We thus say that we have arrived at a best feasible solution in the sense that there is nothing we can do to increase our profit M. How do we know that if we take a different approach we will not do better? We do not know, but we will before long. The essence of this solution is that if we make 1 keg ($n = 1$) of nuts, 4 kegs of taper pins, and no bolts, then given the constraints, we will maximize profit at $11. In so doing the only slack time will be 1 hour on the small lathe ($S_2 = 1$).

The variables at the top of a tableau are called *nonbasic variables* and those at the right are called *basic variables*. Because of these names, a solution to a system of linear equations in which all the nonbasic variables are zero is called a *basic solution*. All the solutions in this section are basic solutions.

Exercises

For each tableau in Exercises 1–3, identify:
(a) The set of nonbasic variables.
(b) The set of basic variables.
(c) The basic solution.

1.

x	y	z	-1	
2	3	4	2	$= -u$
1	2	6	5	$= -v$
-2	1	-3	-3	$= M$

2.

x_1	x_2	-1	
1	2	4	$= -y_1$
3	0	11	$= -y_2$
2	1	7	$= -y_3$
-4	1	-3	$= M$

3.

x_1	x_2	x_3	-1	
3	1	2	4	$= -y_1$
1	0	2	5	$= -y_2$
2	-1	2	3	$= -y_3$
-4	5	-6	-7	$= M$

4. For the tableau in Exercise 1:
(a) Identify the column which contains the nonbasic variable that should be increased in order to increase M.
(b) Identify the row of the basic variable that should be made zero.
(c) Use the solve–substitute–simplify procedure to exchange the variables in (a) and (b).

The sequences of tableaus in Exercise 5 and 6 analyze maximum linear programs. For both sequences answer the following questions:

(a) What is the basic solution to the system of constraint equations in each tableau of the sequence?
(b) For each sequence, which basic solution is the best feasible solution?
(c) Should an additional exchange of variables be made? Explain.

5.

x_1	x_2	-1	
1	2	6	$= -y_1$
3	1	8	$= -y_2$
20	30	0	$= M$

x_1	y_1	-1	
$\frac{1}{2}$	$\frac{1}{2}$	3	$= -x_2$
$\frac{5}{2}$	$-\frac{1}{2}$	5	$= -y_2$
5	-15	-90	$= M$

y_2	y_1	-1	
$-\frac{1}{5}$	$\frac{3}{5}$	2	$= -x_2$
$\frac{2}{5}$	$-\frac{1}{5}$	2	$= -x_1$
-2	-14	-100	$= M$

6.

x_1	x_2	-1	
3	4	96	$= -y_1$
5	6	150	$= -y_2$
120	156	0	$= M$

6. (cont.)

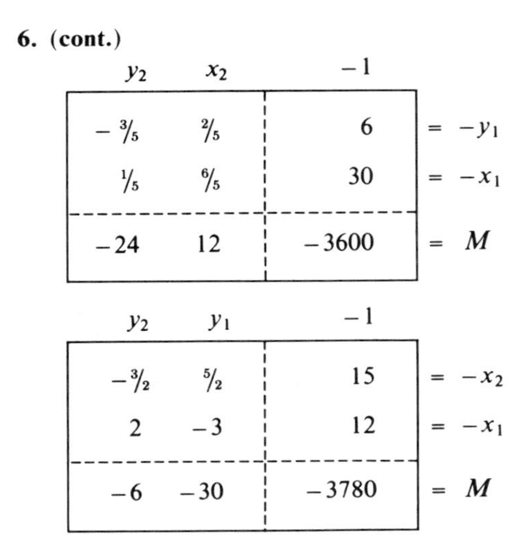

y_2	x_2	-1	
$-3/5$	$2/5$	6	$= -y_1$
$1/5$	$6/5$	30	$= -x_1$
-24	12	-3600	$= M$

y_2	y_1	-1	
$-3/2$	$5/2$	15	$= -x_2$
2	-3	12	$= -x_1$
-6	-30	-3780	$= M$

1.3 Dual Programs and the Duality Theorem

In addition to the maximum linear programs we have studied there are minimum linear programs. These are programs in which the objective function is to be minimized, subject to certain constraints. Minimum linear programs arise from linear programming problems of the following type.

A hospital dietician is faced with the problem of preparing dinner portions for patients which consist of two basic foods purchased in 10-gram packages. Each dinner portion is to contain at least 20 units of vitamin A, 30 units of vitamin B, and 40 units of vitamin C. Each 10-gram package of the first food costs 8 cents and has 2, 1, and 3 units of vitamins A, B, and C, respectively. Each 10-gram package of the second food costs 9 cents and contains 4, 1, and 2 units of vitamins A, B, and C, respectively. How many 10-gram packages of each food should go into one dinner portion so that the minimum vitamin requirements are satisfied and the cost is minimized?

This linear programming problem is translated into a minimum linear program as follows. Let v_1 represent the number of 10-gram packages of food I, and v_2 the number of 10-gram packages of food II. Then $2v_1$ is the vitamin A content per portion of food I and $4v_2$ is the vitamin A content per portion of food II. The total vitamin A output is thus $2v_1 + 4v_2$, which must total at least 20. Let T_1 be the surplus by which $2v_1 + 4v_2$ exceeds 20. (T_1 is called a *surplus variable.*) Then $2v_1 + 4v_2 = 20 + T_1$ with $T_1 \geq 0$, or

$$2v_1 + 4v_2 - 20 = T_1 \tag{1}$$

Similarly,

$$(2) \qquad 1v_1 + 1v_2 - 30 = T_2$$

where $T_2 \geq 0$ is the surplus variable for vitamin B.

$$(3) \qquad 3v_1 + 2v_2 - 40 = T_3$$

is the equation of constraint for vitamin C, where $T_3 \geq 0$. Letting m represent total cost we have

$$8v_1 + 9v_2 = m$$

The minimum linear program is now written as:

> Find nonnegative numbers v_1, v_2, T_1, T_2, and T_3 which minimize
>
> $$8v_1 + 9v_2 = m$$
>
> subject to
>
> $$(1) \qquad 2v_1 + 4v_2 - 20 = T_1$$
> $$(2) \qquad v_1 + v_2 - 30 = T_2$$
> $$(3) \qquad 3v_1 + 2v_2 - 40 = T_3$$

The surplus variables, in contrast to the slack variables of maximum linear programs, are written without negative signs. This is so because a minimum problem calls for adding the surplus variables on the right side of the equation rather than adding them on the left as is done with the slack variables of maximum linear programs. We use a tableau format for minimum linear programs which is different from that of maximum linear programs. Equation (1), for instance, is written in column form as follows:

v_1	2
v_2	4
-1	20
	$= T_1$

The nonbasic variables are written at the left and the basic variables at the bottom. The minimum linear program is represented in tableau form as

v_1	2	1	3	8
v_2	4	1	2	9
-1	20	30	40	0
	$= T_1$	$= T_2$	$= T_3$	$= m$

where the requirement once again is that all variables are nonnegative.

Suppose, without knowing the problem from which it arose, we were faced with the maximum linear program: find nonnegative numbers x_1, x_2, x_3, S_1, *and* S_2 which will maximize

$$20x_1 + 30x_2 + 40x_3 = M$$

subject to

$$(1) \qquad 2x_1 + x_2 + 3x_3 - 8 = -S_1$$

$$(2) \qquad 4x_1 + x_2 + 2x_3 - 9 = -S_2$$

which in tableau form is

x_1	x_2	x_3	-1	
2	1	3	8	$= -S_1$
4	1	2	9	$= -S_2$
20	30	40	0	$= M$

It is obvious that the inside of this tableau and that of the minimum linear program are identical. We therefore write these two linear programs in one *dual tableau:*

	x_1	x_2	x_3	-1	
v_1	2	1	3	8	$= -S_1$
v_2	4	1	2	9	$= -S_2$
-1	20	30	40	0	$= M$
	$= T_1$	$= T_2$	$= T_3$	$= m$	

We agree when reading the maximum linear program to ignore the variables at the left and bottom and when reading the minimum linear program to ignore the variables at the top and right. Two such linear programs are called *dual linear programs*. Obviously, every linear program has a dual linear program. For instance, adding the variables at the left and bottom gives us the minimum linear program dual to our original maximum linear program and allows us to write the following dual tableau:

	n	t	b	-1	
v_1	4	2	2	12	$= -S_1$
v_2	1	0	1	2	$= -S_2$
v_3	0	1	3	4	$= -S_3$
-1	3	2	2.5	0	$= M$
	$= T_1$	$= T_2$	$= T_3$	$= m$	

We are now ready to consider the theory that leads to the fundamental theorem, called the *duality theorem*. In what follows, each of the programs has six variables. The maximum linear program has two constraint equations and the minimum linear program has three constraint equations. The discussion is easily extendable to programs with any finite number of variables and constraint equations. Let a pair of dual programs be represented as:

	x_1	x_2	x_3	-1	
v_1	a_1	a_2	a_3	a_4	$= -y_1$
v_2	b_1	b_2	b_3	b_4	$= -y_2$
-1	c_1	c_2	c_3	c_4	$= M$
	$= u_1$	$= u_2$	$= u_3$	$= m$	

We begin by evaluating the expression

$$E = u_1x_1 + u_2x_2 + u_3x_3 - m \tag{1}$$

Substituting for u_1, u_2, u_3, and m their values as found from the columns of the dual tableau yields

$$E = (a_1v_1 + b_1v_2 - c_1)x_1 + (a_2v_1 + b_2v_2 - c_2)x_2$$
$$+ (a_3v_1 + b_3v_2 - c_3)x_3 - (a_4v_1 + b_4v_2 - c_4)$$

Multiply and collect the coefficients of v_1, v_2, and v_3. Then

$$E = (a_1x_1 + a_2x_2 + a_3x_3 - a_4)v_1 + (b_1x_1 + b_2x_2 + b_3x_3 - b_4)v_2$$
$$- (c_1x_1 + c_2x_2 + c_3x_3 - c_4)$$

We see from the rows of the dual tableau that the expressions in the respective parentheses are equal to $-y_1$, $-y_2$, and M. Therefore,

$$E = -v_1y_1 - v_2y_2 - M \tag{2}$$

Setting the value of E in (1) equal to the value of E in (2) yields

$$u_1x_1 + u_2x_2 + u_3x_3 - m = -v_1y_1 - v_2y_2 - M$$

Finally,

$$\boxed{u_1x_1 + u_2x_2 + u_3x_3 + v_1y_1 + v_2y_2 = m - M}$$

This is called the *key equation* or the *duality equation* of the pair of dual programs. It is true for any solution $(x_1, x_2, x_3, y_1, y_2)$ to the maximum program and any solution $(v_1, v_2, u_1, u_2, u_3)$ to the minimum program of the tableau (not necessarily feasible solutions). For feasible solutions none of the variables can be negative. Thus the left member of the equation cannot be negative. Therefore, $m - M \geq 0$ or $m \geq M$. This startling result says that for any feasible solutions to the programs in a dual pair, the value of the objective variable m of the minimum program cannot be less than the value of the objective variable M of the maximum program. When M is increased and/or m is decreased, both desirable, the gap between them closes. The best that can be done is to eliminate that gap, that is, arrive at the condition $m = M$. When this happens, if at all, then the feasible solutions to both programs are best, and the values of both programs are optimal.

This reasoning is valid for any pair of dual programs with suitable modifications for the number of variables and constraint equations. Hence

THE DUALITY THEOREM. Let M be the value in the objective equation corresponding to a feasible solution of the maximum linear program. Let m be the value in the objective equation corresponding to a feasible solution of the dual minimum linear program. If $m = M$, then this common value is optimal and the corresponding feasible solutions are best feasible solutions.

Note that the theorem does not state that the best feasible solutions are unique. It is quite possible that there is more than one set of best feasible solutions to the dual programs, each such set leading to the optimal value $m = M$. (In other words, "there is more than one way to skin a cat.")

Another interesting conclusion flows from the duality equation when $m = M$. For then

$$u_1x_1 + u_2x_2 + u_3x_3 + v_1y_1 + v_2y_2 = 0$$

Since each variable in this equation is nonnegative, each term on the left is either zero or positive. But neither can any term be positive, for then their sum would not be zero. Therefore, each term is zero and thus at least one factor in each term is zero. For instance, u_1x_1 is zero, whence either u_1 or x_1 is zero. Observe that u_1 and x_1 are at opposite ends of a column in the dual tableau, and that v_1 and y_1 are at opposite ends of a row. *In fact, in each pair of opposites at the ends of a row or column one of the variables must be zero in a best feasible solution.* This fact sheds some light on why the best feasible solutions we have obtained have many zeros. It also leads to a method for solving a linear program when the best feasible solution to its dual program is known. As an illustrative example, recall that a best feasible solution to the maximum linear program of Section 1.1 is asserted to be $(n, t, b, S_1, S_2, S_3) = (1, 4, 0, 0, 1, 0)$, with an optimal value of $M = 11$. We show this result with the dual minimum linear program in the following tableau:

	1	4	0	−1	
v_1	4	2	2	12	= −0
v_2	1	0	1	2	= −1
v_3	0	1	3	4	= −0
−1	3	2	2.5	0	= 11
	$= T_1$	$= T_2$	$= T_3$	$= m$	

Since $n = 1, t = 4$, and $S_2 = 1$, the key equation forces $T_1 = 0, T_2 = 0$, and $v_2 = 0$. The first column of the tableau is thus

v_1	4
0	1
v_3	0
−1	3
	= 0

which yields $v_1 = \frac{3}{4}$. Then the second column of the tableau, which yields $v_3 = \frac{1}{2}$, is

$\frac{3}{4}$	2
0	0
v_3	1
-1	2
	$= 0$

Then the third column of the tableau is

$\frac{3}{4}$	2
0	1
$\frac{1}{2}$	3
-1	2.5
	$= T_3$

which yields $T_3 = \frac{1}{2}$. The fourth column yields $m = 11$.

Therefore, knowing a best feasible solution to the maximum linear program allowed us to deduce $(v_1, v_2, v_3, T_1, T_2, T_3) = (\frac{3}{4}, 0, \frac{1}{2}, 0, 0, \frac{1}{2})$ as a best feasible solution to the dual minimum linear program. Of course, then $m = M = 11$.

This example suggests a converse of the duality theorem. If there exists a best feasible solution for one program in a dual pair, then there exist best feasible solutions for both programs with $m = M$. The proof of this converse is long and will not be given.

Exercises

1. A best feasible solution to the maximum program in

	x_1	x_2	-1	
v_1	4	1	5	$= -y_1$
v_2	2	3	6	$= -y_2$
-1	8	9	0	$= M$
	$= u_1$	$= u_2$	$= m$	

is

$$(x_1, x_2, y_1, y_2) = (9/10, 14/10, 0, 0)$$

Find a best feasible solution to the minimum program and the optimal values of both objective equations.

2. A best feasible solution to the minimum program in

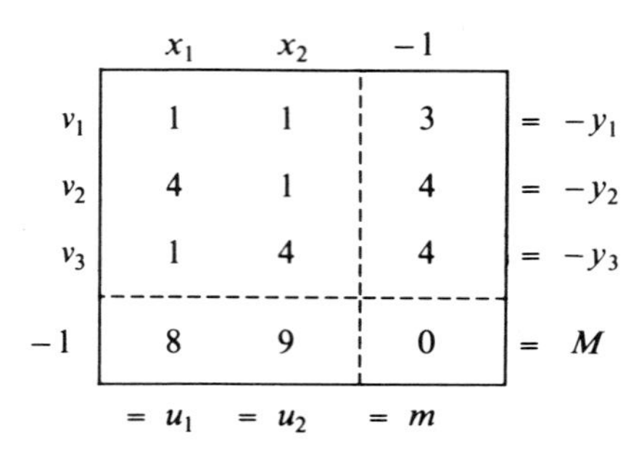

	x_1	x_2	-1	
v_1	1	1	3	$= -y_1$
v_2	4	1	4	$= -y_2$
v_3	1	4	4	$= -y_3$
-1	8	9	0	$= M$
	$= u_1$	$= u_2$	$= m$	

is

$$(v_1, v_2, v_3, u_1, u_2) = (0, 23/15, 28/15, 0, 0)$$

Find a best feasible solution to the maximum program and the optimal values of both objective equations.

3. A maximum linear program has five constraint equations which have six nonbasic variables. How many constraint equations and nonbasic variables does its dual minimum program have?

4. Let

	x_1	x_2	x_3	-1	
	a_1	a_2	a_3	a_4	$= -S_1$
P_1:	b_1	b_2	b_3	b_4	$= -S_2$
	c_1	c_2	c_3	c_4	$= M$

represent a maximum linear program. After changing signs inside the tableau for all entries in the first three rows, replacing x_i with y_i, S_i with T_i, and M with m, and then writing all rows as columns, in the same order, the result is

P_2:

y_1	$-a_1$	$-b_1$	c_1
y_2	$-a_2$	$-b_2$	c_2
y_3	$-a_3$	$-b_3$	c_3
-1	$-a_4$	$-b_4$	c_4
	$= T_1$	$= T_2$	$= m$

P_1 and P_2 are maximum and minimum programs with the same equations.

1.4 Pivot Exchange

In Section 1.2 we used the solve–substitute–simplify procedure to exchange the roles of a basic and nonbasic variable. In this section we develop the operation (computational procedure) of *pivot exchange* which applies solve–substitute–simplify directly to tableaus. Pivot exchange is fundamental to linear algebra and will recur constantly throughout the book. You are well advised to master it.

The equation

(1)

x	y	-1	
3	4	9	$= -z$

has x and y as nonbasic variables and z as the basic variable. If we exchange y and z, that is, make z nonbasic and y basic, we obtain

(2)

x	z	-1	
$3/4$	$1/4$	$9/4$	$= -y$

We can see this from the conventional form for equations by which (1) is written as

$$3x + 4y - 9 = -z$$

We can then solve for $-y$ as follows:

$$3x + z - 9 = -4y$$

$$\tfrac{3}{4}x + \tfrac{1}{4}z - \tfrac{9}{4} = -y$$

which is (2). Comparing (1) and (2) as tableaus we notice that the 4, which is in the column of the nonbasic variable y and the row of the basic variable z (which were exchanged), is replaced by its reciprocal $\frac{1}{4}$, and every other number is divided by that 4. The 4 is pivotal in the exchange of variables.

All the variables, basic and nonbasic, actually act as placeholders for numbers. Recall that a set of numbers that satisfy an equation is a solution of that equation. If we replace the placeholder x by 2 and y by 3 in (1) the result is

(3)

2	3	−1	
3	4	9	$= -z$

We deduce that the placeholder $z = -9$, so we have

(4)

2	3	−1	
3	4	9	$= -(-9)$

If we take $x = 2$, $y = 3$, and $z = -9$ in (2), we obtain

(5)

2	−9	−1	
$\frac{3}{4}$	$\frac{1}{4}$	$\frac{9}{4}$	$= -3$

which is true.

It is no surprise that a solution of (1) is a solution of (2) since the pivot exchange from (1) to (2) merely exchanges a basic and nonbasic variable of an equation. On the other hand, suppose that we select $x = -4$ and $z = 5$. Then from (2) we see that $y = 4$. That is,

(6)

−4	5	−1	
$\frac{3}{4}$	$\frac{1}{4}$	$\frac{9}{4}$	$= -4$

The solution of (2), $x = -4$, $y = 4$, and $z = 5$, is also a solution of (1), as is verified in (7).

(7)

−4	4	−1	
3	4	9	$= -5$

That a solution of (2) is a solution of (1) is a first indication that pivot exchange is reversible, a matter we explore further in the following examples.

EXAMPLE 1

Pivot-exchange x_1 and y in

(1)

x_1	x_2	x_3	-1	
2*	3	-5	9	$= -y$

As a guess from our preceding work we should replace 2 by its reciprocal $\frac{1}{2}$ and all other numbers inside the tableau should be divided by 2. We have starred 2 to identify it as the *pivot* of the exchange.

Carrying out the operation we obtain

(2)

y	x_2	x_3	-1	
$\frac{1}{2}$*	$\frac{3}{2}$	$-\frac{5}{2}$	$\frac{9}{2}$	$= -x_1$

To check we start with

$$2x_1 + 3x_2 - 5x_3 - 9 = -y$$

and proceed to

$$y + 3x_2 - 5x_3 - 9 = -2x_1$$

and

$$\tfrac{1}{2}y + \tfrac{3}{2}x_2 - \tfrac{5}{2}x_3 - \tfrac{9}{2} = -x_1$$

This confirms our guess.

Now pivot-exchange y and x_1 in (2). Again as a guess from the preceding work we should replace $\frac{1}{2}$ by its reciprocal and all other numbers inside the tableau should be divided by $\frac{1}{2}$. We have starred $\frac{1}{2}$ to identify it as the pivot of the exchange. Carrying out the operation yields

(3)

x_1	x_2	x_3	-1	
2	3	-5	9	$= -y$

The reader is advised to check as above. The indication that pivot exchange is reversible is now quite strong.

EXAMPLE 2

Pivot-exchange x_2 and y in

(1)

$$\begin{array}{cccc|l} x_1 & x_2 & x_3 & -1 & \\ \hline a & b^* & c & d & = -y \end{array}$$

where $b \neq 0$ is a constant as are a, c, and d. (Why must we insist that $b \neq 0$?)

By the operation

(2)

$$\begin{array}{cccc|l} x_1 & y & x_3 & -1 & \\ \hline \frac{a}{b} & \frac{1}{b} & \frac{c}{b} & \frac{d}{b} & = -x_2 \end{array}$$

To check:

$$\begin{aligned} ax_1 + bx_2 + cx_3 - d &= -y \\ ax_1 + y + cx_3 - d &= -bx_2 \\ \frac{a}{b}x_1 + \frac{1}{b}y + \frac{c}{b}x_3 - \frac{d}{b} &= -x_2 \end{aligned}$$

Now pivot-exchange y and x_2 in (2). The result is

(3)

$$\begin{array}{cccc|l} x_1 & x_2 & x_3 & -1 & \\ \hline a & b & c & d & = -y \end{array}$$

EXAMPLE 3

Pivot-exchange x_2 and y_1 in

$$\begin{array}{cccc|l} x_1 & x_2 & x_3 & -1 & \\ \hline a & b^* & c & d & = -y_1 \\ e & f & g & h & = -y_2 \end{array}$$

where a, b, c, d, e, f, g, and h are constants with $b \neq 0$. (Since b is to be the pivot, we have starred it.)

Row 1 represents the equation in which the variables exchange and as such row 1 will pivot-exchange exactly as in Example 2, yielding

x_1	y_1	x_3	-1	
$\frac{a}{b}$	$\frac{1}{b}$	$\frac{c}{b}$	$\frac{d}{b}$	$= -x_2$
?	?	?	?	$= -y_2$

What about the second row? That is where we need the substitute–simplify procedure. The new first row is

$$\frac{a}{b}x_1 + \frac{1}{b}y_1 + \frac{c}{b}x_3 - \frac{d}{b} = -x_2 \tag{1}$$

The old second row is

$$ex_1 + fx_2 + gx_3 - h = -y_2 \tag{2}$$

Now substituting the negative of the left side of (1) for x_2 in (2) yields

$$ex_1 + f\left(-\frac{a}{b}x_1 - \frac{1}{b}y_1 - \frac{c}{b}x_3 + \frac{d}{b}\right) + gx_3 - h = -y_2$$

Simplifying terms gives

$$\left(e - \frac{af}{b}\right)x_1 - \frac{f}{b}y_1 + \left(g - \frac{cf}{b}\right)x_3 - \left(h - \frac{df}{b}\right) = -y_2$$

which when placed in the tableau yields, as a final result,

x_1	y_1	x_3	-1	
$\frac{a}{b}$	$\frac{1}{b}$	$\frac{c}{b}$	$\frac{d}{b}$	$= -x_2$
$e - \frac{af}{b}$	$-\frac{f}{b}$	$g - \frac{cf}{b}$	$h - \frac{df}{b}$	$= -y_2$

EXAMPLE 4

Pivot-exchange x_2 and y_1 in

x_1	x_2	x_3	-1	
a	b^*	c	d	$= -y_1$
e	f	g	h	$= -y_2$
m	n	p	q	$= -y_3$

From Example 3 we see that the first two rows are given by

x_1	y_1	x_3	-1	
$\frac{a}{b}$	$\frac{1}{b}$	$\frac{c}{b}$	$\frac{d}{b}$	$= -x_2$
$e - \frac{af}{b}$	$-\frac{f}{b}$	$g - \frac{cf}{b}$	$h - \frac{df}{b}$	$= -y_2$
?	?	?	?	$= -y_3$

A moment's reflection shows that the result in the third row is obtained by the same method that produced the second row: substitution of the value of x_2. The only difference is that the third row has the letters m, n, p, and q, respectively, instead of the letters e, f, g, and h of the second row. The final tableau is thus

x_1	y_1	x_3	-1	
$\frac{a}{b}$	$\frac{1}{b}$	$\frac{c}{b}$	$\frac{d}{b}$	$= -x_2$
$e - \frac{af}{b}$	$-\frac{f}{b}$	$g - \frac{cf}{b}$	$h - \frac{df}{b}$	$= -y_2$
$m - \frac{an}{b}$	$-\frac{n}{b}$	$p - \frac{cn}{b}$	$q - \frac{dn}{b}$	$= -y_3$

EXAMPLE 5

Pivot-exchange y and R in

(1)

x	y	z	-1	
10	2*	4	6	$= -R$
1	8	0	3	$= -S$
5	4	7	9	$= -T$

Following the procedure outlined in Example 4, we replace the pivot by its reciprocal $\frac{1}{2}$ and divide all other numbers in the pivot row by the pivot. Thus $\frac{10}{2} = 5$, $\frac{4}{2} = 2$, and $\frac{6}{2} = 3$. The first row is

x	R	z	-1	
5	$\frac{1}{2}$	2	3	$= -y$
?	?	?	?	$= -S$
?	?	?	?	$= -T$

To get the new numbers in the column of the pivot, following the guidelines of Example 4, we take the negative of the original column numbers divided by the pivot. Thus $-\frac{8}{2} = -4$ and $-\frac{4}{2} = -2$. This yields

x	R	z	-1	
5	$\frac{1}{2}$	2	3	$= -y$
?	-4	?	?	$= -S$
?	-2	?	?	$= -T$

To establish a procedure for obtaining the other numbers not in the row or column of the pivot we develop the *rectangle technique*. Let us start by computing the number to replace 1. Think of 1 and the pivot 2 as being corners of a rectangle. Then the other corners are 10 and 8. Multiply the other corners $(10 \cdot 8)$, divide by the pivot $[(10 \cdot 8)/2 = 40]$, and subtract from the 1$(1 - 40 = -39)$. This number replaces 1. We write: 1 goes to -39.

Similarly, 5 and the pivot 2 are corners. The other corners are 10 and 4. Thus the replacement is $5 - (10 \cdot 4)/2 = -15$.

$$5 \text{ goes to } -15$$

$$0 \text{ goes to } 0 - \frac{8 \cdot 4}{2} = -16$$

$$7 \text{ goes to } 7 - \frac{4 \cdot 4}{2} = -1$$

$$3 \text{ goes to } 3 - \frac{8 \cdot 6}{2} = -21$$

$$9 \text{ goes to } 9 - \frac{4 \cdot 6}{2} = -3$$

The final tableau is

(2)

x	R	z	-1	
5	$\frac{1}{2}$*	2	3	$= -y$
-39	-4	-16	-21	$= -S$
-15	-2	-1	-3	$= -T$

To see that pivot exchange is reversible for this example we proceed to carry out the operation on (2) exchanging R and y and therefore using $\frac{1}{2}$ as the pivot. Replacing the pivot $\frac{1}{2}$ by its reciprocal and as usual dividing each other number in the first row by $\frac{1}{2}$, we get the first row of (3). Following the guidelines of Example 4 and the first part of this example the replacements for the other numbers in the pivot column are the negatives of these numbers divided by the pivot. Replacements for all other numbers that are not in the row or column of $\frac{1}{2}$ are determined by the rectangle technique. Thus -39 is replaced by

$$-39 - \left[\frac{5(-4)}{\frac{1}{2}}\right] = 1$$

Carrying out the details of the rectangle technique gives

-39 goes to 1

-15 goes to 5

-16 goes to 0

-1 goes to 7

-21 goes to 3

-3 goes to 9

The result (3) is identical to (1):

(3)

x	y	z	-1	
10	2	4	6	$= -R$
1	8	0	3	$= -S$
5	4	7	9	$= -T$

As a check we obtain a solution to (1), test this solution in (2), and then obtain a solution to (2) and test it in (1).

Letting $x = 1$, $y = -1$, and $z = 2$, in (1) we see that $R = -10$, $S = 10$, and $T = -6$. Substituting these six values in (2) yields

(4)

1	−10	2	−1	
5	½	2	3	= −(−1)
−39	−4	−16	−21	= −10
−15	−2	−1	−3	= −(−6)

Letting $x = -3$, $R = 6$, and $z = -\frac{1}{2}$ in (2) we see that $y = 16$, $S = -122$, and $T = -\frac{73}{2}$. Substituting these six values in (1) yields

(5)

−3	16	−½	−1	
10	2	4	6	= −6
1	8	0	3	= −(−122)
5	4	7	9	= −(−73/2)

EXAMPLE 6

Pivot-exchange x_1 with y_2 and x_2 with y_1 for

x_1	x_2	-1	
1	2	3	$= -y_1$
−2*	−1	5	$= -y_2$

Using the techniques developed so far and pivoting on −2 as indicated, we exchange x_1 and y_2. This yields

y_2	x_2	-1	
½	3/2*	11/2	$= -y_1$
−½	½	−5/2	$= -x_1$

To pivot-exchange x_2 with y_1, pivot on the $\frac{3}{2}$, yielding

y_2	y_1	-1	
1/3	2/3	11/3	$= -x_2$
−2/3	−1/3	−13/3	$= -x_1$

You should rework this example, pivot-exchanging x_2 with y_1 before x_1 with y_2 to convince yourself that when there is more than one exchange to be made, the order in which you perform the work does not affect the final result.

EXAMPLE 7

Pivot-exchange v_1 with u_2 for

$$\begin{aligned} 10v_1 + v_2 - 5 &= u_1 \\ 2v_1 + 8v_2 - 4 &= u_2 \\ 4v_1 + 0v_2 - 7 &= u_3 \\ 6v_1 + 3v_2 - 9 &= u_4 \end{aligned}$$

In Section 1.3 when writing minimum linear programs in tableau form we established the column tableau. Let us recall that

$$10v_1 + v_2 - 5 = u_1$$

is written as

v_1	10
v_2	1
-1	5
	$= u_1$

Similarly, we write the *columnar tableau* that follows to represent the given four equations:

v_1	10	2	4	6
v_2	1	8	0	3
-1	5	4	7	9
	$= u_1$	$= u_2$	$= u_3$	$= u_4$

You may recognize the inside of this tableau as being identical with the inside of the tableau (1) of Example 5. Suppose, then, we write one *dual tableau* to represent both the *row tableau* of Example 5 and the *columnar tableau* above.

	X	Y	Z	-1	
v_1	10	2*	4	6	$= -R$
v_2	1	8	0	3	$= -S$
-1	5	4	7	9	$= -T$
	$= u_1$	$= u_2$	$= u_3$	$= u_4$	

When reading the rows we ignore the left-hand side and bottom variables. When reading the columns we ignore the top and right-side variables. Notice that the 2 at the intersection of the column of v_1 and the row of u_2, the variables to be exchanged in this example, is the second pivot in Example 5. What if we pivot as in Example 5? Will such pivoting work for both tableaus? To check we reproduce (2) of Example 5, exchanging Y with R and v_1 with u_2 and getting

	X	R	Z	-1	
u_2	5	$\frac{1}{2}$	2	3	$= -Y$
v_2	-39	-4	-16	-21	$= -S$
-1	-15	-2	-1	-3	$= -T$
	$= u_1$	$= v_1$	$= u_3$	$= u_4$	

We know from Example 5 that the rows are correct. To ascertain that the columns are, we apply solve–substitute–simplify to our given four equations. Since we are exchanging v_1 and u_2 we solve the second equation,

$$2v_1 + 8v_2 - 4 = u_2$$

getting

$$\tfrac{1}{2}u_2 - 4v_2 + 2 = v_1 \tag{1}$$

which we see is the second column of the last dual tableau. Now if you substitute (1) for v_1 in the other three equations and rearrange terms, you will see that this last dual tableau is correct. It represents pivot exchange for both systems.

The following observations should be noted.

1. In row tableau format we have written the nonbasic variables at the top and the basic variables at the right. The basic variables have

minus signs. Therefore, when we exchange variables, we are changing signs of the exchanged variables; the nonbasic variable becomes basic and adopts a negative sign, and the basic variable becomes nonbasic and drops the negative sign.

2. In column tableau format both types of variables are without minus signs, and when we exchange we do not change signs of the exchanged variables. (This nonchange of sign is illusory. We are really changing sign internally. Recall that the pivot procedure introduces a minus sign in the column of the pivot. This minus sign is in each term of the equation in columnar form and thus is a sign change.)

We sum up the rules for pivot exchange:

(a) Star the pivot. This pivot, which cannot be zero, lies at the intersection of the row and column of the variables to be exchanged. Exchange the variables. When the tableau is dual, both pairs of variables are exchanged. Change the signs of the variables for the row tableau.

(b) Replace the pivot p by its reciprocal $1/p$.

(c) Replace all other numbers q in the row of the pivot by q/p.

(d) Replace all other numbers r in the column of the pivot by $-r/p$.

(e) Replace all other numbers s which are not in the pivot's row or column by $s - rq/p$, where r and q are the two other corners of the rectangle that has s and the pivot p as corners.

Exercises

1. In each of the following, use pivot exchange to exchange x and z, if possible. Check your work in each exercise by obtaining solutions for each tableau and testing in the other tableau.

(a)

x	y	
2	3	$= z$
4	5	$= w$

(b)

y	x	-1	
1	2	3	$= z$
4	5	6	$= w$

(c)

x	y	-1	
1	2	3	$= w$
5	0	-2	$= z$

(d)

y	x	-1	
1	2	3	$= w$
5	0	-2	$= z$

2. (a) Exchange x_1 and y_2 in

x_1	x_2	x_3	-1	
-6	0	2	2	$= -y_1$
3	2	0	-1	$= -y_2$

(b) Which pairs of variables in the tableau of (a) cannot be exchanged? Is it possible they may be exchanged in another tableau equivalent to it?

3. In the maximum program,

	x_1	x_2	x_3	-1	
v_1	1	2	3	4	$= -y_1$
v_2	5	2	-1	-2	$= -y_2$
v_3	0	4	-2	6	$= -y_3$
-1	8	6	0	0	$= M$
	$= u_1$	$= u_2$	$= u_3$	$= m$	

exchange x_3 and y_2. As a result of this exchange, which variables must be exchanged in the minimum program?

4. In

x_1	x_2	-1	
4	3	5	$= -y_1$
2	1	6	$= -y_2$
0	7	8	$= P$

compare the ease of calculations in exchanging x_2 and y_2 with the difficulty in exchanging x_1 and y_1.

5. In the tableau

	x_1	x_2	-1	
v_1	2	3	4	$= -y_1$
v_2	-1	-1	2	$= -y_2$
-1	20	30	40	$= M$
	$= u_1$	$= u_2$	$= m$	

exchange v_1 and u_1. Show the new tableau with all variables exchanged that must be exchanged.

6. After exchanging x_1 and y_2 in

x_1	x_2	-1	
2	3	-1	$= -y_1$
4	5	0	$= -y_2$

a second pivot exchange is made to reexchange them. What should the final tableau be? Verify your answer by performing the two pivot exchanges and testing with solutions.

7. Show that exchanging x_1 and y_1 in

x_1	x_2	-1	
1	a	c	$= y_1$
0	b	d	$= y_2$

does not alter the inside of the tableau.

8. Show that the exchange of x_1 and y_1 in

x_1	x_2	x_3	-1	
1	0	a	d	$= y_1$
0	1	b	e	$= y_2$
0	0	c	f	$= y_3$

followed by the exchange of x_2 and y_2 does not alter the inside of the tableau.

9. Formulate a generalization based on the results of Exercises 7 and 8.

10. (a) Let $a \neq 0$ be the pivot of a pivot exchange and $b = 0$ an entry in the pivot row. Show that a pivot exchange on a does not alter the column of b.

(b) Let $a \neq 0$ be the pivot of a pivot exchange and $c = 0$ an entry in the pivot column. Show that a pivot exchange on a does not alter the row of c.

11. (a) In

	x_1	x_2	
	2	3	$= y_1$
	4	5	$= y_2$

exchange x_1 and y_1, and follow up with an exchange between x_2 and y_2. Start again and reverse the order. Compare the two final tableaus.

(b) Repeat (a) starting with the exchange between x_1 and y_2, followed by the exchange between x_2 and y_1. Then reverse the order and compare the two tableaus with each other and the two in (a).

(c) For each sequence of exchanges, find the product of the pivots and compare the four products.

1.5 The Simplex Algorithm

Using the key equation of Section 1.3 and pivot exchange, we now illustrate the simplex algorithm (George B. Dantzig, 1947) for solving dual linear programs. Tableau (1) represents the dual linear programs of Section 1.3, for which we already know the optimal feasible solutions.

(1)

	n	t	b	-1	
v_1	4	(2)	2	(12)	$= -S_1$
v_2	1	0	1	2	$= -S_2$
v_3	0	(1*)	3	(4)	$= -S_3$
-1	3	2	2.5	0	$= M$
	$= T_1$	$= T_2$	$= T_3$	$= m$	

Let us start by assuming that $n = t = b = 0$. Therefore, $S_1 = 12$, $S_2 = 2$, $S_3 = 4$, and $M = 0$. Thus

$$(n, t, b, S_1, S_2, S_3) = (0, 0, 0, 12, 2, 4)$$

is a basic feasible solution of the maximum program (that is, both basic and feasible). Now let us try to obtain a basic feasible solution to the minimum program. To do so we must set the nonbasic variables v_1, v_2, and v_3 equal to zero. This gives $m = 0$, and since $M = 0$ we have $m = M$. Unfortunately, when $v_1 = v_2 = v_3 = 0$ we get $T_1 = -3$, $T_2 = -2$, $T_3 = -2.5$, and

$$(v_1, v_2, v_3, T_1, T_2, T_3) = (0, 0, 0, -3, -2, -2.5)$$

is not a basic feasible solution to the minimum program. (It is basic but not feasible since not all variables are nonnegative). Let us analyze this further. We have a feasible solution to the maximum program. We have $m = M$. If we had a feasible solution to the minimum program we would have (by the duality-theorem corollary) optimal solutions. We therefore address ourselves to performing three tasks:

1. Keep the solution to the maximum program basic feasible.
2. Keep $m = M$.
3. Obtain a feasible solution to the minimum program.

How shall we accomplish these tasks? As the tableau now stands, the variables at the left are 0, and thus the basic variables at the bottom are negative since all the numbers in the row of the -1 are positive (except for the 0 in the $m = M$ box). These numbers are called *indicators*. If we can convert all the indicators to negative numbers or zero we shall have accomplished (3). Of course, we must do this in such a way as to maintain the conditions in (1) and (2).

Let us concentrate on converting the indicator 2 to a negative number. To do this we pivot in the column of the indicator 2. If we pivot on a positive number the pivot-exchange rule for columns will convert 2 to a negative number. There are two positive numbers in the column of the indicator 2, namely 1 and 2. (Pivoting on 0 is not allowed.) We select the pivot by the *ratio procedure*. Form the ratios $^{12}/_2$ and $^4/_1$. These ratios are formed by the numbers in the -1 column and the potential pivots, both numbers in a ratio being in the same row. In (1) these numbers are shown linked together: $^{12}/_2 = 6$ and $^4/_1 = 4$. Pivot on the 1 since it is part of the smaller ratio. Is this necessary? Yes. Pivoting on the 2 causes the 4 in the last column to become negative, and that means that we have failed to maintain condition (1).

Suppose, in general, that there is a choice of two pivots (all positive), a and b, with corresponding numbers c and d in the -1 column; then the ratios are c/a and d/b with all numbers positive. Suppose that $c/a > d/b$ and we pivot on a.

	-1
a^*	c
b	d

Then d is replaced by $d - bc/a$. From $c/a > d/b$ we get, multiplying both sides of the inequality by b, that

$$\frac{bc}{a} > \frac{bd}{b}$$

or

$$\frac{bc}{a} > d$$

Thus

$$0 > d - \frac{bc}{a}$$

This means that d is replaced by a negative number. However, pivoting on b keeps c positive,

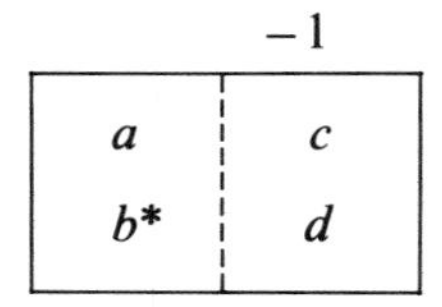

since c is replaced by $c - ad/b$ and multiplying both sides of

$$\frac{c}{a} > \frac{d}{b}$$

by a gives

$$\frac{ac}{a} > \frac{ad}{b} \qquad \text{or} \qquad c > \frac{ad}{b}$$

Thus $c - ad/b > 0$, as desired. We return to (1). Pivoting on the 1 we get

(2)

	n	S_3	b	-1	
v_1	(4*)	-2	-4	(4)	$= -S_1$
v_2	(1)	0	1	(2)	$= -S_2$
T_2	0	1	3	4	$= -t$
-1	3	-2	-3.5	-8	$= M$
	$= T_1$	$= v_3$	$= T_3$	$= m$	

Setting the nonbasic variables (top and left) equal to zero gives a basic feasible solution to the maximum program ($S_1 = 4, S_2 = 2, T = 4$). We have $m = M = 8$ (an improvement), but the basic solution to the minimum program is still not feasible ($T_1 = -3, v_3 = 2, T_3 = 3.5$). Since two of our indicators are now negative, we continue by pivoting in the column of the positive indicator 3. The relevant ratios are $4/4 = 1$ and $2/1 = 2$. Therefore, we pivot on the 4. The result is

(3)

	S_1	S_3	b	-1	
T_1	$1/4$	$-1/2$	-1	1	$= -n$
v_2	$-1/4$	$1/2$	2	1	$= -S_2$
T_2	0	1	3	4	$= -t$
-1	$-3/4$	$-1/2$	$-1/2$	-11	$= M$
	$= v_1$	$= v_3$	$= T_3$	$= m$	

All indicators are negative. We have finished. Setting all nonbasic variables equal to zero we get the basic feasible solution,

$$(n, t, b, S_1, S_2, S_3) = (1, 4, 0, 0, 1, 0)$$

for the maximum linear program and the basic feasible solution

$$(v_1, v_2, v_3, T_1, T_2, T_3) = (3/4, 0, 1/2, 0, 0, 1/2)$$

for the dual minimum linear program. Since $m = M = 11$, these solutions are optimal.

EXAMPLE 1

Solve the dual linear programs in T_1 of Figure 1.1.

All indicators are positive (9, 12, 15) in T_1, so any column will do for the pivot. We select the column of the 15 to dispel the possible illusion that the column to select is the one with the lowest indicator. The relevant ratios are $10/1 = 10$, $12/1 = 12$, and $14/5$ of which the last is the smallest. Therefore, pivot on 5. The result is T_2.

The positive indicators in T_2 are 6 and 9. We pivot in the column of the 9. The relevant ratios are

$$\frac{36/5}{9/5} = 4, \quad \frac{46/5}{14/5} = 23/7, \qquad \text{and} \qquad \frac{14/5}{1/5} = 14$$

We pivot on $14/5$. The result is T_3. The only positive indicator is $3/14$. By

T_1

	x_1	x_2	x_3	-1	
v_1	2	2	1	10	$= -y_1$
v_2	2	3	1	12	$= -y_2$
v_3	1	1	5*	14	$= -y_3$
-1	9	12	15	0	$= M$
	$= u_1$	$= u_2$	$= u_3$	$= m$	

T_2

	x_1	x_2	y_3	-1	
v_1	$9/5$	$9/5$	$-1/5$	$36/5$	$= -y_1$
v_2	$9/5$	$14/5$*	$-1/5$	$46/5$	$= -y_2$
u_3	$1/5$	$1/5$	$1/5$	$14/5$	$= -x_3$
-1	6	9	-3	-42	$= M$
	$= u_1$	$= u_2$	$= v_3$	$= m$	

T_3

	x_1	y_2	y_3	-1	
v_1	$9/14$*	$-9/14$	$-1/14$	$9/7$	$= -y_1$
u_2	$9/14$	$5/14$	$-1/14$	$23/7$	$= -x_2$
u_3	$1/14$	$-1/14$	$13/70$	$15/7$	$= -x_3$
-1	$3/14$	$-45/14$	$-33/14$	$-501/7$	$= M$
	$= u_1$	$= v_2$	$= v_3$	$= m$	

T_4

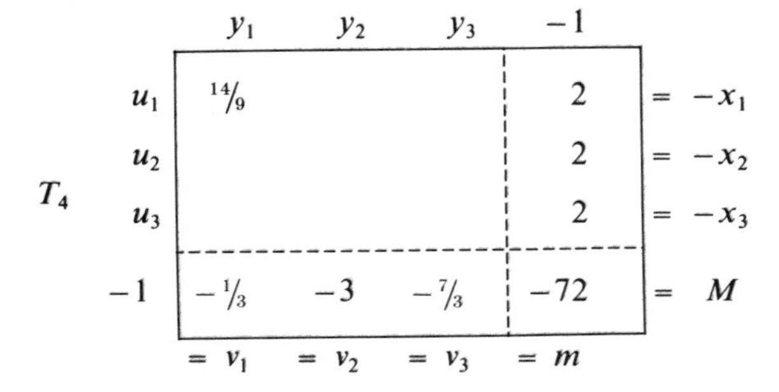

	y_1	y_2	y_3	-1	
u_1	$14/9$			2	$= -x_1$
u_2				2	$= -x_2$
u_3				2	$= -x_3$
-1	$-1/3$	-3	$-7/3$	-72	$= M$
	$= v_1$	$= v_2$	$= v_3$	$= m$	

Figure 1.1

the ratio rule we pivot on $^9/_{14}$ indicated in T_3. Hopefully when we pivot $^3/_{14}$ will be converted to a negative, and all other indicators will stay negative. If that is so, we will have finished, and by setting the nonbasic variables equal to zero, we shall obtain basic feasible solutions to the dual programs. Since $m = M$, these solutions will be optimal. The first numbers to check when pivoting are the indicators. If they are all negative, then calculate the numbers in the last column. The rest of the tableau is not needed. This is done and illustrated in T_4. (This is so because when the nonbasic variables are set equal to zero, all the numbers on the inside of the dashed lines of the tableau are multiplied by zero when solving for the basic variables.) Setting the nonbasic variables at the top of T_4 equal to zero we get $x_1 = 2$, $x_2 = 2$, $x_3 = 2$, and $M = 72$. Thus an optimal solution to the maximum linear program is

$$(x_1, x_2, x_3, y_1, y_2, y_3) = (2, 2, 2, 0, 0, 0)$$

and this gives the optimal value of 72 to the M in the objective equation

$$9x_1 + 12x_2 + 15x_3 = M$$

Setting the nonbasic variables at the left equal to zero gives the optimal solution to the dual minimum linear program as

$$(v_1, v_2, v_3, u_1, u_2, u_3) = (^1/_3, 3, ^7/_3, 0, 0, 0)$$

and the optimal value of m in the minimum objective equation

$$10v_1 + 12v_2 + 14v_3 = m$$

is also 72.

Notice that for this dual program all slack and surplus variables are zero. This means that any problem which has this program as its mathematical model must meet all conditions (constraints) exactly. There is no surplus and no slack.

EXAMPLE 2

Use the simplex algorithm to solve the dual linear programs of T_1 of Figure 1.2.

All indicators are positive. We select the column of 3 to obtain the pivot. By the ratio rule, since $^2/_3$ is less than $^4/_2$, we pivot on 3. As suggested in Example 1, we obtain the new indicators first. Since all are nonpositive, we have finished. Setting all nonbasic variables equal to zero yields for the

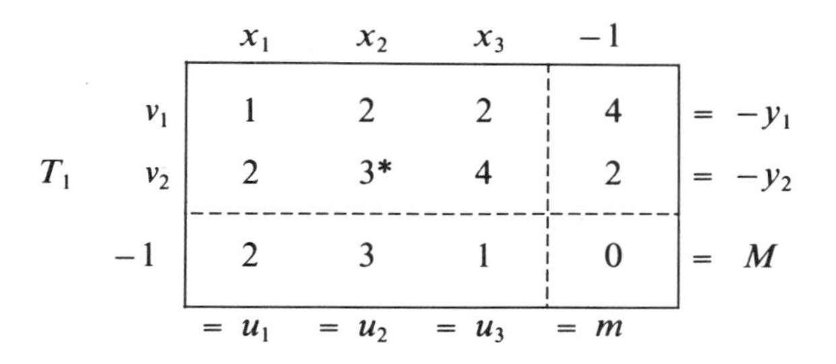

		x_1	x_2	x_3	-1	
T_1	v_1	1	2	2	4	$= -y_1$
	v_2	2	3*	4	2	$= -y_2$
	-1	2	3	1	0	$= M$
		$= u_1$	$= u_2$	$= u_3$	$= m$	

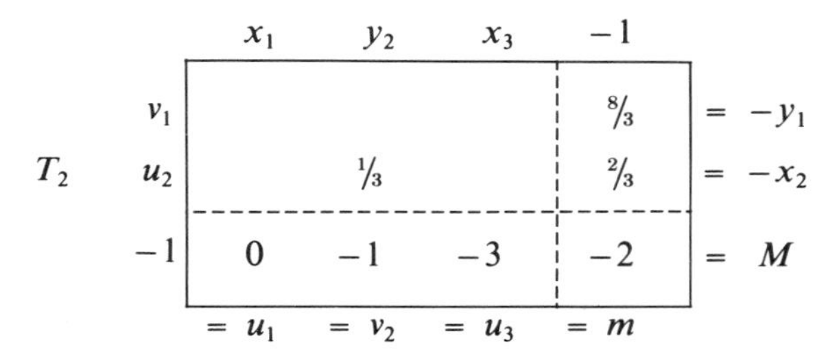

		x_1	y_2	x_3	-1	
T_2	v_1				$8/3$	$= -y_1$
	u_2		$1/3$		$2/3$	$= -x_2$
	-1	0	-1	-3	-2	$= M$
		$= u_1$	$= v_2$	$= u_3$	$= m$	

Figure 1.2

maximum program,

$$(x_1, x_2, x_3, y_1, y_2) = (0, 2/3, 0, 8/3, 0)$$

and for the minimum program,

$$(v_1, v_2, u_1, u_2, u_3) = (0, 1, 0, 0, 3)$$

and $m = M = 2$.

EXAMPLE 3

Use the simplex algorithm to solve the dual linear programs of T_1 in Figure 1.3.

As indicated, we pivot on the 1 of row 1, column 2. This is the only possible pivot in the column since we do not pivot on negative numbers. Now if there are pivots in T_2, they must be in the column of the indicator, 3, since there are no other positive indicators. But in this column there are only negative numbers, which means we do not pivot. What is to be done? Nothing. If in all columns whose indicators are positive there are no pivot elements available, then there are no optimal solutions to either of the dual linear programs, and in fact there is no feasible solution to at least one of

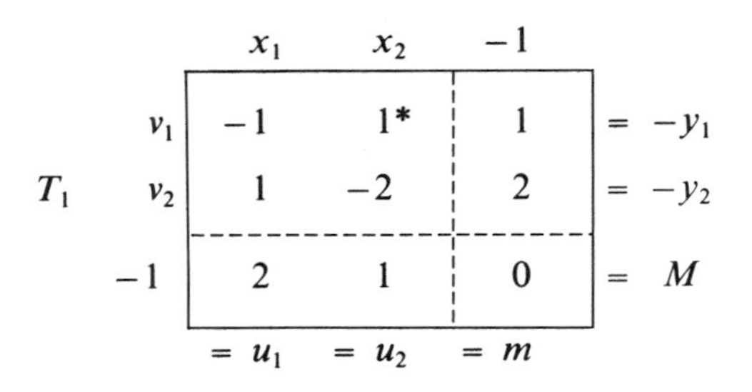

		x_1	x_2	-1	
	v_1	-1	1^*	1	$= -y_1$
T_1	v_2	1	-2	2	$= -y_2$
	-1	2	1	0	$= M$
		$= u_1$	$= u_2$	$= m$	

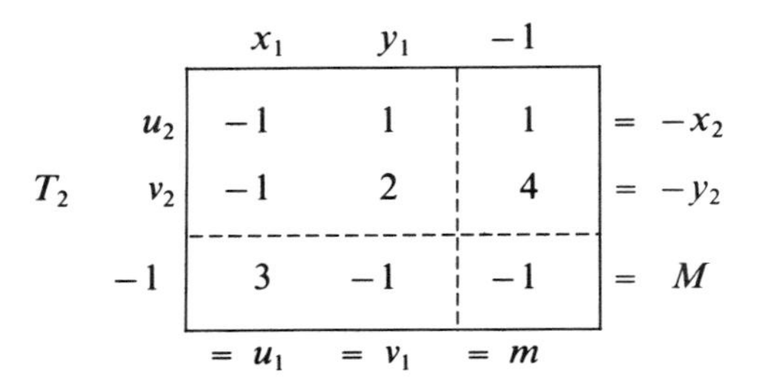

		x_1	y_1	-1	
	u_2	-1	1	1	$= -x_2$
T_2	v_2	-1	2	4	$= -y_2$
	-1	3	-1	-1	$= M$
		$= u_1$	$= v_1$	$= m$	

Figure 1.3

them. Since in this case the maximum program has a basic feasible solution (see T_1)

$$(x_1, x_2, y_1, y_2) = (0, 0, 1, 2)$$

the minimum program has no feasible solution. That this is true is clear from the first column of T_2, which we rewrite here as

$$-u_2 - v_2 - 3 = u_1 \tag{1}$$

Clearly, if u_2 and v_2 are nonnegative, then u_1 must be negative. Therefore, this equation cannot have a nonnegative solution. It is now clear that for all positive indicators, if the other elements of the column are negative or zero, we end up in a situation like (1), which means there are no feasible solutions.

EXAMPLE 4

In all previous examples we have (without saying it) assumed the existence of a basic feasible solution for the maximum linear program. This means that we have assumed no negative numbers in the last column (except possibly for the M value). What happens if we have dual linear programs where such a negative exists, as in T_1 of Figure 1.4?

The answer is to get rid of it. In Figure 1.4 we illustrate a simple case of how to do this.

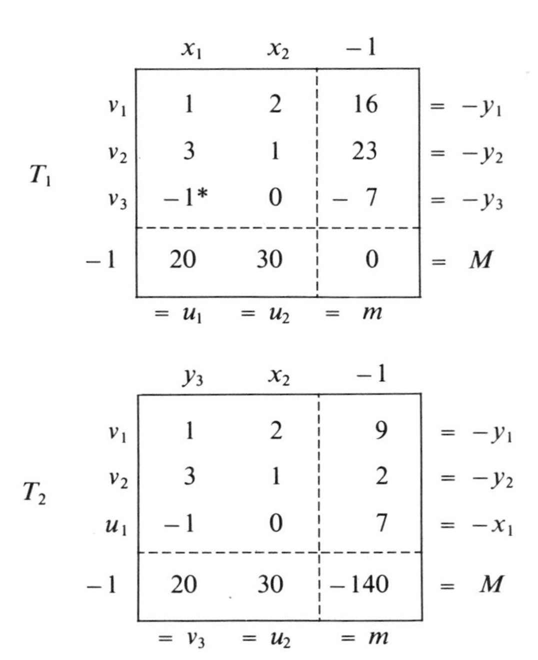

Figure 1.4

There is a general procedure that the reader may want to try and discover. Notice that once the column is all nonnegative, you can then proceed to use the regular simplex algorithm, which we now summarize.

1. Put the dual linear programs in dual tableau format.
2. Select as pivot column any column with a positive indicator.
3. Pivot on a number in that column which is part of the smallest ratio, using the last column (constraint constants) as the ratio numerators. (Negative numbers and 0 cannot be pivots.)
4. Get the indicator numbers first, when pivoting. If they are all nonpositive, then get the last column and proceed to step 5. If any indicator is still positive, fill in the rest of the tableau and then repeat steps 1–4 for some other column with a positive indicator. (*Warning:* Although we have not illustrated it, indicators can become negative and then go back to positive. If this happens in an exercise, keep working until they are all nonpositive.)
5. When all indicators are nonpositive (negative or 0), stop. Assign all nonbasic variables 0 (top and left). Read off the basic variables (right column and bottom row), collect the data in the form of basic feasible optimal solutions to the dual linear programs, and record the optimal value ($m = M$).

Exercises

Using the simplex algorithm, solve each of the following dual linear programs. If only one of the programs is stated, then state the other.

1. (a)

	x	y	-1	
u	3	2	5	$= -X$
v	1	4	6	$= -Y$
-1	6	2	0	$= M$
	$= U$	$= V$	$= m$	

(b)

	a	b	c	-1	
p	3	0	1	3	$= -d$
q	2	1	4	12	$= -e$
-1	4	2	3	0	$= M$
	$= r$	$= s$	$= t$	$= m$	

2. Find nonnegative numbers x_1, x_2, S_1, S_2, and S_3 which will maximize

$$3x_1 + 2x_2 = M$$

subject to

$$\begin{aligned} x_1 + 3x_2 + S_1 &= 8 \\ 2x_1 + x_2 + S_2 &= 7 \\ 3x_1 - x_2 + S_3 &= 5 \end{aligned}$$

3. Find nonnegative numbers a_1, a_2, a_3, T_1, and T_2 which will minimize

$$3a_1 + 5a_2 + 2a_3 = m$$

subject to

$$\begin{aligned} 2a_1 + 3a_2 + a_3 &= 8 + T_1 \\ a_1 + a_2 + a_3 &= 2 + T_2 \end{aligned}$$

4. Find nonnegative numbers u_1, u_2, u_3, T_1, T_2, and T_3 which will minimize

$$45u_1 + 50u_2 + 60u_3 = m$$

subject to

$$\begin{aligned} u_1 + 2u_2 + 4u_3 &= 6 + T_1 \\ 3u_1 + 3u_2 + 2u_3 &= 3 + T_2 \\ u_1 + 3u_2 + u_3 &= 2 + T_3 \end{aligned}$$

5. Find nonnegative numbers x_1, x_2, x_3, S_1, S_2, and S_3 which will maximize

$$10x_1 + 12x_2 + 15x_3 = M$$

subject to

$$\begin{aligned} 2x_1 + 2x_2 + x_3 + S_1 &= 9 \\ 2x_1 + 3x_2 + x_3 + S_2 &= 12 \\ x_1 + x_2 + 5x_3 + S_3 &= 15 \end{aligned}$$

6. Solve the problem in Exercise 1 of Section 1.1.

7. Solve the problem in Exercise 2 of Section 1.1.

8. Solve the problem in Exercise 3 of Section 1.1.

9. Solve the problem in Exercise 4 of Section 1.1.

10. Solve the dual linear program with which Section 1.5 starts, taking column 1 as the initial pivot column.

11. Show that

x_1	x_2	1	
-1	1	1	$= -S_1$
1	-2	2	$= -S_2$
2	1	0	$= M$

has no solution.

1.6 A Transportation Problem

A company has three warehouses, W_1, W_2, and W_3, and two retail outlets, R_1 and R_2. The demands at the retail outlets and the stock in the warehouses of one of the company's products are shown in the respective boxes of Figure 1.5. The per ton shipping costs are shown also. The company wishes to

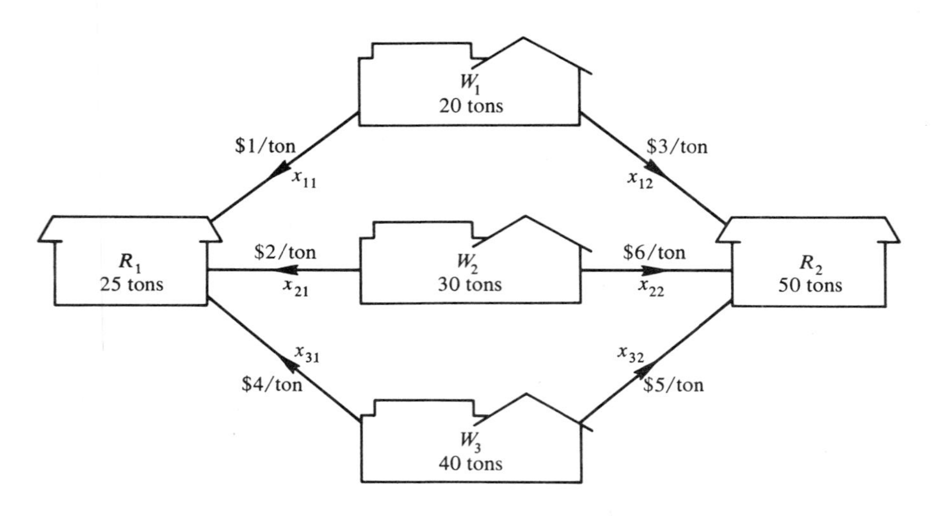

Figure 1.5

(1)

x_{11}	1	0	−1	0	0	1
x_{12}	0	1*	−1	0	0	3
x_{21}	1	0	0	−1	0	2
x_{22}	0	1	0	−1	0	6
x_{31}	1	0	0	0	−1	4
x_{32}	0	1	0	0	−1	5
−1	25	50	−20	−30	−40	0
	$= T_1$	$= T_2$	$= S_1$	$= S_2$	$= S_3$	$= m$

(2)

x_{11}	1	0	−1	0	0	1
T_2	0	1	−1	0	0	3
x_{21}	1	0	0	−1	0	2
x_{22}	0	−1	1	−1	0	3
x_{31}	1	0	0	0	−1	4
x_{32}	0	−1	1*	0	−1	2
−1	25	−50	30	−30	−40	−150
	$= T_1$	$= x_{12}$	$= S_1$	$= S_2$	$= S_3$	$= m$

Figure 1.6

(3)

x_{11}	1	-1	1	0	-1	3
T_2	0	0	1	0	-1	5
x_{21}	1*	0	0	-1	0	2
x_{22}	0	0	-1	-1	1	1
x_{31}	1	0	0	0	-1	4
S_1	0	-1	1	0	-1	2
-1	25	-20	-30	-30	-10	-210
	$= T_1$	$= x_{12}$	$= x_{32}$	$= S_2$	$= S_3$	$= m$

(4) .

x_{11}	-1	-1	1	1	-1	1
T_2	0	0	1	0	-1	5
T_1	1	0	0	-1	0	2
x_{22}	0	0	-1	-1	1	1
x_{31}	-1	0	0	1	-1	2
S_1	0	-1	1	0	-1	2
-1	-25	-20	-30	-5	-10	-260
	$= x_{21}$	$= x_{12}$	$= x_{32}$	$= S_2$	$= S_3$	$= m$

Figure 1.6 (cont.)

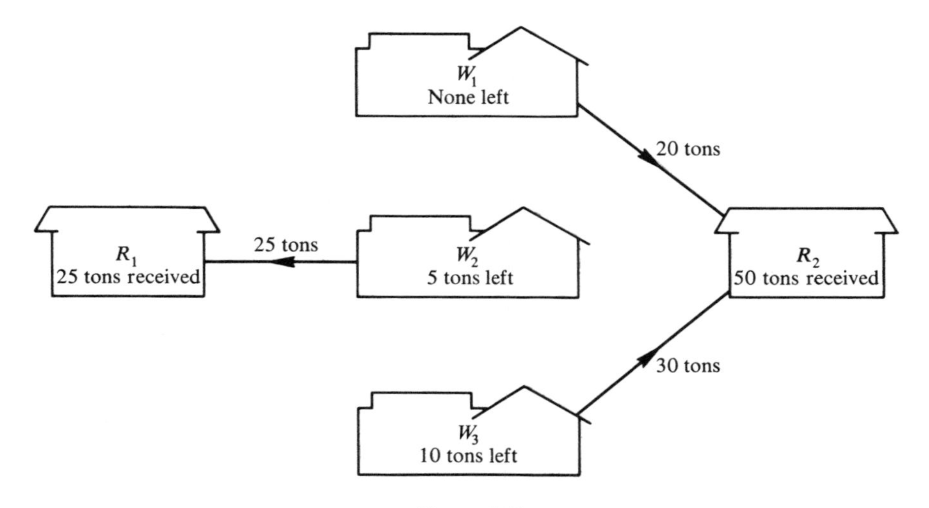

Figure 1.7

determine a shipping schedule which minimizes the total shipping costs, meets the demands of the retail stores, and does not exceed the stock in the warehouses. To start the determination, let x_{ij} represent the tonnage shipped from warehouse W_i to retail store R_j. Thus x_{32} represents the tonnage shipped from W_3 to R_2. Note that the total supply, 90, exceeds the total demand, 75.

The linear program for this problem is: minimize

$$1x_{11} + 3x_{12} + 2x_{21} + 6x_{22} + 4x_{31} + 5x_{32} = m$$

subject to

$$\begin{array}{llll} x_{11} + x_{21} + x_{31} & - 25 & = T_1 & \text{(shipments to } R_1) \\ x_{11} + x_{22} + x_{32} & - 50 & = T_2 & \text{(shipments to } R_2) \\ -x_{11} - x_{12} & + 20 & = S_1 & \text{(shipments from } W_1) \\ -x_{21} - x_{22} & + 30 & = S_2 & \text{(shipments from } W_2) \\ -x_{31} - x_{32} & + 40 & = S_3 & \text{(shipments from } W_3) \end{array}$$

We allow the retail outlets to accept excess shipments, and thus the constraints for the outlets have surplus variables T_1 and T_2. The tableau for the minimum linear program and the tableaus resulting from applying the simplex algorithm are in Figure 1.6. Figure 1.7 is a schematic diagram of the optimal shipping schedule. Transportation problems, in general, have *sparse* tableaus; that is, there are many zeros in the tableaus. Dantzig and others have devised special methods for the speedy solution of these problems.

1.7 Chapter Review

1. Linear programming problems are translated into mathematical models called linear programs. There are two types of linear programs, maximum linear programs and minimum linear programs.
2. Linear programs are written in tableau format. Maximum linear programs are written in rows; minimum linear programs are written in columns. A dual pair of linear programs is housed in a single dual tableau.
3. The pivot-exchange operation is developed. Pivot exchange operates on tableaus. It is reversible and preserves solutions. The rules for pivot exchange are:
   (a) Star the pivot. This pivot, which cannot be zero, lies at the

intersection of the row and column of the variables to be exchanged. Exchange the variables. When the tableau is dual, both pairs of variables are exchanged. Change the sign of the variables for the row tableau.

(b) Replace the pivot p by its reciprocal $1/p$.
(c) Replace all other numbers q in the row of the pivot by q/p.
(d) Replace all other numbers r in the column of the pivot by $-r/p$.
(e) Replace all other numbers s which are not in the pivot's row or column by $s - rq/p$, where r and q are the two other corners of the rectangle that has s and the pivot p as corners.

4. The key equation for the dual tableau

	x_1	x_2	x_3	-1	
v_1	a_1	a_2	a_3	a_4	$= -y_1$
v_2	b_1	b_2	b_3	b_4	$= -y_2$
-1	c_1	c_2	c_3	c_4	$= M$
	$= u_1$	$= u_2$	$= u_3$	$= m$	

is

$$u_1x_1 + u_2x_2 + u_3x_3 + v_1y_1 + v_2y_2 = m - M$$

When the dual tableau represents dual linear programs, the x's, y's, u's, and v's are nonnegative and therefore from the key equation $m \geq M$. When $m = M$ the feasible solutions to the dual linear programs are best. This is summed up in the duality theorem.

THE DUALITY THEOREM. Let M be the value in the objective equation corresponding to a feasible solution of the maximum linear program. Let m be the value in the objective equation corresponding to a feasible solution of the dual minimum linear program. If $m = M$, then this common value is optimal and the corresponding feasible solutions are best feasible solutions.

5. The simplex algorithm for solving dual linear programs is derived. Its rules are:

(a) Put the dual linear programs in dual tableau format.
(b) Select as pivot column any column with a positive indicator.
(c) Pivot on a number in that column which is part of the smallest ratio, using the last column (constraint constants) as the ratio numerators. (Negative numbers and 0 cannot be pivots.)
(d) Get the indicator numbers first, when pivoting. If they are all

nonpositive, then get the last column and proceed to (e). If any indicator is still positive, fill in the rest of the tableau and then repeat (c) and (d) for some other column with a positive indicator. (*Warning:* Although we have not illustrated it, indicators can become negative and then go back to positive. If this happens in an exercise, keep working until they are all nonpositive.)

(e) When all indicators are nonpositive (negative or 0), stop. Assign all nonbasic variables 0 (top and left). Read off the basic variables (right column and bottom row), collect the data in the form of basic feasible optimal solutions to the dual linear programs, and record the optimal value ($m = M$). Check these solutions in the initial tableau.

Review Exercises

1. Solve the following programs.

(a) Find nonnegative numbers x, y, z, S_1, and S_2 which will maximize

$$2x + 3y + 3z = M$$

subject to

$$x + 3y + 4z + S_1 = 30$$

$$x + 5y + 2z + S_2 = 40$$

(b) Find nonnegative numbers x_1, x_2, x_3, S_1, S_2, and S_3 which will maximize

$$4x_1 + 13x_2 + 3x_3 = M$$

subject to

$$x_1 + 3x_2 + 6x_3 + S_1 = 48$$

$$x_1 + 6x_2 + 3x_3 + S_2 = 90$$

$$x_1 + 9x_2 + 10x_3 + S_3 = 137$$

(c) Find nonnegative numbers u_1, u_2, u_3, T_1, T_2, and T_3 which will minimize

$$12u_1 + 6u_2 + 4u_3 = m$$

subject to

$$
\begin{aligned}
2u_1 + u_2 + 2u_3 &= 1 + T_1 \\
u_1 + 2u_2 &= 2 + T_2 \\
2u_1 + u_3 &= 1 + T_3
\end{aligned}
$$

2. A manufacturer makes chemicals A and B using three machines, I, II, and III, in the process. The number of hours each machine must be used for each chemical, and the profit in dollars on each ton of each chemical are given by the table:

	I	II	III	Profit
A	20	15	4	40
B	4	15	24	35

Machine I cannot be used more than 32 hours per week; machine II cannot be used more than 30 hours per week; machine III cannot be used more than 20 hours per week. How many tons of each chemical should be made, per week, to realize the maximum profit? What is the maximum profit?

3. Assume that the following table gives the amounts, in appropriate units, of the calories, protein, and calcium found in 1 pound of bread, meat, potatoes, and milk; also the daily minimum requirements and the cost per pound of the foods listed.

	Bread	Meat	Potatoes	Milk	Daily minimum requirement
Calories	300	500	60	60	750
Protein	20	24	4	8	36
Calcium	100	10	10	50	200
Cost/lb	30	100	20	25	

In what amounts should one buy these foods to obtain a satisfactory daily mix at the least cost?

4. Show, with an example, that if the ratio test is violated in choosing the pivot row, then the basic solution of the resulting system is not feasible.

5. A farmer has 50-pound bags of corn, linseed oil meal, and alfalfa. Each bag of corn has 1 pound of protein, 3 pounds of fat, and 1 pound of mineral matter. Each bag of linseed oil meal contains 2 pounds of protein, 3 pounds of fat, and 3 pounds of mineral matter. Each bag of alfalfa contains 4 pounds of protein, 2 pounds of fat, and 1 pound of mineral matter.

The farmer wants to use x bags of corn, y bags of linseed oil, and z bags of alfalfa to make a mixture that would contain at least 6 pounds of protein, at least 3 pounds of fat, and at least 2 pounds of mineral matter. If a bag of corn cost 50 cents, a bag of linseed oil meal costs 50 cents, and a bag of alfalfa costs 60 cents, how many bags of each should be mixed to satisfy the minimum requirements at the least cost?

6. Two kinds of pills contain the same ingredients: aspirin, bicarbonate, and codeine. The smaller pill has 2 grains of aspirin, 5 grains of bicarbonate, and 1 grain of codeine. The larger pill has 1 grain of aspirin, 8 grains of bicarbonate, and 6 grains of codeine. What is the least number of pills that will provide at least 12 grains of aspirin, 74 grains of bicarbonate, and 28 grains of codeine?

CLA

CHAPTER 2

The Fundamental Pivot-Exchange Algorithm, Tableaus, and Mappings

The simplex algorithm of Chapter 1 used pivot-exchange manipulation of tableaus to solve dual linear programs. The fundamental pivot-exchange algorithm, developed in this chapter, uses pivot-exchange manipulation to "dissect" a given tableau. This algorithm leads to a fully exchanged tableau which splits into blocks, each of which yields explicit information about the characteristics of the initial tableau. We suggest, at this point, a review of Section 1.4.

2.1 The Four-Towns Problems

Suppose that A, B, C, and D are four towns, each of which produces and consumes goods. They are linked by six roads, as shown in Figure 2.1. The arrows in the figure show the directions of the flow of goods from town to town. For instance, A sends goods to B and C and receives goods from D. Payments for shipments are made weekly and are measured in \$1000 units.

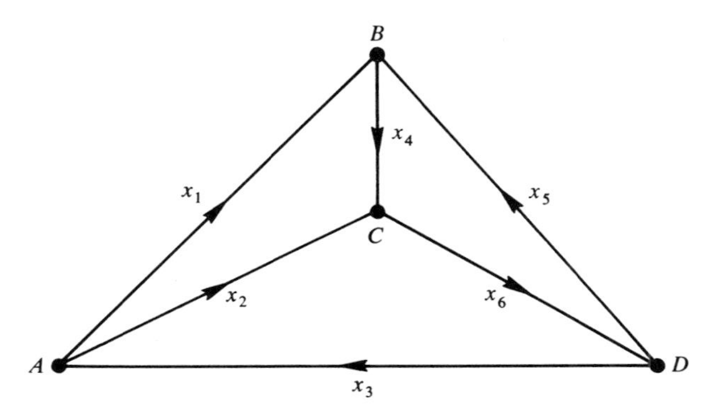

Figure 2.1

Let x_1 be the value of goods sent from A to B, x_2 the value of goods sent from A to C, and x_3 the value of goods sent from D to A. Thus A receives x_1 for its shipments to B, x_2 for its shipments to C, and pays x_3 for shipments received from D. Let y_A be the weekly net receipts (amount received minus amount paid out) for A. Then

$$x_1 + x_2 - x_3 = y_A$$

In tableau form this is

x_1	x_2	x_3	
1	1	-1	$= y_A$

Similarly, x_4 is the value of goods shipped from B to C, and x_5 the value of goods from D to B. (See Figure 2.1.) If y_B are the weekly net receipts for B, then

$$x_4 - (x_1 + x_5) = y_B$$

or, in tableau form,

x_1	x_4	x_5	
-1	1	-1	$= y_B$

To combine the two tableaus into one requires the presence of all variables from x_1 to x_5. This is accomplished by using zeros as coefficients for variables that do not appear in equations. The single tableau for the two equations is

x_1	x_2	x_3	x_4	x_5	
1	1	-1	0	0	$= y_A$
-1	0	0	1	-1	$= y_B$

Let x_6 be the value of weekly shipments from C to D, y_C the weekly receipts for C, and y_D the weekly receipts for D. Then

$$x_6 - (x_2 + x_4) = y_C$$

$$x_3 + x_5 - x_6 = y_D$$

All four equations are written in the one tableau:

(1)

x_1	x_2	x_3	x_4	x_5	x_6	
1	1	-1	0	0	0	$= y_A$
-1	0	0	1	-1	0	$= y_B$
0	-1	0	-1	0	1	$= y_C$
0	0	1	0	1	-1	$= y_D$

The four-towns problems can now be concisely stated.

1. Given the six values of weekly shipments $(x_1, x_2, x_3, x_4, x_5, x_6)$, compute the four values of weekly receipts (y_A, y_B, y_C, y_D).
2. Given a desired four values of weekly receipts (y_A, y_B, y_C, y_D) determine if there are six values of weekly shipments $(x_1, x_2, x_3, x_4, x_5, x_6)$ for which these receipts are possible.

EXAMPLE 1

Suppose that in the four-towns problems $x_1 = 2$, $x_2 = 1$, $x_3 = 3$, $x_4 = 5$, $x_5 = 1$, and $x_6 = 2$.

Replacing the x's by these values, (1) is

(2)

2	1	3	5	1	2	
1	1	-1	0	0	0	$= y_A$
-1	0	0	1	-1	0	$= y_B$
0	-1	0	-1	0	1	$= y_C$
0	0	1	0	1	-1	$= y_D$

From row 1 of (2):

$$(1)(2) + (1)(1) + (-1)(3) + (0)(5) + (0)(1) + (0)(2) = y_A$$
$$0 = y_A$$

From row 2:

$$(-1)(2) + (0)(1) + (0)(3) + (1)(5) + (-1)(1) + (0)(2) = y_B$$
$$2 = y_B$$

From row 3:

$$-4 = y_C$$

From row 4:

$$2 = y_D$$

The six x's in Example 1 are written as the 6-tuple (2, 1, 3, 5, 1, 2). It is understood, from the order in which the numbers are written that the first is the value of x_1, the second is the value of x_2, and so forth. As such, it is called an *ordered* 6-tuple. We write

$$(x_1, x_2, x_3, x_4, x_5, x_6) = (2, 1, 3, 5, 1, 2)$$

The four numbers found for the y's are written as an ordered 4-tuple

$$(y_A, y_B, y_C, y_D) = (0, 2, -4, 2)$$

This means that

$$y_A = 0, \quad y_B = 2, \quad y_C = -4, \quad y_D = 2$$

The 6-tuple (2, 1, 3, 5, 1, 2) yields the 4-tuple (0, 2, −4, 2). This is indicated in symbolic form as

$$(2, 1, 3, 5, 1, 2) \mapsto (0, 2, -4, 2)$$

The symbol "$\mapsto$" is read "*maps to.*" We also call (0, 2, −4, 2) the *image* of (2, 1, 3, 5, 1, 2).

EXAMPLE 2

Suppose in the first four-towns problem that

$$(x_1, x_2, x_3, x_4, x_5, x_6) = (-2, 3, 5, 1, 4, 7)$$

"$x_1 = -2$" means there is a reversal of direction of shipments, from B to A instead of from A to B. The tableau is

-2	3	5	1	4	7	
1	1	-1	0	0	0	$= y_A$
-1	0	0	1	-1	0	$= y_B$
0	-1	0	-1	0	1	$= y_C$
0	0	1	0	1	-1	$= y_D$

The reader is invited to verify with arithmetic computations that $(-2, 3, 5, 1, 4, 7) \mapsto (-4, -1, 3, 2)$. In words, verify that $(-4, -1, 3, 2)$ is the image of $(-2, 3, 5, 1, 4, 7)$. A feature of the first four-towns problem is that each 6-tuple $(x_1, x_2, x_3, x_4, x_5, x_6)$ has exactly one 4-tuple (y_A, y_B, y_C, y_D) as its image, which we indicate by

$$(x_1, x_2, x_3, x_4, x_5, x_6) \mapsto (y_A, y_B, y_C, y_D)$$

In particular, in Examples 1 and 2,

$$(\ 2, 1, 3, 5, 1, 2) \mapsto (\ 0, \ 2, -4, 2)$$
$$(-2, 3, 5, 1, 4, 7) \mapsto (-4, -1, \ 3, 2)$$

We might ask: Suppose we want a 4-tuple of receipts $(3, -6, 1, 1)$. Is there a shipment schedule that allows this? That is, find $(x_1, x_2, x_3, x_4, x_5, x_6)$ (if any) such that

$$(x_1, x_2, x_3, x_4, x_5, x_6) \mapsto (3, -6, 1, 1)$$

For such y values the tableau is

x_1	x_2	x_3	x_4	x_5	x_6	
1	1	-1	0	0	0	$= 3$
-1	0	0	1	-1	0	$= -6$
0	-1	0	-1	0	1	$= 1$
0	0	1	0	1	-1	$= 1$

The nature of this problem is different and more difficult than those of Examples 1 and 2. The variables at the top of the tableau (nonbasic variables), namely $x_1, x_2, \ldots, x_6$, play a different role from the y's at the right of the tableau (basic variables). From our previous work we see that the nonbasic variables can always be assigned numbers and simple arithmetic calculations lead to values for the basic variables. Thus we are led to the realization

(1)

x_1	x_2	x_3	x_4	x_5	x_6		
1*	1	−1	0	0	0	=	y_A
−1	0	0	1	−1	0	=	y_B
0	−1	0	−1	0	1	=	y_C
0	0	1	0	1	−1	=	y_D

(2)

$-y_A$	x_2	x_3	x_4	x_5	x_6		
1	1	−1	0	0	0	=	$-x_1$
1	1*	−1	1	−1	0	=	y_B
0	−1	0	−1	0	1	=	y_C
0	0	1	0	1	−1	=	y_D

(3)

$-y_A$	$-y_B$	x_3	x_4	x_5	x_6		
0	−1	0	−1	1	0	=	$-x_1$
1	1	−1	1	−1	0	=	$-x_2$
1	1	−1	0	−1	1	=	y_C
0	0	1	0	1*	−1	=	y_D

(4)

$-y_A$	$-y_B$	x_3	x_4	$-y_D$	x_6		
0	−1	−1	−1	−1	1	=	$-x_1$
1	1	0	1	1	−1	=	$-x_2$
1	1	0	0	1	0	=	y_C
0	0	1	0	1	−1	=	$-x_5$

Figure 2.2

that to solve this problem the roles of the variables must be exchanged with the y's becoming the nonbasic variables. In Section 1.4 we showed, by examples, that pivot exchange is reversible and that solutions of one tableau are solutions of any other tableau obtained by pivot exchange. Both of these results are true and are used in what follows. To start, if x_1 and y_A are exchanged, then y_A becomes a nonbasic variable and it can be assigned any value we please. Figure 2.2 shows the initial tableau in (1). The pivot exchange on 1* results in (2) and exchanges the roles of x_1 and y_A. Notice that the pivoting inside the tableau is the same as in Chapter 1. Here both exchanged variables now have minus signs. As in Chapter 1, both exchanged variables change sign. In selecting the next pivot, ignore all numbers in the row or column of the previous pivot. Otherwise we would not increase the number of y's that become nonbasic.

The pivot exchange on 1* in (2) results in (3) exchanging the roles of x_2 and y_B. The pivot exchange on 1* in (3) results in (4) exchanging x_5 and y_D. The last nonbasic variable y_C cannot be exchanged with x_3, x_4, or x_6, the only available variables still to be exchanged. This is so because the three possible pivots are all zero, and we cannot pivot on a zero.

Let us try to answer the question: For what value if any, does $(x_1, x_2, x_3, x_4, x_5, x_6) \mapsto (3, -6, 1, 1)$? Replacing the y's with their given values, (4) becomes

(5)

-3	6	x_3	x_4	-1	x_6	
0	−1	−1	−1	−1	1	$= -x_1$
1	1	0	1	1	−1	$= -x_2$
1	1	0	0	1	0	$= 1$
0	0	1	0	1	−1	$= -x_5$

For row 1:

$$(0)(-3) + (-1)(6) + (-1)(x_3) + (-1)(x_4) + (-1)(-1) + (1)(x_6) = -x_1$$

or

$$-6 - x_3 - x_4 + 1 + x_6 = -x_1$$

or

$$-x_3 - x_4 + x_6 - 5 = -x_1$$

Because x_3, x_4, and x_6 continue to be nonbasic variables we can assign any values to them. Each such 3-tuple determines a value of x_1.

By the same reasoning with rows 2 and 4 and the nonbasic variables x_3, x_4, and x_6, we can calculate x_2 and x_5.

But row 3 has a message we cannot disregard. It says that

$$(1)(-3) + (1)(6) + (0)(x_3) + (0)(x_4) + (1)(-1) + (0)(x_6) = 1$$

or $2 = 1$, which is false. This falsehood is independent of the values assigned to x_3, x_4, and x_6 (since each is multiplied by 0). This means that the message of row 3 of (5) is that no solution of (4) can have $y_A = 3$, $y_B = -6$, $y_C = 1$, and $y_D = 1$, regardless of the x's. Therefore, no solution of (1) can have these y values since we know that all solutions of (4) are solutions of (1), and all solutions of (1) are solutions of (4). Thus we must conclude that there is no 6-tuple $(x_1, x_2, x_3, x_4, x_5, x_6)$ that has $(3, -6, 1, 1)$ as an image. In four-towns terms it is impossible to have shipments that allow simultaneous receipts $y_A = 3$, $y_B = -6$, $y_C = 1$, and $y_D = 1$.

EXAMPLE 3

Find (if any) the values of x_1, x_2, x_3, x_4, x_5, and x_6 that produce $(y_A, y_B, y_C, y_D) = (3, -5, 1, 1)$.

In (4) of Figure 2.2 replace all the y's with their assigned values, which yields (6):

(6)

-3	5	x_3	x_4	-1	x_6	
0	-1	-1	-1	-1	1	$= -x_1$
1	1	0	1	1	-1	$= -x_2$
1	1	0	0	1	0	$= 1$
0	0	1	0	1	-1	$= -x_5$

This time let us examine row 3 first.

$$(1)(-3) + (1)(5) + (0)(x_3) + (0)(x_4) + (1)(-1) + (0)(x_6) = 1$$

Therefore, the y's are consistent and we proceed to find values of x's such that

$$(x_1, x_2, x_3, x_4, x_5, x_6) \mapsto (3, -5, 1, 1)$$

The alert reader will see immediately that for each choice of values for x_3, x_4, and x_6 we can find corresponding values of x_1, x_2, and x_5 and thus obtain the desired result. We shall pursue this matter in greater detail in Section 2.4.

Exercises

Exercises 1–5 refer to the tableau of the four-towns problems.

x_1	x_2	x_3	x_4	x_5	x_6	
1	1	-1	0	0	0	$= y_A$
-1	0	0	1	-1	0	$= y_B$
0	-1	0	-1	0	1	$= y_C$
0	0	1	0	1	-1	$= y_D$

1. Find (y_A, y_B, y_C, y_D) if $(x_1, x_2, x_3, x_4, x_5, x_6)$ is
(a) $(0, 0, 0, 0, 0, 0)$.
(b) $(1, 0, 1, 0, 1, 0)$.
(c) $(0, -1, 2, 1, -1, 0)$.
(d) $(1, -1, 1, -1, 1, -1)$.

2. For each 4-tuple in Exercise 1, show that $y_A + y_B + y_C + y_D = 0$. Explain why this is so.

3. There is a breakdown of transportation between towns A and B. Write a tableau (with five columns) that defines the new situation.

4. Town A is closed to all transportation in or out. What tableau defines this situation?

5. Interpret the statement $(0, 0, 0, 0, 0, 0) \mapsto (0, 0, 0, 0)$.

6. Let $(2, 1, -1, 3, 4, 6) \mapsto (y_A, y_B, y_C, y_D)$ and $(4, 2, -2, 6, 8, 12) \mapsto (y_A, y_B, y_C, y_D)$. Observe that each x in one 6-tuple is twice the corresponding x in the other. Find and compare the corresponding y's.

7. Variables x_1, x_2, x_3 and y_1, y_2 are related in accordance with the equations in the tableau

x_1	x_2	x_3	
2	1	-1	$= y_1$
3	4	-3	$= y_2$

(a) Which are the nonbasic variables? Which are the basic variables?
(b) Find (y_1, y_2) if $(x_1, x_2, x_3) =$
(i) $(1, 2, -1)$.
(ii) $(0, 5, -5)$.
(iii) $(0, 0, 0)$.

(c) Try to find a value (x_1, x_2, x_3) for which $(x_1, x_2, x_3) \mapsto (0, 0)$. Do you think there are values of (x_1, x_2, x_3) other than the one you found?

8. Variables u, v and x, y, z are related as shown in

u	v	
-1	2	$= x$
-2	3	$= y$
-3	4	$= z$

(a) Which are the nonbasic variables? Which are the basic variables?
(b) Find (x, y, z) if (u, v) is
  (i) $(3, 2)$.
  (ii) $(-2, 5)$.
  (iii) $(0, 0)$.
(c) Try to find a value of (u, v) for which $(u, v) \mapsto (0, 0, 0)$. Do you think there are values of (u, v) other than the one you found?

9. Using the data in Exercise 8, show that tripling each u and v also triples each x, y, and z. What happens to the values of x, y, and z when two (u, v) pairs are added together?

2.2 The Fundamental Pivot-Exchange Algorithm

Recall that a tableau presents a system of linear equations. It may be that for some values of the variables each equation in the system is true. In this case we say that these values of the variables *satisfy* the system. It is possible for other values of the variables to fail to satisfy at least one equation in the system. Then the system is *not satisfied.* For example, the system

(1)

x_1	x_2	
2	3	$= y_1$
-2	5	$= y_2$

is satisfied by

$$(x_1, x_2, y_1, y_2) = (2, 1, 7, 1)$$

since

$$(2) \qquad \begin{array}{cc} 2 & 1 \\ \hline \multicolumn{1}{|c}{2} & \multicolumn{1}{c|}{3} \\ \multicolumn{1}{|c}{-2} & \multicolumn{1}{c|}{5} \\ \hline \end{array} \begin{array}{l} \\ = 7 \quad \checkmark \\ = 1 \quad \checkmark \end{array}$$

But the system is not satisfied by

$$(x_1, x_2, y_1, y_2) = (2, 1, 3, 1)$$

since the first equation in (3) is false.

$$(3) \qquad \begin{array}{cc} 2 & 1 \\ \hline \multicolumn{1}{|c}{2} & \multicolumn{1}{c|}{3} \\ \multicolumn{1}{|c}{-2} & \multicolumn{1}{c|}{5} \\ \hline \end{array} \begin{array}{l} \\ = 3 \\ = 1 \end{array}$$

The set of all 4-tuples (x_1, x_2, y_1, y_2) that satisfy the system (1) is called the system's *solution set.*

Pivot exchange does not affect a solution set. For instance, if we pivot on 3 in (1) the result is

$$(4) \qquad \begin{array}{cc} x_1 & -y_1 \\ \hline \multicolumn{1}{|c}{2/3} & \multicolumn{1}{c|}{1/3} \\ \multicolumn{1}{|c}{-16/3} & \multicolumn{1}{c|}{-5/3} \\ \hline \end{array} \begin{array}{l} \\ = -x_2 \\ = \quad y_2 \end{array}$$

$$(x_1, x_2, y_1, y_2) = (2, 1, 7, 1)$$

also satisfies (4):

$$(5) \qquad \begin{array}{cc} 2 & -7 \\ \hline \multicolumn{1}{|c}{2/3} & \multicolumn{1}{c|}{1/3} \\ \multicolumn{1}{|c}{-16/3} & \multicolumn{1}{c|}{-5/3} \\ \hline \end{array} \begin{array}{l} \\ = -1 \quad \checkmark \\ = \quad 1 \quad \checkmark \end{array}$$

On the other hand,

$$(x_1, x_2, y_1, y_2) = (2, 1, 3, 1)$$

fails to satisfy (4). Indeed in

(6)

2	-1	
$2/3$	$1/3$	$= -1$
$-16/3$	$-5/3$	$= 1$

the second equation, $(-16/3)(2) + (-5/3)(-1) = 1$, is false. Similarly, we know that solutions of (4) are solutions of (1) and nonsolutions of (4) are nonsolutions of (1). These facts about solution sets obtained by pivot exchange are a result of the reversibility of the pivot-exchange procedure.

In Chapter 1 pivot exchange is used as part of the simplex algorithm. That algorithm prescribes the choice of pivots, and this in turn determines which pair of nonbasic and basic variables are exchanged. In this chapter, no longer concerned with linear programming and the simplex algorithm, we have greater freedom in the choice of pivots.

The fact that pivot exchange preserves solution sets does not depend on the choice of pivot. Thus we have the freedom of choosing as pivot any entry (other than zero) in a tableau and can be certain that the solution set is not altered. Moreover, it is usually possible to choose the sequence of pivots in more than one way. For instance, suppose that we wish to exchange both x's with both y's in

x_1	x_2	
2	3	$= y_1$
-2	5	$= y_2$

expressing the x's in terms of the y's. In the initial tableau (1) we arbitrarily choose 3 as pivot.

(1)

x_1	x_2	
2	3^*	$= y_1$
-2	5	$= y_2$

This exchanges x_2 and y_1 and yields (2).

(2)

x_1	$-y_1$	
$2/3$	$1/3$	$= -x_2$
$-16/3^*$	$-5/3$	$= y_2$

x_1 and y_2 remain to be exchanged. $-16/3$ must be the pivot in (2). Why? Pivoting on $-16/3$ gives (3).

(3)

$-y_2$	$-y_1$	
$1/8$	$1/8$	$= -x_2$
$-3/16$	$5/16$	$= -x_1$

Now suppose we had chosen 2 as the pivot in (1), resulting in (4).

(1)

x_1	x_2	
2*	3	$= y_1$
-2	5	$= y_2$

(4)

$-y_1$	x_2	
$1/2$	$3/2$	$= -x_1$
1	8*	$= y_2$

The pivot in (4) must be 8, yielding the resulting tableau (5). Now compare (3) and (5).

(5)

$-y_1$	$-y_2$	
$5/16$	$-3/16$	$= -x_1$
$1/8$	$1/8$	$= -x_2$

The variables in (3) and (5) appear in different orders. To facilitate the comparison arrange the variables in (3) so that they have the same order as those in (5). This is done as follows:

(a) Interchange the columns of (3), giving

(6)

$-y_1$	$-y_2$	
$1/8$	$1/8$	$= -x_2$
$5/16$	$-3/16$	$= -x_1$

This interchange of columns does not change the system of equations, merely the order in which the nonbasic variables are written.

(b) Now interchange rows, giving

(7)

$-y_1$	$-y_2$	
$5/16$	$-3/16$	$= -x_1$
$1/8$	$1/8$	$= -x_2$

The interchange of rows does not change the system, only the order of the equations.

Thus from (3) we obtain (7), which is the same as (5), which shows that we have a choice as to the sequences of pivot exchanges.

To get from (1) to (2) we pivoted on the number in row 1, column 2. If we now pivot on $1/3$, the number in (2) in the same location, x_2 and y_1 exchange places, restoring them to their original roles. The result is the initial tableau (1).

(2)

x_1	$-y_1$	
$2/3$	$1/3^*$	$= -x_2$
$-16/3$	$-5/3$	$= y_2$

(1)

x_1	x_2	
2	3	$= y_1$
-2	5	$= y_2$

If in (2) we pivot on $2/3$, exchanging x_1 and x_2, we get no closer to the desired result, in which all the x's are to be basic variables and all y's are to be nonbasic. The best thing we can say for this exchange (x_1 and x_2) is that it marks time.

Summarizing:

1. Pivot exchange is reversible.
2. Pivot exchange preserves solution sets.
3. It is possible to arrive at the same end result using different sequences of pivot exchanges.
4. Do not exchange any variables more than once.

EXAMPLE 1

Exchange as many x's with y's as possible in (1).

(1)

x_1	x_2	x_3	
1*	2	3	$= y_1$
2	1	-1	$= y_2$
-1	1	-2	$= y_3$

We can choose any of nine numbers as pivots in (1). The pivot 1* is chosen because the resulting calculations are simple. The result is (2).

(2)

$-y_1$	x_2	x_3	
1	2	3	$= -x_1$
-2	-3	-7	$= y_2$
1	3	1*	$= y_3$

There are four locations in (2) from which to choose the second pivot. Which are they? We use 1* to get

(3)

$-y_1$	x_2	$-y_3$	
-2	-7	-3	$= -x_1$
5	18*	7	$= y_2$
1	3	1	$= -x_3$

The choice of the pivot 18 in (3) is forced. Explain why.

(4)

$-y_1$	$-y_2$	$-y_3$	
$-1/18$	$7/18$	$-5/18$	$= -x_1$
$5/18$	$1/18$	$7/18$	$= -x_2$
$1/6$	$-1/6$	$-1/6$	$= -x_3$

All variables in (4) are now exchanged. The x's are expressed in terms of the y's. Tableau (4) is called a *fully exchanged* form of (1).

EXAMPLE 2

Exchange as many x's and y's as possible in (1).

(1)

x_1	x_2	x_3	
1*	2	3	$= y_1$
5	6	7	$= y_2$

In (1) 1 is the pivot. This removes row 1 and column 1 as sources of subsequent pivots.

(2)

$-y_1$	x_2	x_3	
1	2	3	$= -x_1$
-5	-4	-8*	$= y_2$

In (2) the pivot is -8. (We could have selected -4.) This removes row 2 and column 3 as the source of subsequent pivots.

(3)

$-y_1$	x_2	$-y_2$	
$-7/8$	$1/2$	$3/8$	$= -x_1$
$5/8$	$1/2$	$-1/8$	$= -x_3$

No pivot exchanges are made in (3) since it is a *fully exchanged* tableau of (1).

Only two of the x's can be made basic variables. Why?

Another possible sequence is as follows.

(1)

x_1	x_2	x_3	
1	2*	3	$= y_1$
5	6	7	$= y_2$

(4)

x_1	$-y_1$	x_3	
$1/2$	$1/2$	$3/2$	$= -x_2$
2*	-3	-2	$= y_1$

(5)

$-y_2$	$-y_1$	x_3	
$-1/4$	$5/4$	2	$= -x_2$
$1/2$	$-3/2$	-1	$= -x_1$

The solution sets of (3) and (5) are the same, even though the tableaus bear little resemblance to each other. This factor can be used to derive a check for the fundamental pivot-exchange algorithm. Choose values for x_1, x_2, and x_3, arbitrarily, say $(x_1, x_2, x_3) = (-2, 0, 1)$ and in (1) calculate $(y_1, y_2) = (1, -3)$. Then $(x_1, x_2, x_3, y_1, y_2) = (-2, 0, 1, 1, -3)$ is a solution of (1). Now check it in (3):

-1	0	3		
$-7/8$	$1/2$	$3/8$	$= 2$	✓
$5/8$	$1/2$	$-1/8$	$= -1$	✓

and also in (5),

3	-1	1		
$-1/4$	$5/4$	2	$= 0$	✓
$1/2$	$-3/2$	-1	$= 2$	✓

EXAMPLE 3

Exchange as many x's with y's as possible in (1):

(1)

x_1	x_2	
-3	1	$= y_1$
7	-1^*	$= y_2$
32	-5	$= y_3$

(2)

x_1	$-y_2$	
4^*	1	$= y_1$
-7	-1	$= -x_2$
-3	-5	$= y_3$

(3)

	$-y_1$	$-y_2$	
	$\frac{1}{4}$	$\frac{1}{4}$	$= -x_1$
	$\frac{7}{4}$	$\frac{3}{4}$	$= -x_2$
	$\frac{3}{4}$	$-\frac{17}{4}$	$= y_3$

Let $(x_1, x_2) = (1, 2)$. From (1) $(y_1, y_2, y_3) = (-1, 5, 22)$. Thus $(1, 2, -1, 5, 22)$ is a solution of (1). Checking in (3) yields

1	-5	
$\frac{1}{4}$	$\frac{1}{4}$	$= -1$
$\frac{7}{4}$	$\frac{3}{4}$	$= -2$
$\frac{3}{4}$	$-\frac{17}{4}$	$= 22$

which is true.

EXAMPLE 4

Obtain a fully exchanged tableau for

(1)

x_1	x_2	x_3	
2	3	-1	$= y_1$
1*	2	-1	$= y_2$
4	7	-3	$= y_3$

Pivoting as indicated yields

(2)

$-y_2$	x_2	x_3	
-2	-1	1*	$= y_1$
1	2	-1	$= -x_1$
-4	-1	1	$= y_3$

In (3) the only possible pivot position is occupied by 0. Thus (3) is a fully exchanged tableau for (1).

(3)

$-y_2$	x_2	$-y_1$	
-2	-1	1	$= -x_3$
-1	1	1	$= -x_1$
-2	0	-1	$= y_3$

No matter how we select pivots in trying to obtain a fully exchanged tableau, the solution set is unaltered. But, one may ask, how many pivot exchanges are needed to obtain a fully exchanged tableau? Is that number of exchanges the same for every choice of pivots? The answer to the second question is: yes. If pivots are chosen so that no two are in the same row or column, the number of exchanges is the same. We shall prove this assertion in Section 2.5. Meanwhile we assume it to be true and call that number of exchanges the *pivot rank* of the initial tableau. The pivot rank of tableau (1) in Example 1 is 3. The pivot rank of each of the initial tableaus in Examples 2, 3, and 4 is 2.

EXAMPLE 5

Find a fully exchanged tableau for and the pivot rank of

(1)

w	x	y	z	
4	-4	5	2	$= P$
1	5	2^*	-4	$= Q$
-6	-10	-9	8	$= R$
6	-2	8	0	$= S$
-3	-9	-5	7	$= T$

Pivoting as indicated yields

(2)

w	x	$-Q$	z	
$3/2$	$-33/2$	$-5/2$	12	$= P$
$1/2$	$5/2$	$1/2$	-2	$= -y$
$-3/2$	$25/2$	$9/2$	-10	$= R$
2	-22	-4	16	$= S$
$-1/2$	$7/2$	$5/2$	-3^*	$= T$

(3)

w	x	$-Q$	$-T$		
$-1/2$	$-5/2$	$15/2$	4	$=$	P
$5/6$	$1/6$	$-7/6$	$-2/3$	$=$	$-y$
$1/6$	$5/6$	$-23/6$	$-10/3$	$=$	R
$-2/3$*	$-10/3$	$28/3$	$16/3$	$=$	S
$1/6$	$-7/6$	$-5/6$	$-1/3$	$=$	$-z$

(4)

$-S$	x	$-Q$	$-T$		
$-3/4$	0	$1/2$	0	$=$	P
$5/4$	-4	$21/2$	6	$=$	$-y$
$1/4$	0	$-3/2$	-2	$=$	R
$-3/2$	5	-14	-8	$=$	$-w$
$1/4$	-2	$3/2$	1	$=$	$-z$

In (4) a pivot would have to be selected from row 1 or 3 and column 2. But the entries in row 1, column 2, and in row 3, column 2, are both zero. No further exchanges are possible. A fully exchanged tableau for (1) is (4) and the pivot rank of (1) is 3.

We sum up this section with a listing of the procedure to carry out the fundamental pivot-exchange algorithm.

1. Select as pivot any nonzero number in the inside of the initial tableau, exchange the corresponding basic and nonbasic variables, and carry out the pivot procedure.
2. If there are other pairs of variables to exchange, select the next pivot. This pivot comes from the intersection of a row and a column of variables not previously exchanged. Proceed as in step 1.
3. Repeat step 2 as often as possible.

The final tableau is called a *fully exchanged tableau* and the number of pivots is the pivot rank of the initial tableau.

The fundamental pivot-exchange algorithm has many uses. These uses emerge as we continue. In the next section it is used to help resolve the second of the four-towns problems.

Exercises

1. Let

(1)

x_1	x_2	
1	2	$= y_1$
3	2	$= y_2$

be a system of linear equations.

(a) Show that $(x_1, x_2, y_1, y_2) = (2, -1, 0, 4)$ is a solution of (1).

(b) Which pivot requires the exchange of x_1 and y_1? Perform the exchange and call the new system (2).

(c) Check the solution in (a) in the tableau (2).

(d) In (2), exchange x_2 and y_2 resulting in (3). Check the solution in (a) in (3).

2. Using (1) in Exercise 1 and checking the solution at each pivot exchange:

(a) Exchange x_1 and y_2.

(b) Then exchange x_2 and y_1.

(c) Show that the fully exchanged tableaus in Exercises 1 and 2 are the same.

(d) What is the pivot rank of (1)?

3. Consider

(1)

x_1	x_2	
1	2	$= y_1$
2	-1	$= y_2$
3	4	$= y_3$

(a) Show that $(x_1, x_2, y_1, y_2, y_3) = (1, -2, -3, 4, -5)$ is a solution of (1).

(b) Find a fully exchanged tableau of (1) in which y_3 remains a basic variable. Check each pivot exchange with the solution in (a).

(c) Find a fully exchanged tableau of (1) for which y_2 is a basic variable. Check each pivot exchange with the solution in (a).

(d) Find a fully exchanged tableau of (1) for which y_1 is a basic variable.

(e) What is the pivot rank of (1)?

4. Consider

(1)

x_1	x_2	x_3	
1	2	-1	$= y_1$
2	1	2	$= y_2$

(a) Find the solution of (1) for which $(x_1, x_2, x_3) = (2, -1, 3)$.
(b) Find a fully exchanged tableau of (1).

5. Show that

x	y	
1	−2	$= u$
−3	6	$= v$
−12	24	$= w$

has pivot rank 1.

6. Find the pivot rank of

x_1	x_2	x_3	
1	2	3	$= y_1$
2	3	−1	$= y_2$
3	5	2	$= y_3$

7. Show that the numbers inside the tableau

x_1	x_2	x_3	x_4	
1	0	0	2	$= y_1$
0	1	0	3	$= y_2$
0	0	1	4	$= y_3$

are unaltered by pivot exchange on any of the 1's.

8. Let

(1)

x_1	x_2	x_3	x_4	x_5	
1	0	0	0	a	$= y_1$
0	1	0	0	b	$= y_2$
0	0	1	0	c	$= y_3$
0	0	0	1	d	$= y_4$

be a system of linear equations. Show that there is a fully exchanged tableau of (1) whose entries are the same, in each location, as those of (1).

9. A fully exchanged tableau of a system is

$-y_1$	$-y_2$	x_3	
1	2	3	$= -x_1$
2	1	-1	$= -x_2$
3	4	0	$= y_3$

Find the initial tableau of that system (the x's are the initial nonbasic variables).

2.3 Image Sets

In connection with the second four-towns problem, we asked: Which 6-tuple, if any, has an image $(y_A, y_B, y_C, y_D) = (3, -6, 1, 1)$? In other words, which shipping schedule results in weekly net receipts (in \$1000 units) of 3 for A, -6 for B, 1 for C, and 1 for D? In an attempt to answer this question we used the fundamental pivot-exchange algorithm on (1) to arrive at a fully exchanged tableau (2). (See Figure 2.2.)

(1)

x_1	x_2	x_3	x_4	x_5	x_6	
1	1	-1	0	0	0	$= y_A$
-1	0	0	1	-1	0	$= y_B$
0	-1	0	-1	0	1	$= y_C$
0	0	1	0	1	-1	$= y_D$

(2)

$-y_A$	$-y_B$	x_3	x_4	$-y_D$	x_6	
0	-1	-1	-1	-1	1	$= -x_1$
1	1	0	1	1	-1	$= -x_2$
1	1	0	0	1	0	$= y_C$
0	0	1	0	1	-1	$= -x_5$

If y_A is replaced by 3, y_B by -6, y_C by 1, and y_D by 1, the third row reads 2 = 1, which is false. This means that

$$(x_1, x_2, x_3, x_4, x_5, x_6, y_A, y_B, y_C, y_D) = (x_1, x_2, x_3, x_4, x_5, x_6, 3, -6, 1, 1)$$

is not a solution of the system. Another way to say this is: There is no 6-tuple $(x_1, x_2, x_3, x_4, x_5, x_6)$ whose image is $(3, -6, 1, 1)$. Symbolically,

$$(x_1, x_2, x_3, x_4, x_5, x_6) \not\mapsto (3, -6, 1, 1)$$

It is convenient, at this time, to introduce another way to say the same thing. We say that $(3, -6, 1, 1)$ has no *preimage*. Using this language we assert that $(3, -5, 1, 1)$ does have a preimage—in fact, many of them.

Substituting $(3, -5, 1, 1)$ in (2) yields

(3)

-3	5	x_3	x_4	-1	x_6	
0	-1	-1	-1	-1	1	$= -x_2$
1	1	0	1	1	-1	$= -x_2$
1	1	0	0	1	0	$= 1$
0	0	1	0	1	-1	$= -x_5$

Now select any numbers for x_3, x_4, and x_6. This is allowable since these variables are nonbasic. If $x_3 = 1$, $x_4 = 2$, and $x_6 = -1$ we have

(4)

-3	5	1	2	-1	-1	
0	-1	-1	-1	-1	1	$= -x_1$
1	1	0	1	1	-1	$= -x_2$
1	1	0	0	1	0	$= 1$
0	0	1	0	1	-1	$= -x_5$

From (4) we deduce that

$$x_1 = 8, \qquad x_2 = -4, \qquad x_5 = -1$$

Thus

$$(8, -4, 1, 2, -1, -1) \mapsto (3, -5, 1, 1)$$

That

$$(2, \quad 3, 2, -1, \quad 2, \quad 3) \mapsto (3, -5, 1, 1)$$

is seen from (3) by substituting $x_3 = 2$, $x_4 = -1$, and $x_6 = 3$ and then calculating x_1, x_2, and x_5. Therefore, we have two preimages of (3, −5, 1, 1). The collection of all preimages of (3, −5, 1, 1) is called the ***preimage set*** of (3, −5, 1, 1) for the tableau (1). It is advisable to check preimages in (1). Thus

8	−4	1	2	−1	−1			
1	1	−1	0	0	0	=	3	✓
−1	0	0	1	−1	0	=	−5	✓
0	−1	0	−1	0	1	=	1	✓
0	0	1	0	1	−1	=	1	✓

2	3	2	−1	2	3			
1	1	−1	0	0	0	=	3	✓
−1	0	0	1	−1	0	=	−5	✓
0	−1	0	−1	0	1	=	1	✓
0	0	1	0	1	−1	=	1	✓

The decision as to which 4-tuples (y_A, y_B, y_C, y_D) have preimages is made by examining row 3 of (2). y_A, y_B, and y_D can be selected arbitrarily since they are nonbasic variables. y_C is then fixed by row 3. Any other selection of y_C other than the one determined by row 3 will result in a 4-tuple that has no preimage. This is another way of saying that a 4-tuple must satisfy row 3 to be the image of a 6-tuple. The set of all images is called the *image set*. The image set of tableau (1) does not depend on the x's. (y_A, y_B, y_C, y_D) is the image of some 6-tuple provided the y's satisfy row 3 of (2). Since there are zeros in row 3 in all x columns, row 3 is equivalent to

(5)

$-y_A$	$-y_B$	$-y_D$		
1	1	1	=	y_C

Thus any 4-tuple of y's satisfying (5) is in the image set of (1).

To describe precisely an image set it is useful to use symbols. In the case of the above tableau the image set is

$\{(y_A, y_B, y_C, y_D):$

$-y_A$	$-y_B$	$-y_D$	
1	1	1	$= y_C\}$

Letting { } be read "the set of all" and ":" read "such that," the symbolism is read "The set of all ordered 4-tuples (y_A, y_B, y_C, y_D) such that

$-y_A$	$-y_B$	$-y_D$	
1	1	1	$= y_C$

EXAMPLE 1

Find the image set of

(1)

x_1	x_2	x_3	
1	0	1	$= y_1$
0	1	-1	$= y_2$
2	2	0	$= y_3$

To obtain a fully exchanged tableau of (1), pivot as indicated:

(1)

x_1	x_2	x_3	
1*	0	1	$= y_1$
0	1	-1	$= y_2$
2	2	0	$= y_3$

(2)

$-y_1$	x_2	x_3	
1	0	1	$= -x_1$
0	1*	-1	$= y_2$
-2	2	-2	$= y_3$

(3)

$-y_1$	$-y_2$	x_3	
1	0	1	$= -x_1$
0	1	-1	$= -x_2$
-2	-2	0	$= y_3$

(4)

$-y_1$	$-y_2$	
-2	-2	$= y_3$

Tableau (3) is fully exchanged. Row 3 defines the relationship that must exist among the y's which are components of the 3-tuples (y_1, y_2, y_3) of the image set of (1). The auxiliary tableau (4) shows this relationship without the x_3 column. The x_3 column does not affect (4) since there is a 0 in the y_3 row where it meets the x_3 column. The image set of (1), in set language rather than in the form (4), is

	$-y_1$	$-y_2$	
$\{(y_1, y_2, y_3)$:	-2	-2	$= y_3\}$

y_1, and y_2 are the nonbasic variables in (3) and (4) and they determine y_3. Some 3-tuples in the image set are (1, 1, 4), (2, −5, −6), and (3, −1, 4), as can be verified in (4).

It is instructive to solve the problem in Example 1 choosing other pivots. For instance, pivoting as indicated below starting with the initial tableau (1) yields the fully exchanged tableau (6).

(1)

x_1	x_2	x_3	
1	0	1*	$= y_1$
0	1	−1	$= y_2$
2	2	0	$= y_3$

(5)

x_1	x_2	$-y_1$	
1	0	1	$= -x_3$
1	1	1	$= y_2$
2*	2	0	$= y_3$

(6)

$-y_3$	x_2	$-y_1$	
$-\frac{1}{2}$	−1	1	$= -x_3$
$-\frac{1}{2}$	0	1	$= y_2$
$\frac{1}{2}$	1	0	$= -x_1$

(7)

$-y_3$	$-y_1$	
$-\frac{1}{2}$	1	$= y_2$

The second row of (6) rewritten as (7) displays the relationship among the y's that is necessary for (y_1, y_2, y_3) to qualify as a member of the image set of the initial tableau (1). Tableaus (7) and (4) are clearly equivalent, and thus (as we knew) the selection of pivots does not affect the image set.

From (4) it is easy to determine if a triple (y_1, y_2, y_3) is a member of the image set. For instance: If $(y_1, y_2, y_3) = (2, -\frac{1}{2}, 3)$, we have

-2	$\frac{1}{2}$	
-2	-2	$= 3$

which is true. Therefore $(2, -\frac{1}{2}, 3)$ is in the image set of (1).

If $(y_1, y_2, y_3) = (2, 1, 3)$, then we have

-2	-1	
-2	-2	$= 3$

which is false. Therefore, $(2, 1, 3)$ is not in the image set of (1).

Compare the pivot rank of (1) and the number of nonbasic variables in (4).

EXAMPLE 2

Find the image set of

(1)

x_1	x_2	x_3	
1	-4	3	$= y_1$
-2	8	-6	$= y_2$
2	-8	6	$= y_3$

A fully exchanged tableau is obtained from the initial tableau (1), as indicated.

(1)

x_1	x_2	x_3		
1*	−4	3	=	y_1
−2	8	−6	=	y_2
2	−8	6	=	y_3

(2)

$-y_1$	x_2	x_3		
1	−4	3	=	$-x_1$
2	0	0	=	y_2
−2	0	0	=	y_3

Rows 2 and 3 of (2) give the conditions that must be satisfied by (y_1, y_2, y_3) to qualify for membership in the image set of (1). They are displayed in the auxiliary tableau

(3)

$-y_1$		
2	=	$-y_2$
−2	=	y_3

The image set of (1) in set language is

$$\left\{(y_1, y_2, y_3): \begin{array}{c|cc} -y_1 & & \\ \hline 2 & = & -y_2 \\ -2 & = & -y_3 \end{array}\right\}$$

When y_1, the nonbasic variable, is assigned a value, y_2 and y_3 are determined. Compare the pivot rank of (1) and the number of nonbasic variables in (3).

EXAMPLE 3

Find the image set of

(1)

x_1	x_2		
3	1	=	y_1
1	−1	=	y_2
2	0	=	y_3
0	3	=	y_4

Pivoting as indicated yields the fully exchanged tableau (3).

(1)

	x_1	x_2		
	3	1*	=	y_1
	1	-1	=	y_2
	2	0	=	y_3
	0	3	=	y_4

(2)

	x_1	$-y_1$		
	3	1	=	$-x_2$
	4	1	=	y_2
	2*	0	=	y_3
	-9	-3	=	y_4

(3)

	$-y_3$	$-y_1$		
	$-3/2$	1	=	$-x_2$
	-2	1	=	y_2
	$1/2$	0	=	$-x_1$
	$9/2$	-3	=	y_4

The auxiliary tableau that must be satisfied for (y_1, y_2, y_3, y_4) to be in the image is

(4)

	$-y_3$	$-y_1$		
	-2	1	=	y_2
	$9/2$	-3	=	y_4

Arbitrary choices for y_1 and y_3 lead to members of the image set of (1). Compare the pivot rank of the initial tableau (1) and the number of nonbasic variables in (4).

EXAMPLE 4

Find the image set of

	x_1	x_2	
	2	3	$= y_1$
	4	3	$= y_2$

Pivot as indicated:

(1)

x_1	x_2	
2*	3	$= y_1$
4	3	$= y_2$

(2)

$-y_1$	x_2	
$\frac{1}{2}$	$\frac{3}{2}$	$= -x_1$
-2	-3*	$= y_2$

(3)

$-y_1$	$-y_2$	
$-\frac{1}{2}$	$\frac{1}{2}$	$= -x_1$
$\frac{2}{3}$	$-\frac{1}{3}$	$= -x_2$

All y's in the fully exchanged tableau (3) are nonbasic variables. Therefore, for any values of y_1 and y_2, (y_1, y_2) has a preimage. All (y_1, y_2) are images, and thus the image set of (1) is the set of all ordered pairs of real numbers. This set is denoted R^2 (read R two).

EXAMPLE 5

Find the image set of

(1)

w	x	y	z	
4	-4	5	2	$= P$
1	5	2*	-4	$= Q$
-6	-10	-9	8	$= R$
6	-2	8	0	$= S$
-3	-9	-5	7	$= T$

(See Example 5, Section 2.2.)
A fully exchanged tableau is

(2)

$-S$	x	$-Q$	$-T$	
$-3/4$	0	$1/2$	0	$= P$
$5/4$	-4	$21/2$	6	$= -y$
$1/4$	0	$-3/2$	-2	$= R$
$-3/2$	5	-14	-8	$= -w$
$1/4$	-2	$3/2$	1	$= -z$

The image set is determined by the equations that relate P, Q, R, S, and T. These are

(3)

$-S$	$-Q$	$-T$	
$-3/4$	$1/2$	0	$= P$
$1/4$	$-3/2$	-2	$= R$

The image set is

$\left\{ (P, Q, R, S, T) : \right.$

$-S$	$-Q$	$-T$	
$-3/4$	$1/2$	0	$= P$
$1/4$	$-3/2$	2	$= R$

$\left. \right\}$

Note the number of nonbasic variables and the number of equations in (5) and the pivot rank of the initial tableau.

The following theorem summarizes the results of the five examples.

THEOREM 2.1. If a tableau has m basic variables and pivot rank r, then its image set is defined by $m - r$ equations having $m - r$ basic variables and r nonbasic variables.

If $m = r$, the image set is the set of all m-tuples. We denote this set by the symbol R^m.

Proof: After exchanging r distinct pairs of variables x and y, there are $m - r$ y's at the right and r y's at the top. Moreover, only zeros appear in the fully exchanged version of the initial tableau, where the columns of the remaining x's cross the rows of the remaining y's. (Otherwise, the pivot rank of the initial tableau would be greater than r.) These zeros eliminate the x's from the equations expressing the $m - r$ y's at the right in terms of the r y's at the top. The image set of the initial tableau is defined by these $m - r$ equations. They have $m - r$ basic variables and r nonbasic variables. If $m = r$, then all y's are at the top and the image set is the set of all m-tuples.

Exercises

Find the image set for each of the tableaus given in Exercises 1–10.

1.

x	y	
2	3	$= u$
4	6	$= v$

2.

x	y	z	
2	3	4	$= u$
6	9	12	$= v$

3.

x	y	
2	3	$= z$
1	2	$= w$

4.

x_1	x_2	x_3	
1	3	1	$= y_1$
−1	2	3	$= y_2$
1	8	5	$= y_3$

5.

x_1	x_2	
1	2	$= y_1$
0	3	$= y_2$
0	4	$= y_3$

6.

x_1	x_2	
1	2	$= y_1$
−1	2	$= y_2$
2	3	$= y_3$
3	4	$= y_4$

7.

x_1	x_2	x_3	
1	0	0	$= y_1$
0	2	0	$= y_2$
0	0	3	$= y_3$
1	2	3	$= y_4$

8.

x_1	x_2	x_3	
0	1	1	$= y_1$
2	0	2	$= y_2$
3	3	0	$= y_3$

9.

x_1	x_2	x_3	x_4	
1	2	3	4	$= y_1$
4	3	2	1	$= y_2$

10.

x_1	x_2	x_3	
1	$2/3$	$-1/3$	$= y_1$
$2/3$	$1/3$	$7/9$	$= y_2$
-3	0	1	$= y_3$

11. A fully exchanged tableau of an initial tableau is

$-y_1$	x_2	$-y_3$	
1	2	3	$= -x_1$
4	0	-5	$= y_2$
6	7	8	$= -x_3$

(a) Describe the image set of the initial tableau in set language.
(b) Determine if each of the following is in the image set of the tableau: (0, 0, 0), (1, 1, 1), (2, 2, 2).

12. A fully exchanged tableau of an initial tableau is

x_1	$-y_3$	$-y_1$	x_4	
2	1	0	5	$= -x_3$
0	3	-2	0	$= y_2$
3	0	4	2	$= -x_2$
0	4	-1	0	$= y_4$

(a) Describe the image set of the initial tableau.
(b) Find three members of the image set.
(c) Find a 4-tuple that is not in the image set.

13. Given any tableau with n nonbasic variables and m basic variables, show that the zero m-tuple is in the image set of the tableau.

2.4 Kernel Sets

In Section 2.1, in connection with the four-towns problems, we saw that $(y_A, y_B, y_C, y_D) = (3, -5, 1, 1)$ has a preimage, in fact many of them, and promised to pursue this further. We now keep that promise. Again we rely on a fully reduced tableau of that problem, in search of preimages of

(3, −5, 1, 1). The initial tableau (see Figure 2.2) is

(1)

x_1	x_2	x_3	x_4	x_5	x_6	
1	1	−1	0	0	0	$= y_A$
−1	0	0	1	−1	0	$= y_B$
0	−1	0	−1	0	1	$= y_C$
0	0	1	0	1	−1	$= y_D$

A fully reduced tableau is

(2)

$-y_A$	$-y_B$	x_3	x_4	$-y_D$	x_6	
0	−1	−1	−1	−1	1	$= -x_1$
1	1	0	1	1	−1	$= -x_2$
1	1	0	0	1	0	$= y_C$
0	0	1	0	1	−1	$= -x_5$

When y_A is replaced by 3, y_B by −5, y_C by 1, and y_D by 1, we get

(3)

−3	5	x_3	x_4	−1	x_6	
0	−1	−1	−1	−1	1	$= -x_1$
1	1	0	1	1	−1	$= -x_2$
1	1	0	0	1	0	$= 1$
0	0	1	0	1	−1	$= -x_5$

The first equation of (3),

$$-5 - x_3 - x_4 + 1 + x_6 = -x_1$$

can be simplified to

$$-x_3 - x_4 + x_6 - 4 = -x_1$$

The second can be simplified to

$$x_4 - x_6 + 1 = -x_2$$

The fourth can be simplified to

$$x_3 - x_6 - 1 = -x_5$$

These constitute the auxiliary tableau

(4)

x_3	x_4	x_6	1	
−1	−1	1	−4	$= -x_1$
0	1	−1	1	$= -x_2$
1	0	−1	−1	$= -x_5$

The solution set of (4) satisfies the initial tableau (1) when $(y_A, y_B, y_C, y_D) = (3, -5, 1, 1)$. That is, every 6-tuple that satisfies (4) maps to $(3, -5, 1, 1)$. Therefore, the set of preimages of $(3, -5, 1, 1)$ is the set of all $(x_1, x_2, x_3, x_4, x_5, x_6)$ that satisfy (4). Mathematicians are interested in finding the preimage set of $(0, 0, 0, 0)$. There is a special name for this preimage set. If T is a tableau with n nonbasic variables and m basic variables, then the preimage set of the zero m-tuple is called the *kernel set* of T. Figure 2.3 pictorially represents the kernel and image sets of a tableau T.

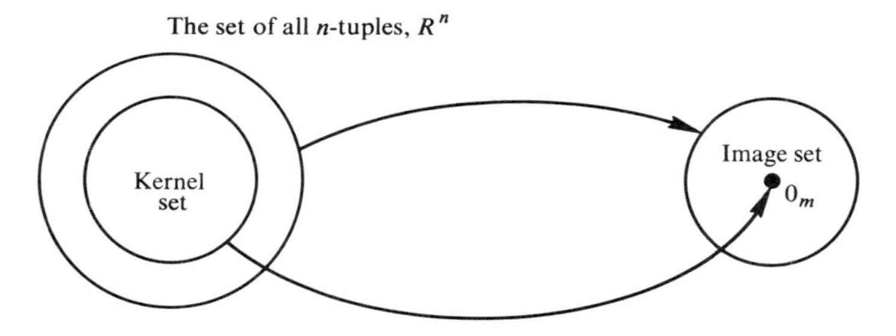

Figure 2.3

Economists studying the four-towns problems are interested in the kernel set because it translates into shipping schedules for which $y_A = y_B = y_C = y_D = 0$. (Each town breaks even.)

To find the kernel of the four-towns tableau, replace y_A by 0, y_B by 0, y_C by 0, and y_D by 0 in the fully exchanged tableau (2). The result is

(5)

0	0	x_3	x_4	0	x_6	
0	−1	−1	−1	−1	1	$= -x_1$
1	1	0	1	1	−1	$= -x_2$
1	1	0	0	1	0	$= 0$
0	0	1	0	1	−1	$= -x_5$

Deleting the columns that are labeled 0 and row 3, which has 0's in all x positions, leaves the auxiliary tableau

(6)

x_3	x_4	x_6	
-1	-1	1	$= -x_1$
0	1	-1	$= -x_2$
1	0	-1	$= -x_5$

The kernel of the four-towns tableau is the set of all 6-tuples $(x_1, x_2, x_3, x_4, x_5, x_6)$ satisfying (6). If $x_3 = 2$, $x_4 = 1$, and $x_6 = 1$ we get

(7)

2	1	1	
-1	-1	1	$= -x_1$
0	1	-1	$= -x_2$
1	0	-1	$= -x_5$

from which $x_1 = 2$, $x_2 = 0$, $x_5 = -1$. Thus $(2, 0, 2, 1, -1, 1)$ is in the kernel. Substituting this 6-tuple in (1) gives $y_A = 0$, $y_B = 0$, $y_C = 0$, $y_D = 0$, which checks the arithmetic. The auxiliary tableau for this kernel set has three equations, with three nonbasic variables. The pivot rank of the initial tableau is 3. In the examples that follow, relations among these numbers unfold.

EXAMPLE 1

Find the kernel set of

(1)

x_1	x_2	x_3	
1	0	1	$= y_1$
0	1	-1	$= y_2$
2	2	0	$= y_3$

This is the tableau of Example 1 of Section 2.3. A fully exchanged tableau, found after two pivot exchanges, is

(2)

$-y_1$	$-y_2$	x_3	
1	0	1	$= -x_1$
0	1	-1	$= -x_2$
-2	-2	0	$= \ y_3$

To find the kernel set let $(y_1, y_2, y_3) = (0, 0, 0)$. This yields

(3)

0	0	x_3	
1	0	1	$= -x_1$
0	1	-1	$= -x_2$
-2	-2	0	$= \ 0$

The kernel set of (1) is the solution set of the auxiliary tableau

(4)

x_3	
1	$= -x_1$
-1	$= -x_2$

This tableau reads $x_1 = -x_3$ and $x_2 = x_3$. Replacing x_1 with $-x_3$, x_2 with x_3, y_1 with 0, y_2 with 0, and y_3 with 0 in the initial tableau yields

$-x_3$	x_3	x_3		
1	0	1	$= 0$	✓
0	1	-1	$= 0$	✓
2	2	0	$= 0$	✓

which is true for all (x_1, x_2, x_3). This is a check that (3) defines the kernel of the initial tableau (1). From (4) $(x_1, x_2, x_3) = (-1, 1, 1)$ and $(x_1, x_2, x_3) = (2, -2, -2)$ are in the kernel. The reader should check by substituting these 3-tuples into (1).

EXAMPLE 2

Find the kernel set of

(1)

x_1	x_2	x_3	
1*	−4	3	$= y_1$
−2	8	−6	$= y_2$
2	−8	6	$= y_3$

(See Example 2 of Section 2.3.)
A fully exchanged tableau

(2)

$-y_1$	x_2	x_3	
1	−4	3	$= -x_1$
2	0	0	$= y_2$
−2	0	0	$= y_3$

is obtained by one pivot exchange. Let $(y_1, y_2, y_3) = (0, 0, 0)$. The auxiliary tableau is

(3)

x_2	x_3	
−4	3	$= -x_1$

The kernel is

$\{(x_1, x_2, x_3)$:

x_2	x_3	
−4	3	$= -x_1\}$

If $x_2 = 3, x_3 = -7$, then from (3) we see that $(33, 3, -7)$ is in the kernel.

EXAMPLE 3

Find the kernel set of the tableau

(1)

x_1	x_2	
3	1	$= y_1$
1	−1	$= y_2$
2	0	$= y_3$
0	3	$= y_4$

(See Example 3, Section 2.3.)

A fully exchanged tableau is found after two pivot exchanges:

(2)

$-y_3$	$-y_1$	
$-3/2$	1	$= -x_2$
-2	1	$= y_2$
$1/2$	0	$= -x_1$
$9/2$	-3	$= y_4$

Let $(y_1, y_2, y_3, y_4) = (0, 0, 0, 0)$. The kernel consists of $(x_1, x_2) = (0, 0)$.

EXAMPLE 4

Find the kernel set of the tableau

(1)

w	x	y	z	
4	-4	5	2	$= P$
1	5	2	-4	$= Q$
-6	-10	-9	8	$= R$
6	-2	8	0	$= S$
-3	-9	-5	7	$= T$

(See Example 5 of Section 2.2.)

After three pivots a fully exchanged tableau is

(2)

$-S$	x	$-Q$	$-T$	
$-3/4$	0	$1/2$	0	$= P$
$5/4$	-4	$21/2$	6	$= -y$
$1/4$	0	$-3/2$	-2	$= R$
$-3/2$	5	-14	-8	$= -w$
$1/4$	-2	$3/2$	1	$= -z$

The kernel is the preimage set of $(P, Q, R, S, T) = (0, 0, 0, 0, 0)$. In the usual manner, deleting the columns headed S, Q, T, and rows P and R, we are left with

(3)

x	
-4	$= -y$
5	$= -w$
-2	$= -z$

which defines the kernel. The kernel set is

$$\left\{(x, y, z, w): \begin{array}{|c|l} \multicolumn{1}{c}{x} & \\ \hline -4 & = -y \\ 5 & = -w \\ -2 & = -z \\ \hline \end{array}\right\}$$

Check: Replace y, w, and z in the initial tableau by $4x$, $-5x$, and $2x$, respectively, and each of P, Q, R, S, and T by 0.

$-5x$	x	$4x$	$2x$		
4	-4	5	2	$= 0$	✓
1	5	2	-4	$= 0$	✓
-6	-10	-9	8	$= 0$	✓
6	-2	8	0	$= 0$	✓
-3	-9	-5	7	$= 0$	✓

Each row is true for all values of x.

The following theorem summarizes the results.

THEOREM 2.2. Let T be a tableau with n nonbasic variables and m basic variables and pivot rank r. Then the kernel set is defined by r equations having r basic variables and $n - r$ nonbasic variables. If $n = r$, the kernel contains only the zero n-tuple $(0, 0, \ldots, 0)$.

Proof: After exchanging r distinct pairs of variables x and y there are r x's at the right of the tableau and $(n - r)$ x's remaining at the top. At this stage, setting all y's at the top equal to zero results in zero values for the y's remaining at the right. Therefore, $(y_1, y_2, \ldots, y_m)$, equal to the zero m-tuple, implies a relation between the x's at the right and the x's remaining at the top. These relations are the r equations containing

r basic variables and $(n - r)$ nonbasic variables. If $n = r$, there are no nonbasic x's at the top and the kernel contains only the zero n-tuple.

Exercises

Find the kernel set of each of the tableaus in Exercises 1–13.

1.

x	y	
3	5	$= z$
1	2	$= w$

2.

x	y	
1	3	$= z$
2	6	$= w$

3.

x_1	x_2	x_3	
2	1	3	$= y_1$
3	0	2	$= y_2$

4.

x_1	x_2	
1	2	$= y_1$
3	8	$= y_2$
4	1	$= y_3$

5.

x_1	x_2	x_3	
2	-4	-6	$= y_1$
-1	2	3	$= y_2$
1	-2	-3	$= y_3$

6.

x_1	x_2	x_3	
1	-2	4	$= y_1$
3	-5	2	$= y_2$
-1	2	-4	$= y_3$

7.

x_1	x_2	x_3	
1	0	3	$= y_1$
3	1	4	$= y_2$
-2	3	-21	$= y_3$

8.

x_1	x_2	x_3	
1	2	4	$= y_1$
3	5	7	$= y_2$
6	8	9	$= y_3$

9.

x_1	x_2	x_3	
1	2	2	$= y_1$
2	1	4	$= y_2$
-1	-1	2	$= y_3$
2	2	4	$= y_4$

10.

x_1	x_2	x_3	x_4	
2	1	0	3	$= y_1$
1	4	2	−2	$= y_2$

11.

x_1	x_2	
1	5	$= y_1$
2	6	$= y_2$
3	7	$= y_3$
4	8	$= y_4$

12.

x_1	x_2	
1	2	$= y_1$

13.

x_1	
2	$= y_1$
3	$= y_2$
4	$= y_3$

14. Let T be a tableau with n nonbasic variables and n basic variables. Prove each of the following:

(a) The image set of T is R^n, the set of all n-tuples, if its kernel contains only the zero n-tuple.

(b) The kernel of T contains only the zero n-tuple if the image set of T is R^n.

(c) If the image set of T is defined by $n - r$ equations with r nonbasic variables, then the kernel of T is defined by r equations with $n - r$ nonbasic variables.

2.5 Uniqueness of Pivot Rank

There is a nice device by which one displays the image set and the kernel set of a tableau. We illustrate with our old friend

(1)

x_1	x_2	x_3	x_4	
4	−4	5	2	$= y_1$
1	5	2	−4	$= y_2$
−6	−10	−9	8	$= y_3$
6	−2	8	0	$= y_4$
−3	−9	−5	7	$= y_5$

After three pivot exchanges we obtain a fully exchanged tableau:

(2)

$-y_4$	x_2	$-y_2$	$-y_5$		
$-3/4$	0	$1/2$	0	=	y_1
$5/4$	-4	$21/2$	6	=	$-x_3$
$1/4$	0	$-3/2$	-2	=	y_3
$-3/2$	5	-14	-8	=	$-x_1$
$1/4$	-2	$3/2$	1	=	$-x_4$

Rearranging rows, which we know does not affect solutions, yields

(3)

$-y_4$	x_2	$-y_2$	$-y_5$		
$-3/2$	5	-14	-8	=	$-x_1$
$5/4$	-4	$21/2$	6	=	$-x_3$
$1/4$	-2	$3/2$	1	=	$-x_4$
$-3/4$	0	$1/2$	0	=	y_1
$1/4$	0	$-3/2$	-2	=	y_3

Now rearranging columns (also allowable) yields

(4)

$-y_2$	$-y_4$	$-y_5$	x_2		
-14	$-3/2$	-8	5	=	$-x_1$
$21/2$	$5/4$	6	-4	=	$-x_3$
$3/2$	$1/4$	1	-2	=	$-x_4$
$1/2$	$-3/4$	0	0	=	y_1
$-3/2$	$1/4$	-2	0	=	y_3

The advantage of (4) is the simplicity of presentation of our discoveries. The dashed lines are for emphasis. We list these discoveries in terms of (4).

1. The kernel set of the initial tableau (1) is defined by the last column and the first three rows.
2. The image set is defined by the first three columns and the last two rows.

Any fully exchanged tableau of any initial tableau can be put into this format:

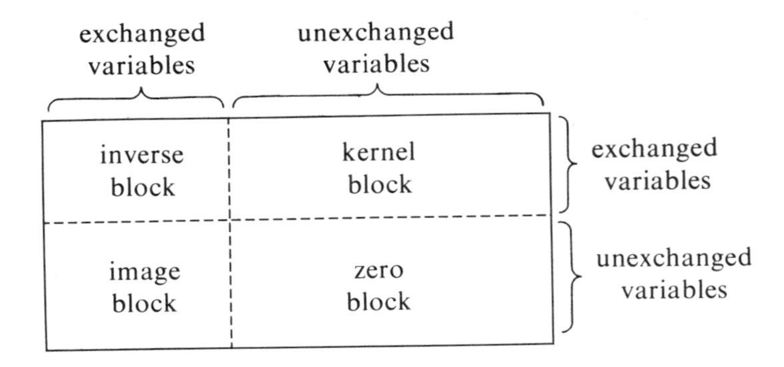

The detailed procedure for any initial tableau with n nonbasic and m basic variables is as follows:

1. Start with an initial tableau.
2. Using the fundamental pivot-exchange algorithm, obtain a fully exchanged tableau.
3. Rearrange the rows of the fully exchanged tableau, setting the unexchanged variables at the bottom.
4. Rearrange the columns of the fully exchanged tableau with the unexchanged variables set at the right.
5. Read the kernel and image sets of the initial tableau as indicated in the above schema.
6. The block marked "inverse block" indicates a portion of the initial tableau which is invertible. This is discussed in Section 2.6.

From the schematic diagram we see that

(a) If none of the original m basic variables are left at the right, the image set is R^m, the set of all m-tuples.
(b) If none of the original n nonbasic variables are left on top, the kernel set consists only of the n-tuple of all zeros.
(c) The case when both (a) and (b) are true is the subject of Section 2.6.

The schema is useful in demonstrating (as promised) that every tableau T has a unique pivot rank. We demonstrate with an initial tableau that has seven nonbasic and six basic variables. We write the numbers on the inside of the tableau only when necessary.

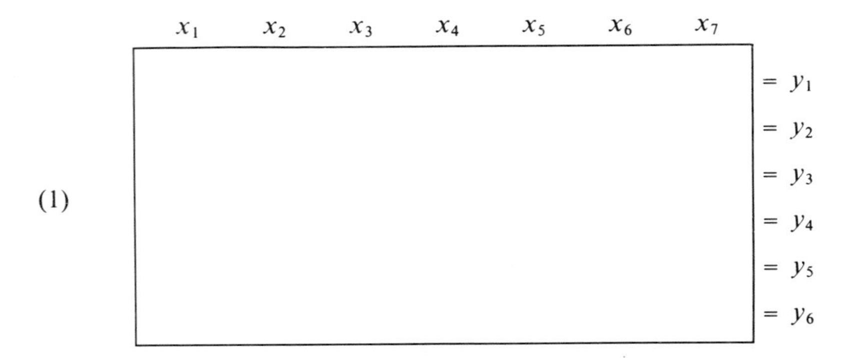

Suppose that there is no unique pivot rank for (1). This means that by different pivot selections it is possible to derive different fully exchanged tableaus from (1) using different numbers of pivots. One thing is certain. If the number of pivots to get fully exchanged tableaus differs, then one of these numbers is largest. Let us assume that this largest number of pivots is four and that the fully exchanged tableau (2) is a result of four pivots. Also assume that (3) is another fully exchanged tableau of (1) obtained with three pivots.

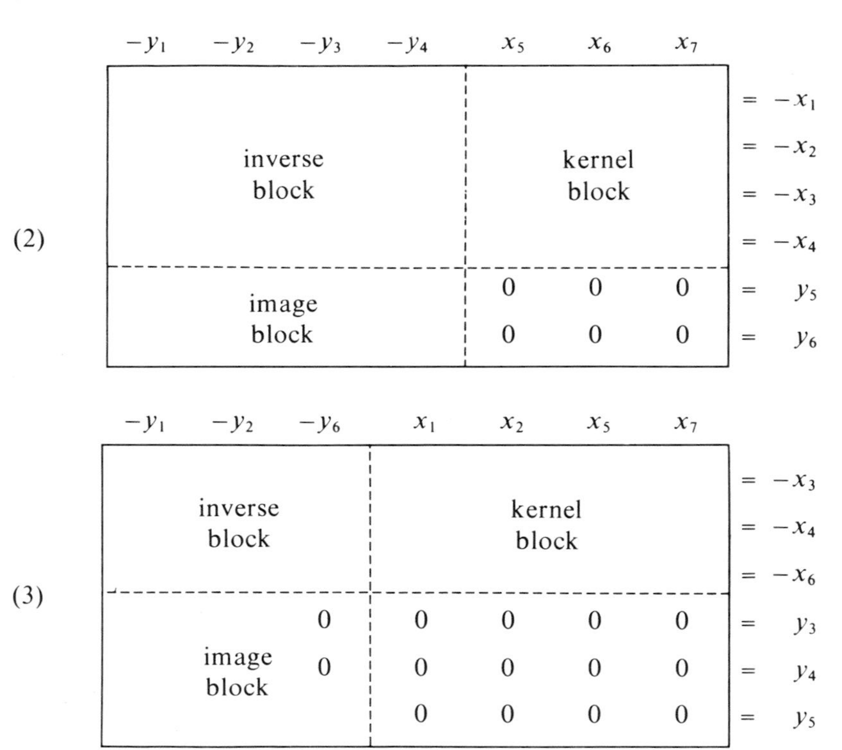

We notice that in (2) y_3 and y_4 are exchanged and y_6 is not exchanged, but in (3) y_6 is exchanged whereas y_3 and y_4 are not. In (3) try to exchange y_6 with y_3 or y_4. If this is impossible to do, then (3) has the two zeros shown in its image block. If, in (3), y_6 can exchange with one of y_3 or y_4 (say y_3), then the indicated zero is not there and we obtain (4) from (3). We pay no attention to the x's because they do not enter into the argument.

(4)

$-y_1$	$-y_2$	$-y_3$	x_1	x_2	x_5	x_7	
							$= -x_3$
	inverse block			kernel block			$= -x_4$
							$= -x_6$
			0	0	0	0	$= y_6$
	image block		0	0	0	0	$= y_4$
			0	0	0	0	$= y_5$

Now, since (2), (3), and (4) are all obtained from (1) by pivot exchange they all have the same solution set as (1) and therefore (2), (3), and (4) have the same solution set. We know that one way to obtain a solution is to select any numbers for the nonbasic variables and then calculate the values for the basic variables. Since in (2) y_1, y_2, y_3, and y_4 appear as nonbasic variables, there is a solution to (2) [and therefore to (3) and (4)] in which $y_1 = 0$, $y_2 = 0$, $y_3 = 0$, and $y_4 = 1$. We proceed to show that such a solution of (2) cannot be a solution of (3) or (4). If it is a solution of (3) we have

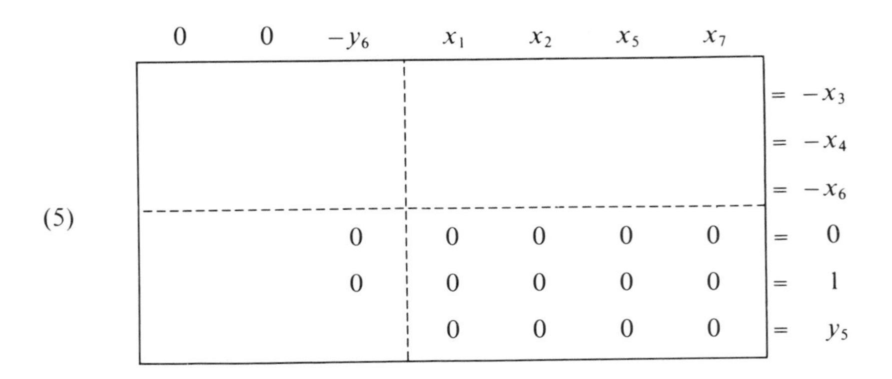

But clearly row 5 of (5) is not true. Similarly, if $y_1 = 0$, $y_2 = 0$, $y_3 = 0$, and $y_4 = 1$ is part of a solution to (4), we obtain (6) from (4):

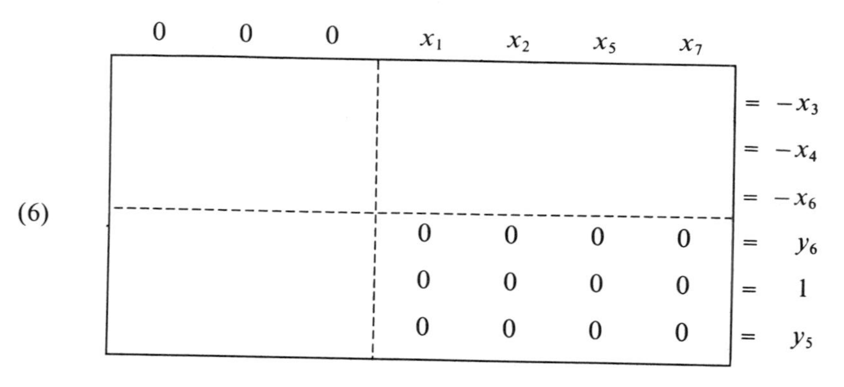

Again row 5 is not true. This demonstrated difficulty arises because we assumed a fully exchanged tableau [that is, (3)] obtained with fewer pivots than the agreed-on maximum number. Therefore, such an assumption forces a contradiction of the fact that pivot exchange preserves solution sets. We therefore conclude that pivot rank is unique. The ideas used in the above demonstration are all that are needed to prove the uniqueness of pivot rank for a tableau of any size.

Exercises

Obtain a schema for each of Exercises 1–13. (See exercises, Section 2.4).

1.

x	y	
3	5	$= z$
1	2	$= w$

2.

x	y	
1	3	$= z$
2	6	$= w$

3.

x_1	x_2	x_3	
2	1	3	$= y_1$
3	0	2	$= y_2$

4.

x_1	x_2	
1	2	$= y_1$
3	8	$= y_2$
4	1	$= y_3$

5.

x_1	x_2	x_3	
2	-4	-6	$= y_1$
-1	2	3	$= y_2$
1	-2	-3	$= y_3$

6.

x_1	x_2	x_3	
1	-2	4	$= y_1$
3	-5	2	$= y_2$
-1	2	-4	$= y_3$

7.

x_1	x_2	x_3	
1	0	3	$= y_1$
3	1	4	$= y_2$
−2	3	−21	$= y_3$

8.

x_1	x_2	x_3	
1	2	4	$= y_1$
3	5	7	$= y_2$
6	8	9	$= y_3$

9.

x_1	x_2	x_3	
1	2	2	$= y_1$
2	1	4	$= y_2$
−1	−1	2	$= y_3$
2	2	4	$= y_4$

10.

x_1	x_2	x_3	x_4	
2	1	0	3	$= y_1$
1	4	2	−2	$= y_2$

11.

x_1	x_2	
1	5	$= y_1$
2	6	$= y_2$
3	7	$= y_3$
4	8	$= y_4$

12.

x_1	x_2	
1	2	$= y_1$

13.

x_1	
2	$= y_1$
3	$= y_2$
4	$= y_3$

14. Obtain a schema for the four-towns tableau.

2.6 *Invertible Tableaus*

The reader may recall examples of $m \times n$ tableaus (that is, tableaus with m basic and n nonbasic variables) whose image sets are R^m and others whose kernels are the zero n-tuple. Of special interest is the tableau that has both of these properties.

EXAMPLE 1

Find the image set and kernel set of

(1)

x_1	x_2	
1	2	$= y_1$
2	−1	$= y_2$

Pivoting as indicated,

(1)

x_1	x_2	
1*	2	$= y_1$
2	−1	$= y_2$

(2)

$-y_1$	x_2	
1	2	$= -x_1$
−2	−5*	$= y_2$

(3)

$-y_1$	$-y_2$	
$1/5$	$2/5$	$= -x_1$
$2/5$	$-1/5$	$= -x_2$

From (3), the fully exchanged tableau, we see that for every value of (y_1, y_2) there is one value of (x_1, x_2), and that the only preimage in (1) of $(y_1, y_2) = (0, 0)$ is $(x_1, x_2) = (0, 0)$.

From the first observation follows the conclusion that the image set of (1) is R^2, the set of all ordered pairs (2-tuples); from the second, that the kernel of (1) contains only (0, 0). Both conclusions follow from the fact that all x's and all y's have exchanged their original roles as nonbasic and basic variables in the fully exchanged tableau (3). This implies that for each (y_1, y_2) in the image set there is exactly one (x_1, x_2).

In essence the initial tableau (1) has been inverted to the fully exchanged tableau (3). That is, from (3) we see that corresponding to any (y_1, y_2) there is exactly one (x_1, x_2). From (1) we obtain

$$(2, 3) \mapsto (8, 1)$$
$$(-1, 1) \mapsto (1, -3)$$
$$(0, 0) \mapsto (0, 0)$$
$$(1, -1) \mapsto (-1, 3)$$

This means that (2, 3) is a preimage of (8, 1), (−1, 1) is a preimage of (1, −3), and so on. Now viewing (3) as an initial tableau yields

$$\begin{aligned}
(8, 1) &\mapsto (2, 3)\\
(1, -3) &\mapsto (-1, 1)\\
(0, 0) &\mapsto (0, 0)\\
(-1, 3) &\mapsto (1, -1)
\end{aligned}$$

Thus the tableaus (1) and (3) are related to each other by the simple expedient of "inverting" the arrow. Thus (1) and (3) are called *inverses* of each other.

EXAMPLE 2

Let (1) be a given initial tableau

(1)

x_1	x_2	
1*	2	$= y_1$
2	4	$= y_2$

In one pivot exchange we obtain the fully exchanged tableau

(2)

$-y_1$	x_2	
1	2	$= -x_1$
−2	0	$= y_2$

The image set of (1) is determined by $2y_1 = y_2$ and the kernel by $2x_2 = -x_1$. Not all x's and y's have been exchanged in the fully exchanged tableau. This implies that some ordered pairs are not images and those ordered pairs that are images have more than one preimage.

For instance, $(y_1, y_2) = (1, 3)$, by the last row of (2) has no preimage. That is, $(x_1, x_2) \mapsto (1, 3)$ is false for all (x_1, x_2). On the other hand, $(y_1, y_2) = (1, 2)$ is in the image set. Substituting (1, 2) in (2) yields

(3)

−1	x_2	
1	2	$= -x_1$
−2	0	$= 2$

The preimage set of (1, 2) is determined by row 1. In set language it is

$$\{(x_1, x_2) : \begin{array}{cc} -1 & x_2 \\ \hline 1 & 2 \end{array} = -x_1\}$$

Some elements of this set are (1, 0), (−1, 1), and (7, −3). Symbolically

$$(1, \quad 0) \mapsto (1, 2)$$
$$(-1, \quad 1) \mapsto (1, 2)$$
$$(7, -3) \mapsto (1, 2)$$

In Example 1 we found an inverse of (1), namely (3), because we were able to exchange all variables, and therefore each pair (y_1, y_2) has exactly one preimage. Since we cannot do this for (1) of Example 2, it has no inverse tableau. It is clear from these examples that a tableau T with $(x_1, x_2, \ldots, x_n) \mapsto (y_1, y_2, \ldots, y_m)$ can be inverted only if in the fully exchanged tableau all x's and y's are exchanged. This is possible only if

(a) $m = n$.
(b) the pivot rank of the initial tableau is also n.

If this is the case, the tableau with

$$(y_1, y_2, \ldots, y_n) \mapsto (x_1, x_2, \ldots, x_n)$$

is called the *inverse tableau* of T. It is designated T^{-1}.

EXAMPLE 3

Find T^{-1} if T is a tableau

x_1	x_2	x_3	
a	b	c	$= y_1$
d	e	f	$= y_2$

There is no T^{-1} for T regardless of the inside of the tableau. This is so because T has three nonbasic variables and two basic variables, and therefore not all variables can be exchanged.

EXAMPLE 4

Find the inverse tableau of

(1)

x_1	x_2	x_3	
1*	2	3	$= y_1$
2	1	-1	$= y_2$
-1	1	-2	$= y_3$

Using the fundamental pivot-exchange algorithm yields

(2)

$-y_1$	x_2	x_3	
1	2	3	$= -x_1$
-2	-3	-7	$= y_2$
1	3	1*	$= y_3$

(3)

$-y_1$	x_2	$-y_3$	
-2	-7	-3	$= -x_1$
5	18*	7	$= y_2$
1	3	1	$= -x_3$

(4)

$-y_1$	$-y_2$	$-y_3$	
$-1/18$	$7/18$	$-5/18$	$= -x_1$
$5/18$	$1/18$	$7/18$	$= -x_2$
$1/6$	$-1/6$	$-1/6$	$= -x_3$

Tableau (4) is the inverse of (1) and of course (1) is the inverse of (4). Why?

To check, assign a value to (x_1, x_2, x_3), say (1, 2, 3) and calculate (y_1, y_2, y_3) in (1):

(5)

1	2	3	
1	2	3	$= y_1$
2	1	-1	$= y_2$
-1	1	-2	$= y_3$

Calculating yields $(y_1, y_2, y_3) = (14, 1, -5)$. Thus $(1, 2, 3) \mapsto (14, 1, -5)$.

Now substituting these values in (4) gives:

(6)

14	1	−5		
$-\frac{1}{18}$	$\frac{7}{18}$	$-\frac{5}{18}$	$= -1$	✓
$\frac{5}{18}$	$\frac{1}{18}$	$\frac{7}{18}$	$= -2$	✓
$\frac{1}{6}$	$-\frac{1}{6}$	$-\frac{1}{6}$	$= -3$	✓

which is true.

EXAMPLE 5

If S is tableau (1), find S^{-1}.

(1)

x_1	x_2	x_3	
1*	2	3	$= y_1$
2	1	−1	$= y_2$
0	−3	−7	$= y_3$

Pivoting as indicated,

(2)

$-y_1$	x_2	x_3	
1	2	3	$= -x_1$
−2	−3*	−7	$= y_2$
0	−3	−7	$= y_3$

(3)

$-y_1$	$-y_2$	x_3	
$-\frac{1}{3}$	$\frac{2}{3}$	$-\frac{5}{3}$	$= -x_1$
$\frac{2}{3}$	$-\frac{1}{3}$	$\frac{7}{3}$	$= -x_2$
2	−1	0	$= y_3$

Tableau (3) is a fully exchanged tableau of (1). Since x_3 and y_3 cannot be exchanged, we conclude that S^{-1} does not exist.

EXAMPLE 6

Find the inverse tableau of

(1)

x_1	x_2	
0	1*	$= y_1$
2	3	$= y_2$

Using the fundamental pivot-exchange algorithm yields

(2)

x_1	$-y_1$	
0	1	$= -x_2$
2*	-3	$= y_2$

(3)

$-y_2$	$-y_1$	
0	1	$= -x_2$
$\frac{1}{2}$	$-\frac{3}{2}$	$= -x_1$

Alternatively, if we pivot in (1) on 3, we arrive at the fully exchanged tableau

(4)

$-y_1$	$-y_2$	
$-\frac{3}{2}$	$\frac{1}{2}$	$= -x_1$
1	0	$= -x_2$

Clearly (4) is also obtainable from (3) by interchanging rows and columns.

The results of our examples are generalized in the following theorem.

THEOREM 2.3. Let T be an $m \times n$ tableau. Then T has an inverse tableau T^{-1} if and only if

1. $m = n$.
2. the pivot rank of the initial tableau is n. (All variables are exchanged using the fundamental pivot-exchange algorithm.)

COROLLARY If the inverse of T is T^{-1}, then the inverse of T^{-1} is T. Symbolically, $(T^{-1})^{-1} = T$.

This corollary implies that an invertible tableau has exactly one inverse. This fact has been clear all along since pivot exchange is reversible. The tableau T

(1)

x_1	x_2	x_3	x_4	
4	−4	5	2	$= y_1$
1	5	2	−4	$= y_2$
−6	−10	−9	8	$= y_3$
6	−2	8	0	$= y_4$
−3	−9	−5	7	$= y_5$

has a fully exchanged tableau with rearranged rows and columns:

(2)

$-y_2$	$-y_4$	$-y_5$	x_2	
−14	$-3/2$	−8	5	$= -x_1$
$21/2$	$5/4$	6	−4	$= -x_3$
$3/2$	$1/4$	1	−2	$= -x_4$
$1/2$	$-3/4$	0	0	$= y_1$
$-3/2$	$1/4$	−2	0	$= y_3$

Subdividing (2) into blocks (as in Section 2.5) means that the kernel of T is described by the upper-right-hand block and the image set of T by the lower-left-hand block. The upper-left-hand block is designated inverse block because if the tableau T is "properly restricted," this block represents the inverse tableau of the "restricted tableau." To see this, abstract this block to obtain

(3)

$-y_2$	$-y_4$	$-y_5$	
−14	$-3/2$	−8	$= -x_1$
$21/2$	$5/4$	6	$= -x_3$
$3/2$	$1/4$	1	$= -x_4$

Applying the fundamental pivot-exchange algorithm to (3) yields

(4)

x_1	x_3	x_4	
1	2	-4	$= y_2$
6	8	0	$= y_4$
-3	-5	7	$= y_5$

Tableau (4) is seen to be a "restricted" portion of T obtained from the rows of y_2, y_4, y_5 and from the columns of x_1, x_3, and x_4 of (1).

Exercises

Find the inverse tableau, if any, of each of the tableaus in Exercises 1–7.

1.

x	y	
2	3	$= z$
4	5	$= w$

2.

x	y	
3	2	$= u$
1	1	$= v$

3.

x	
4	$= y$

4.

x_1	x_2	x_3	
1	3	1	$= y_1$
-1	2	3	$= y_2$
1	8	5	$= y_3$

5.

x_1	x_2	x_3	x_4	
1	0	0	0	$= y_1$
0	1	0	0	$= y_2$
0	0	1	0	$= y_3$
0	0	0	1	$= y_4$

6.

x_1	x_2	x_3	x_4	
1	0	0	0	$= y_1$
0	2	0	0	$= y_2$
0	0	3	0	$= y_3$
0	0	0	4	$= y_4$

7.

x_1	x_2	x_3	
1	2	3	$= y_1$
0	1	2	$= y_2$
0	0	1	$= y_3$

8. Given the tableau T,

x_1	x_2	x_3	
1	2	4	$= y_1$
3	8	10	$= y_2$
-3	-5	-7	$= y_3$

(a) Find the inverse of T (if any).

(b) If there is one, show that it is the same for two different sequences of pivot exchanges.

9. It is claimed that the inverse of

(1)

x_1	x_2	x_3	
3	1	0	$= y_1$
1	2	-3	$= y_2$
1	3	1	$= y_3$

is

(2)

y_1	y_2	y_3	
$5/11$	$-1/11$	$-3/11$	$= x_1$
$-4/11$	$3/11$	$9/11$	$= x_2$
$-1/11$	$-2/11$	$5/11$	$= x_3$

Verify this assertion by a sequence of pivot exchanges:
(a) on (1) to derive (2); (b) on (2) to derive (1).

[*Caution:* Observe that the variables in (2) have no minus signs.]

10. Find the inverse, if any, of

x_1	x_2	x_3	
1	1	0	$= y_1$
0	1	1	$= y_2$
1	0	1	$= y_3$

11. Show that the inverse of

x_1	x_2	x_3	
0	1	0	$= y_1$
0	0	1	$= y_2$
1	0	0	$= y_3$

is

y_1	y_2	y_3	
0	0	1	$= x_1$
1	0	0	$= x_2$
0	1	0	$= x_3$

12. Show that the inverse of

x_1	x_2	x_3	
2	0	0	$= y_1$
0	3	0	$= y_2$
0	0	4	$= y_3$

is

y_1	y_2	y_3	
$1/2$	0	0	$= x_1$
0	$1/3$	0	$= x_2$
0	0	$1/4$	$= x_3$

Note: In Exercises 13–15 a "proof" for the case of a 3×3 tableau is sufficient to show what a general proof would look like.

13. The entries of a tableau that are at the intersection of the first column, first row; second column, second row; and so on, are called *diagonal entries*. All other entries are called *nondiagonal entries*. A tableau is called *diagonal* if all its nondiagonal entries are zeros. (See Exercise 12.) Prove that if a tableau is invertible and diagonal, then the inverse tableau is also diagonal.

14. A tableau is called *upper triangular* if all entries below the diagonal entries are zeros. Prove that if a tableau is invertible and upper triangular, then the inverse tableau is also upper triangular.

15. A tableau is called *lower triangular* if all entries above the diagonal entries are zeros. Prove that if a tableau is invertible and lower triangular, then the inverse tableau is also lower triangular.

2.7 Tableaus with Constant Columns

So far all tableaus in Chapter 2 have nonbasic variables at the top and basic variables on the right. This state of affairs is distinguished from that of Chapter 1, where the tableaus of linear programs have "-1" at the top, in addition to the nonbasic variables. We now introduce constant columns into the tableaus of this chapter and discuss image sets and preimage sets of such tableaus, reserving the discussion of kernel for Chapter 7. To see how these tableaus arise, suppose, for instance, that each of the centers A, B, C, and D in the four-towns problem is given a constant weekly amount by a super-agency. To be specific, let A, B, C, and D be given the weekly constant sums of 3, 5, 0, and -2 \$1000 bills. (-2 means that D gives up \$2000 each week.) These constants are shown in the "1" column of the modified tableau:

x_1	x_2	x_3	x_4	x_5	1	
1	1	-1	0	0	3	$= y_A$
-1	0	0	1	1	5	$= y_B$
0	-1	0	-1	0	0	$= y_C$
0	0	1	0	-1	-2	$= y_D$

EXAMPLE 1

Let (1) be a tableau with a constant column headed by "1."

(1)

x_1	x_2	x_3	1	
1*	2	3	4	$= y_1$
−3	−5	−4	−9	$= y_2$
−1	−2	−1	−2	$= y_3$

The fundamental pivot-exchange algorithm applied to (1) yields

(2)

$-y_1$	x_2	x_3	1	
1	2	3	4	$= -x_1$
3	1*	5	3	$= y_2$
1	0	2	2	$= y_3$

(3)

$-y_1$	$-y_2$	x_3	1	
−5	−2	−7	−2	$= -x_1$
3	1	5	3	$= -x_2$
1	0	2*	2	$= y_3$

(4)

$-y_1$	$-y_2$	$-y_3$	1	
$-3/2$	−2	$7/2$	5	$= -x_1$
$1/2$	1	$-5/2$	−2	$= -x_2$
$1/2$	0	$1/2$	1	$= -x_3$

From (4), a fully exchanged tableau of (1), we see that every value of (y_1, y_2, y_3) produces exactly one value of (x_1, x_2, x_3). Hence (1) is invertible and the tableaus defined by (1) and (4) are inverses of each other. This implies that the image set of (1) is R^3, as is the image set of (4). Observe, in this example, how Theorem 2.3 applies to tableaus with constant columns.

To verify the arithmetic, let $(x_1, x_2, x_3) = (0, 0, 0)$. Then from (1), $(y_1, y_2, y_3) = (4, -9, 2)$. Substituting these values in (4) yields

(5)

-4	9	2	1		
$-\frac{3}{2}$	-2	$\frac{7}{2}$	5	$= 0$	✓
$\frac{1}{2}$	1	$-\frac{5}{2}$	-2	$= 0$	✓
$\frac{1}{2}$	0	$\frac{1}{2}$	1	$= 0$	✓

which is true.

EXAMPLE 2

Let (1) be the tableau

(1)

x_1	x_2	x_3	1	
1^*	2	3	4	$= y_1$
-3	-5	-4	-9	$= y_2$
2	3	1	2	$= y_3$

Pivoting as indicated yields

(2)

$-y_1$	x_2	x_3	1	
1	2	3	4	$= -x_1$
3	1^*	5	3	$= y_2$
-2	-1	-5	-6	$= y_3$

(3)

$-y_1$	$-y_2$	x_3	1	
-5	-2	-7	-2	$= -x_1$
3	1	5	3	$= -x_2$
1	1	0	-3	$= y_3$

Tableau (3) is fully exchanged. The image set of (1) is determined by

(4)

$-y_1$	$-y_2$	1	
1	1	-3	$= y_3$

This system has two nonbasic variables and one equation. The pivot rank of (1) is 2. Note that Theorem 2.1 applies to this tableau, which has a constant column. To find the preimage set of a member of the image set say $(y_1, y_2, y_3) = (1, 0, -4)$, substitute into (3), yielding

(5)

-1	0	x_3	1	
-5	-2	-7	-2	$= -x_1$
3	1	5	3	$= -x_2$
1	1	0	-3	$= -4$

The preimage set of $(1, 0, -4)$ is the solution set of

-1	x_3	1	
-5	-7	-2	$= -x_1$
3	5	3	$= -x_2$

or

x_3	1	
-7	3	$= -x_1$
5	0	$= -x_2$

EXAMPLE 3

Let (1) be the tableau

(1)

w	x	y	z	1	
4	-4	5	2	-3	$= P$
1	5	2	-4	0	$= Q$
-6	-10	-9	8	3	$= R$
6	-2	8	0	-12	$= S$
-3	-9	-5	7	2	$= T$

(This is a modification of the tableau of Example 5 of Section 2.2.)

After three pivot exchanges (1) has as a fully exchanged tableau

(2)

$-S$	x	$-Q$	$-T$	1	
$-3/4$	0	$1/2$	0	6	$= P$
$5/4$	-4	$21/2$	6	-3	$= -y$
$1/4$	0	$-3/2$	-2	-4	$= R$
$-3/2$	5	-14	-8	2	$= -w$
$1/4$	-2	$3/2$	1	-1	$= -z$

The image set of (1) is obtained from

(3)

$-S$	$-Q$	$-T$	1	
$-3/4$	$1/2$	0	6	$= P$
$1/4$	$-3/2$	-2	-4	$= R$

If $(Q, S, T) = (0, 0, 0)$, then $(P, R) = (6, -4)$ and $(P, Q, R, S, T) = (6, 0, -4, 0, 0)$ is a member of the image set. To find the set of preimages of $(6, 0, -4, 0, 0)$, substitute in (2), omitting rows P and R and the columns headed 0. The result is

(4)

x	1	
-4	-3	$= -y$
5	2	$= -w$
-2	-1	$= -z$

Exercises

Find the inverse, if any, of each of the tableaus shown in Exercises 1–8. The initial pivot is indicated.

1.

x_1	x_2	1	
4	2*	2	$= y_1$
5	3	3	$= y_2$

2.

x_1	x_2	1	
2	1*	2	$= y_1$
4	0	4	$= y_2$

3.

x_1	x_2	x_3	1	
2	1*	3	4	$= y_1$
1	0	4	3	$= y_2$
3	4	2	1	$= y_3$

4.

x_1	x_2	x_3	1	
1*	0	0	a	$= y_1$
0	2	0	b	$= y_2$
0	0	3	c	$= y_3$

5.

x_1	x_2	x_3	1	
0	1*	1	a	$= y_1$
1	0	1	a	$= y_2$
1	1	0	a	$= y_3$

6.

x_1	x_2	x_3	1	
1*	2	3	4	$= y_1$
0	1	2	3	$= y_2$
0	0	1	2	$= y_3$

7.

x_1	x_2	x_3	1	
1*	−4	3	1	$= y_1$
−2	8	−6	2	$= y_2$
2	−8	6	3	$= y_3$

8.

x_1	x_2	1	
3	1*	1	$= y_1$
1	−1	2	$= y_2$
2	0	3	$= y_3$

9. Find the image set in
(a) Exercise 2.
(b) Exercise 7.
(c) Exercise 5.
(d) Exercise 8.

10. Find the image set of

x_1	x_2	1	
3	1*	1	$= y_1$
1	−1	2	$= y_2$
2	0	3	$= y_3$
0	3	4	$= y_4$

11. Given the tableau in Exercise 8. Find the preimage set of $(y_1, y_2, y_3) = (1, 2, 3)$.

12. Given the tableau in Exercise 10. Find the preimage set of $(y_1, y_2, y_3, y_4) = (1, 2, 3, 4)$.

13. Show that Theorem 2.1 is also valid for tableaus with constant columns.

14. Show that Theorem 2.3 is also valid for tableaus with constant columns.

15. Given the tableau in Exercise 10. Show that $(y_1, y_2, y_3, y_4) = (0, 0, 0, 0)$ has no preimage. How does this fact induce caution in trying to apply Theorem 2.2 to tableaus with constant columns?

2.8 Mappings

In this section we provide a different perspective in our study of tableaus. We introduce the mathematical concept of mapping and show how tableaus relate to mappings.

The word "mapping" in ordinary English usage suggests the drawing to scale of a picture of a terrain, possibly preserving such relations as direction between points and ratio of distances between points. Mathematicians use the word somewhat differently, although there is some connection with standard English usage. Basically they use the word "mapping" to mean that there are two sets of objects and a rule that assigns each element in the first set to exactly one element in the second set. A mapping is shown schematically in Figure 2.4.

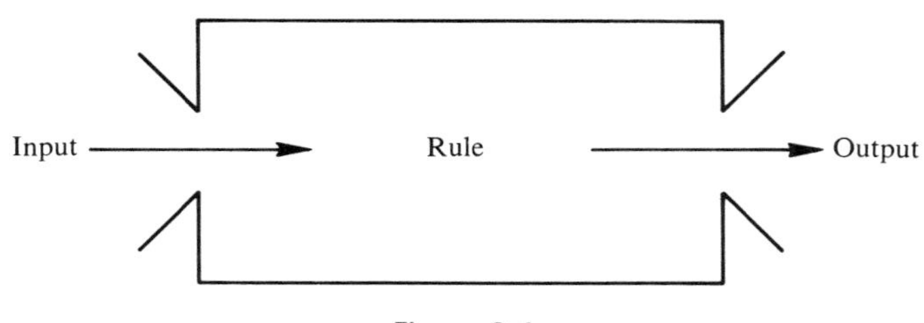

Figure 2.4

The four-towns problem illustrates a mathematical mapping. It has three requisite components:

1. A first set, called the *domain* (set of inputs).
2. A rule.
3. A second set, called the *codomain* (set of outputs).

The first set in the four-towns mapping is the set of all 6-tuples $(x_1, x_2, x_3, x_4, x_5, x_6)$, where the x's represent the value of weekly shipments from town to town. The second set consists of all 4-tuples (Y_A, Y_B, Y_C, Y_D), where the Y's are weekly net receipts (payments received minus payments sent).

The rule that assigns each 6-tuple to exactly one 4-tuple is the system of linear equations that makes up the four-towns tableau.

A word about the kinds of sets serving as domains and codomains of the mappings we discuss in this book. They consist of sets of n-tuples. Each n-tuple consists of real numbers that are positive, negative, zero, rational, or irrational. The set of all real numbers is denoted R. The set of ordered pairs of numbers (2-tuples) is represented by R^2 (read "R two"). Some of its members are (2, 1), $(\sqrt{3}, 8)$, and $(-2, 0)$. The set of ordered triples (3-tuples) is represented by R^3 (read "R three"). Some of its members are (0, 0, 0), $(1, 3, -8)$, and $(3, 4, -\frac{1}{2})$.

The set of ordered quadruples (4-tuples) is denoted R^4. Three of its members are

$$(0, 1, 2, -3) \qquad (-1, 3, \pi, 1) \qquad (0, 0, 0, 0)$$

Any member of R^4 can be represented by (x_1, x_2, x_3, x_4), where the x's can take on any real value.

Each of the numbers in an ordered n-tuple is called a *component* of the

n-tuple. The set of all ordered n-tuples is designated R^n, where n is any positive integer. A member of R^n is designated $(x_1, x_2, \ldots, x_n)$. The n-tuple whose components all are zeros is called the *zero n-tuple.* The zero pair is (0, 0), the zero triple is (0, 0, 0), and so on.

Mappings are named by letters. The four-towns mapping is called M. Symbolically write

$$M : R^6 \to R^4$$

which we read as "the mapping M from R six to R four." We have R^6 as the domain of M and R^4 as the codomain. The domain supplies all the inputs. The codomain contains all the outputs.

The arrow $\to$ is different from $\mapsto$. The arrow $\to$ relates the domain and codomain. The arrow $\mapsto$ relates a specific element of the domain to a specific element of the codomain.

The four-towns mapping M is represented by the tableau

x_1	x_2	x_3	x_4	x_5	x_6	
1	1	−1	0	0	0	$= y_A$
−1	0	0	1	−1	0	$= y_B$
0	−1	0	−1	0	1	$= y_C$
0	0	1	0	1	−1	$= y_D$

In the four-towns mapping M if $(x_1, x_2, x_3, x_4, x_5, x_6) = (2, 1, 3, 5, 1, 2)$, then $(Y_A, Y_B, Y_C, Y_D) = (0, 2, -4, 2)$. This is stated in a variety of ways.

1. $(2, 1, 3, 5, 1, 2) \mapsto (0, 2, -4, 2)$ (M is understood).
2. M *maps* (2, 1, 3, 5, 1, 2) *to* (0, 2, −4, 2).
3. The *image* of (2, 1, 3, 5, 1, 2) *under* M is (0, 2, −4, 2).
4. A *preimage* of (0, 2, −4, 2) *under* M is (2, 1, 3, 5, 1, 2).
5. $M((2, 1, 3, 5, 1, 2)) = (0, 2, -4, 2)$. [Read M of (2, 1, 3, 5, 1, 2) equals (0, 2, −4, 2).]

EXAMPLE 1

The mapping S defined by

(1)

x	y	z	
2	3	−1	$= u$
1	−1	2	$= v$

is also described by

$$S: R^3 \to R^2 \qquad \text{with rule} \begin{cases} 2x + 3y - z = u \\ x - y + 2z = v \end{cases}$$

or

$$(x, y, z) \mapsto (2x + 3y - z, x - y + 2z) = (u, v)$$

If $(x, y, z) = (2, 3, -4)$, the tableau becomes

$$\begin{array}{ccc} 2 & 3 & -4 \\ \hline \multicolumn{1}{|c}{2} & 3 & \multicolumn{1}{c|}{-1} \\ \multicolumn{1}{|c}{1} & -1 & \multicolumn{1}{c|}{2} \\ \hline \end{array} \begin{array}{l} \\ = u \\ = v \end{array} \tag{2}$$

which yields $(u, v) = (17, -9)$.

The following statements are equivalent.

1. The image of $(2, 3, -4)$ under S is $(17, -9)$.
2. $(2, 3, -4) \mapsto (17, -9)$.
3. $(17, -9)$ is a preimage of $(2, 3, -4)$.
4. S maps $(2, 3, -4)$ to $(17, -9)$.
5. $S(2, 3, -4) = (17, -9)$.

EXAMPLE 2

$$\begin{array}{cc} x & y \\ \hline \multicolumn{1}{|c}{3} & \multicolumn{1}{c|}{2} \\ \hline \end{array} = z$$

is the tableau of the mapping, $M: R^2 \mapsto R^1$ with rule $3x + 2y = z$. The reader is asked to verify that

$$\begin{array}{rcr} (0, 0) & \mapsto & 0 \\ (-2, 3) & \mapsto & 0 \\ (2, -3) & \mapsto & 0 \\ (1, 1) & \mapsto & 5 \\ (-1, 1) & \mapsto & -1 \\ (-3;, 4) & \mapsto & -1 \end{array}$$

Every element of the domain is assigned exactly one image. But an image in the codomain such as 0 or -1 has more than one preimage.

EXAMPLE 3

Let $F: R^4 \to R^5$ be a mapping with tableau

w	x	y	z	
4	-4	5	2	$= P$
1	5	2	-4	$= Q$
-6	-10	-9	8	$= R$
6	-2	8	0	$= S$
-3	-9	-5	7	$= T$

Given any domain member, say $(1, -2, 3, 4)$ its image is obtained from the tableau

1	-2	3	4	
4	-4	5	2	$= P$
1	5	2	-4	$= Q$
-6	-10	-9	8	$= R$
6	-2	8	0	$= S$
-3	-9	5	7	$= T$

Calculations show that $P = 35$, $Q = -19$, $R = 19$, $S = 34$, and $T = 58$. Thus $(1, -2, 3, 4) \mapsto (35, -19, 19, 34, 58)$. Given any element of the domain, its image is ascertained from the rule. It is not clear whether or not a preimage exists for a given codomain element and, if so, are there many?

EXAMPLE 4

Figure 2.5 lists five mappings:

(a) $F: R^3 \to R^3$.
(b) $G: R \to R^5$.
(c) $H: R^5 \to R$.
(d) $J: R^2 \to R^3$.
(e) $K: R^3 \to R^2$.

In Figure 2.5 the name for the mapping is attached to the tableau of the mapping.

(a)

	x_1	x_2	x_3	
F:	2	−1	−1	$= y_1$
	−4	5	0	$= y_2$
	−2	−2	−1	$= y_3$

(b)

	x_1	
	2	$= y_1$
	3	$= y_2$
G:	−1	$= y_3$
	3	$= y_4$
	2	$= y_5$

(c)

	x_1	x_2	x_3	x_4	x_5	
H:	2	−½	−1	−3	4	$= y$

(d)

	x_1	x_2	
	1	2	$= y_1$
J:	−1	1	$= y_2$
	−3	3	$= y_3$

(e)

	x	y	z	
K:	5	−2	−4	$= v$
	2	3	1	$= w$

Figure 2.5

In summary, for the mappings $R^n \to R^m$ there are two sets R^m and R^n (m and n are not necessarily distinct); R^n is the *domain* of the mapping and R^m the *codomain.* The *rule* that assigns to each member of R^n a unique member of R^m consists of homogeneous linear equations (that is, equations of the form $3x_1 + 2x_2 + x_3 = y_1$ as contrasted with nonhomogeneous linear equations such as $3x_1 + 2x_2 + x_3 + 6 = y_1$; nonhomogeneous linear equations differ from homogeneous linear equations in that they have a constant term).

Exercises

1. Given the mapping

L:

x_1	x_2	
3	−1	$= y_1$
1	0	$= y_2$
0	2	$= y_3$

(a) What are its domain and codomain?
(b) What is its rule?
(c) Find the image of each of the following: (0, 0), (1, 1), (−1, 1), (2, 3), (−2, 3).
(d) Using the results in (c), find a preimage of (0, 0, 0) and (3,2,6).
(e) Complete the following:
(i) $(0, 1) \mapsto$ __________.
(ii) $L : R^2 \rightarrow$ __________.
(iii) $L(1, 0) =$ __________.
(iv) A preimage of __________ is (2, 0).
(v) L maps (3, −2) to __________.

2. Given the mapping

M:

x	y	
3	2	$= z$

(a) What are its domain and codomain?
(b) What is its rule?
(c) Find the image of each of the following: (0, 0), (1, 1), (−1, 1), (2, 3), (−2, 3).
(d) Complete the following:
(i) $(0, 1) \mapsto$ __________.
(ii) M __________ $\rightarrow R$.
(iii) $(x, y) \mapsto$ __________.
(iv) $M(2, -1) =$ __________.
(v) A preimage of __________ is (4, 3).
(vi) L maps (10, 20) to __________.

3. Describe each of the following by a statement of the form __________: $R^n \rightarrow R^m$, with __________ $\mapsto$ __________.

(a)

H:

x_1	x_2	
1	0	$= y_1$
0	1	$= y_2$
2	3	$= y_3$

(b)

J:

x_1	x_2	
1	0	$= y_1$
0	1	$= y_2$
1	1	$= y_3$

(c)

	x_1	x_2	x_3	
M:	1	2	3	$= y$

(d)

	x	
N:	1	$= y_1$
	2	$= y_2$
	3	$= y_3$
	4	$= y_4$

(e)

	x_1	x_2	x_3	
J:	1	0	0	$= y_1$
	0	1	0	$= y_2$
	0	0	1	$= y_3$

4. Which of the following are not mappings with rules consisting of homogeneous linear equations?

(a) $M: R^2 \to R^2, (x, y) \mapsto (x^2, y^2)$.
(b) $M: R \to R^2, (x) \mapsto (x, \sqrt{x})$.
(c) $M: R \to R, (x) \mapsto (2x + 3)$.
(d) $M: R^3 \to R, (x_1, x_2, x_3) \mapsto (x_1 + x_2 + x_3)$.
(e) $M: R^3 \to R^3, (x_1, x_2, x_3) \mapsto (x_1 + 1, x_2 + 2, x_3 + 3)$.

5. Let $M: R^3 \to R^2$ with tableau

x_1	x_2	x_3	
2	-3	4	$= y_1$
-4	1	3	$= y_2$

(a) Find $M(1, 2, 3)$ and $M(3, 6, 9)$.
(b) Find $M(a_1, a_2, a_3)$ and $M(3a_1, 3a_2, 3a_3)$.
(c) Show that the components of $M(3a_1, 3a_2, 3a_3)$ are three times the corresponding components of $M(a_1, a_2, a_3)$.

6. Let $N: R^3 \to R^2$, have as tableau

x_1	x_2	x_3	
3	2	-2	$= y_1$
4	-1	0	$= y_2$

$(1, 2, 5) \mapsto (-3, 2)$ and $(2, 4, -3) \mapsto (20, 4)$. Show that $(1 + 2, 2 + 4, 5 - 3) \mapsto (20 - 3, 4 + 2)$.

7. Using Exercise 6 as a guide, let $(p_1, p_2, p_3) \mapsto (a_1, a_2)$ and $(q_1, q_2, q_3) \mapsto (b_1, b_2)$. Show for N, that $(p_1 + q_1, p_2 + q_2, p_3 + q_3) \mapsto (a_1 + b_1, a_2 + b_2)$.

2.9 Chapter Review

1. The fundamental pivot-exchange algorithm helps to answer the following questions about tableaus with rules that are homogeneous linear equations.
   (a) What is the image set of a tableau?
   (b) What is the kernel set of a tableau?
   (c) Is a tableau invertible?

The answers to these questions are displayed in the following schema:

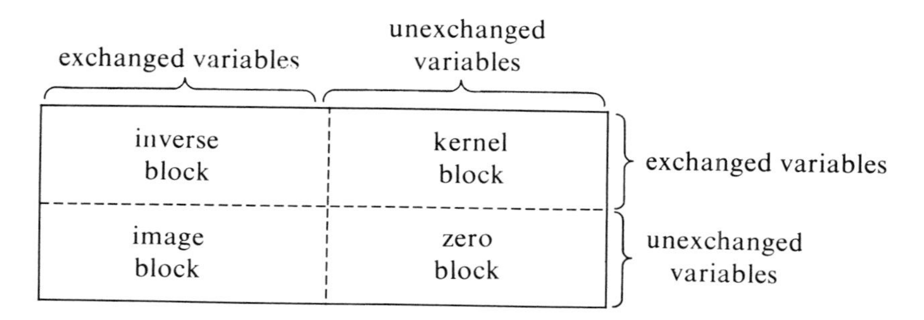

In the schema the block marked inverse block represents the invertible part of the tableau. If there are no other blocks, the tableau is invertible.

2. The uniqueness of pivot rank is demonstrated.
3. The theorems of this chapter are
   (a) *Theorem 2.1:* If a tableau has m basic variables and pivot rank r, then its image set is defined by $m - r$ equations having $m - r$ basic variables and r nonbasic variables.
   (b) *Theorem 2.2:* Let T be a tableau with n nonbasic variables and m basic variables which has pivot rank r. Then the kernel of T is defined by r equations having r basic variables and $n - r$ nonbasic variables. If $n = r$, the kernel contains only the zero n-tuple.
   (c) *Theorem 2.3:* Let T be an $m \times n$ tableau. Then T has an inverse tableau T^{-1} if and only if
      (i) $m = n$.
      (ii) the pivot rank of the initial tableau is n.
4. Theorems 2.1 and 2.3 also apply to tableaus with constant columns (that is, with rules consisting of nonhomogeneous linear equations). The question of kernels for such tableaus is postponed to Chapter 7.
5. The concept of mapping is introduced. Mappings defined by tableaus with rules consisting of homogeneous linear equations are discussed.

Chapter Review Exercises

1. Given the tableau T with constant column

x_1	x_2	1	
2	3	6	$= y_1$
4	5	7	$= y_2$

(a) Show that T is invertible.
(b) Find the preimage of each of (2, 3), (3, 2), (0, 0), (6, 7).

2. Given the mapping $P: R^2 \rightarrow R^3$ such that $(x_1, x_2) \mapsto (x_1 + x_2, x_1 - x_2, x_2)$.
(a) Set up the tableau.
(b) Find the image set of P.
(c) Find the preimage set, if any, of each of (0, 0, 0), (1, 2, 3), and (2, 1, 2).
(d) Find the kernel set of P.

3. Let T be a tableau with constant column such that after one pivot exchange we have

$-y_1$	x_2	1	
1	2	3	$= -x_1$
4	5	6	$= y_2$

Find the initial tableau T.

4. Let $S: R^3 \rightarrow R^2$ be a mapping whose tableau after one pivot exchange is

$-y_1$	x_2	x_3	
2	3	4	$= -x_1$
3	0	0	$= y_2$

(a) What is the image set of S?
(b) What is the kernel set of S?
(c) What is the initial tableau of S?

5. Let T be a tableau with constant column such that (1) results after two pivot exchanges.

(1)

$-y_1$	$-y_2$	x_3	1	
1	−1	3	1	$= -x_1$
2	1	−2	2	$= -x_2$
3	2	0	3	$= y_3$
2	3	0	4	$= y_4$

(a) What is the image set of T?
(b) What is the preimage set of (1, 2, 3, 4)?
(c) What is the initial tableau T?

6. The image set of a tableau with constant column is determined by

$$\begin{cases} y_1 = 2y_2 - 3y_4 + 2 \\ y_3 = y_2 + y_4 + 1 \end{cases}$$

and its kernel is determined by

$$\begin{cases} x_2 = 2x_1 + x_3 \\ x_4 = x_1 - x_3 \end{cases}$$

Do these data determine T? Justify your answer.

7. Given

x_1	x_2	1	
1	2	3	$= y_1$
−1	−2	4	$= y_2$
3	6	1	$= y_3$

verify that (6, 1, 10) is in the image set and then find its preimage set.

CLA

CHAPTER 3

Matrices

3.1 What Is a Matrix?

Some things have an amazing number of uses. The wheel, for example, makes it easier to propel ancient man's cart or modern man's automobile. It is used as a steering device, as gears in a machine, and in a roulette wheel. In mathematics, matrices have many uses. They were first recognized in 1850 by James Joseph Sylvester, used by Sir William Rowan Hamilton in 1853 to generalize numbers, and then used by Arthur Cayley in 1858 in his study of systems of linear equations and functions. Today they are used in the study of physics, economics, heredity, and communication systems, as well as in pure mathematics.

EXAMPLE 1 PAYOFF MATRICES

Joe and Pete play a game in which each tosses a coin. They agree on the following rules: If both coins fall heads, Joe pays Pete 3 cents; if both fall tails, Joe pays Pete 4 cents; if Pete's coin falls heads and Joe's falls tails, Pete pays Joe 2 cents; finally, if Pete's coin falls tails and Joe's heads, Pete pays Joe 5 cents.

These rules become clear when displayed in a data table. The numbers in the table are the amounts Joe pays Pete. Therefore negative numbers mean that Pete pays Joe the absolute value of the negative number.

		Joe	
		H	T
Pete	H	3	−2
	T	−5	4

The array of numbers inside the box is called a *matrix*. It is written

$$\begin{bmatrix} 3 & -2 \\ -5 & 4 \end{bmatrix}$$

In the context of the mathematical subject called game theory, this matrix is called a *payoff matrix*.

EXAMPLE 2 INVENTORY MATRICES

A manufacturing company has three plants each making a different electronic device. Each device requires four distinct parts, call them A, B, C, and D. Each day plant I uses 30 A parts, 43 B parts, 37 C parts, and 16 D parts. Plant II uses 25 A parts, 15 B parts, 30 C parts, and 12 D parts. Plant III uses 61 A parts, 50 B parts, 55 C parts, and 30 D parts.

The data table organizes the information in a memorable form. For any datum just scan the appropriate row and column.

			Plant	
		I	II	III
Part	A	30	25	61
	B	43	15	50
	C	37	30	55
	D	16	12	30

The matrix for this data table is

$$\begin{bmatrix} 30 & 25 & 61 \\ 43 & 15 & 50 \\ 37 & 30 & 55 \\ 16 & 12 & 30 \end{bmatrix}$$

In Example 2 the matrix has 4 rows and 3 columns. It is called a *four-by-three matrix* (written 4×3), or it is said to have *dimension* or *size* 4×3. The payoff matrix of Example 1 has dimension 2×2. In stating the dimension of a matrix the first number stated is the number of rows. In Figure 3.1, (a) is a 3×5 matrix, (b) has dimension 1×4, (c) has dimension 4×1, and (d) is of size 6×3. The notation in (d) for the individual components (entries) of the matrix identifies the location of the entry. Thus a_{23} is the entry in the second row, third column. In (a) $a_{32} = -7$ while $a_{23} = 3$.

(a) $$\begin{bmatrix} 2 & 0 & 0 & 1 & -1 \\ 1 & -5 & 3 & -3 & -2 \\ 0 & -7 & 8 & -6 & 4 \end{bmatrix}$$

(b) $$\begin{bmatrix} a & b & c & d \end{bmatrix}$$

(c) $$\begin{bmatrix} 1.5 \\ 3 \\ 2 \\ -1 \end{bmatrix}$$

(d) $$\begin{bmatrix} a_{11} & a_{12} & a_{13} \\ a_{21} & a_{22} & a_{23} \\ a_{31} & a_{32} & a_{33} \\ a_{41} & a_{42} & a_{43} \\ a_{51} & a_{52} & a_{53} \\ a_{61} & a_{62} & a_{63} \end{bmatrix}$$

Figure 3.1

EXAMPLE 3 AN ADJACENCY MATRIX

Five players are entered in a round-robin chess tournament. The graph shows which players have already played each other. For instance, A and C have played, as have D and E, but D and B have not. This graph is represented in a table in which 1 indicates the players have played their game and 0 means they have not.

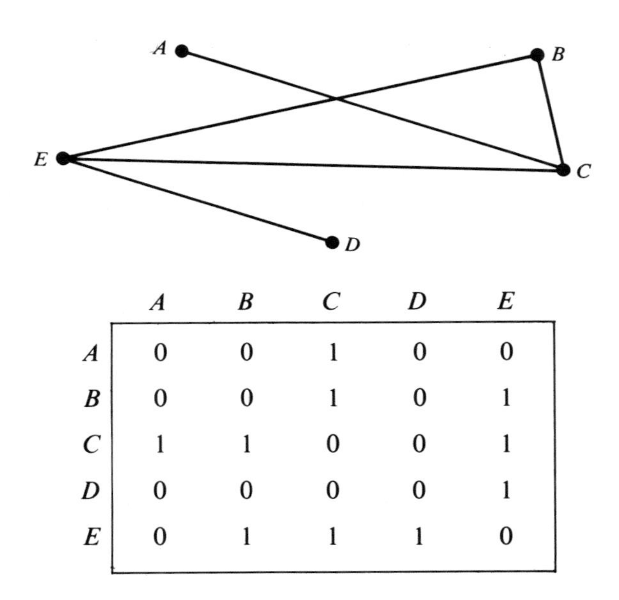

	A	B	C	D	E
A	0	0	1	0	0
B	0	0	1	0	1
C	1	1	0	0	1
D	0	0	0	0	1
E	0	1	1	1	0

The abstraction of this table is a 5×5 matrix called the *adjacency matrix* of the graph.

$$\begin{bmatrix} 0 & 0 & 1 & 0 & 0 \\ 0 & 0 & 1 & 0 & 1 \\ 1 & 1 & 0 & 0 & 1 \\ 0 & 0 & 0 & 0 & 1 \\ 0 & 1 & 1 & 1 & 0 \end{bmatrix}$$

We shall examine this graph in more detail later.

EXAMPLE 4 A BUS NETWORK MATRIX

Four cities A, B, C, and D are connected, if at all, by two-way bus routes as shown:

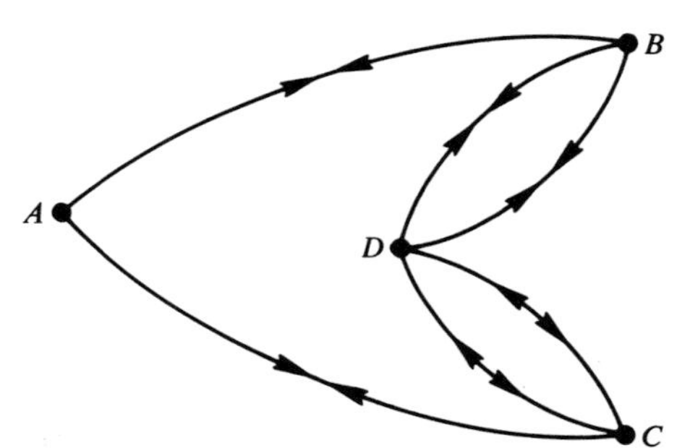

There are three bus routes out of B, one to A and two to D. A complete description of this network is clearer when given in matrix form.

Table

	A	B	C	D
A	0	1	1	0
B	1	0	0	2
C	1	0	0	2
D	0	2	2	0

$$\text{Matrix} \quad \begin{bmatrix} 0 & 1 & 1 & 0 \\ 1 & 0 & 0 & 2 \\ 1 & 0 & 0 & 2 \\ 0 & 2 & 2 & 0 \end{bmatrix}$$

In this bus network matrix $a_{24} = 2$ means that there are two bus routes out of B to D. Observe that for all i, $a_{ii} = 0$. Why is this so?

EXAMPLE 5 COEFFICIENT MATRIX

A system of linear equations can be represented by a matrix. Consider, for example, the tableau representation for the system

x_1	x_2	x_3	
3	2	−3	= 4
1	−1	2	= 5

The inside of the tableau is called the *coefficient matrix* of the system. It is

$$\begin{bmatrix} 3 & 2 & -3 \\ 1 & -1 & 2 \end{bmatrix}$$

The same system of equations is also written as

x_1	x_2	x_3	-1	
3	2	-3	4	$= 0$
1	-1	2	5	$= 0$

The matrix of this tableau is called the *augmented matrix* of the system. It is

$$\begin{bmatrix} 3 & 2 & -3 & 4 \\ 1 & -1 & 2 & 5 \end{bmatrix}$$

EXAMPLE 6 MATRIX OF A TABLEAU

x_1	x_2	
a_{11}	a_{12}	$= y_1$
a_{21}	a_{22}	$= y_2$

The matrix of this tableau is

$$\mathbf{A} = \begin{bmatrix} a_{11} & a_{12} \\ a_{21} & a_{22} \end{bmatrix}$$

Matrices are represented by capital letters **A**, **B**, **C**, etc. (See Figure 3.2.)

(a) $$\mathbf{B} = \begin{bmatrix} 1 & 2 & 3 \\ -4 & a & 2 \\ d & -5 & 0 \end{bmatrix}$$

(b) $$\mathbf{A} = \begin{bmatrix} a_{11} & a_{12} \\ a_{21} & a_{22} \\ a_{31} & a_{32} \end{bmatrix}$$

(c) $$\mathbf{D} = \begin{bmatrix} a_{11} & a_{12} & a_{13} \\ a_{21} & a_{22} & a_{23} \end{bmatrix}$$

(d) $$\mathbf{C} = \begin{bmatrix} c_{11} & c_{12} & c_{13} \\ c_{21} & c_{22} & c_{23} \\ c_{31} & c_{32} & c_{33} \end{bmatrix}$$

Figure 3.2

EXAMPLE 7 MATRICES OF A LINEAR PROGRAM

Consider the maximum linear program defined by the tableau

x_1	x_2	x_3	-1		
a_{11}	a_{12}	a_{13}	b_1	$=$	$-y_1$
a_{21}	a_{22}	a_{23}	b_2	$=$	$-y_2$
c_1	c_2	c_3	f	$=$	M

Its matrix is

$$\begin{bmatrix} a_{11} & a_{12} & a_{13} & b_1 \\ a_{21} & a_{22} & a_{23} & b_2 \\ c_1 & c_2 & c_3 & f \end{bmatrix}$$

This matrix contains submatrices such as

$$\mathbf{A} = \begin{bmatrix} a_{11} & a_{12} & a_{13} \\ a_{21} & a_{22} & a_{23} \end{bmatrix}$$

which is a 2×3 coefficient matrix,

$$\begin{bmatrix} b_1 \\ b_2 \end{bmatrix}$$

a 2×1 matrix containing the constraint constants as elements,

$$[c_1, c_2, c_3]$$

a 1×3 matrix of indicators, and

$$[f]$$

a 1×1 matrix.

Exercises

1. Consider the following matrix:

$$\mathbf{A} = \begin{bmatrix} 0 & 1.4 & 21.5 & 49.0 & 28.0 \\ 1.1 & 52.6 & 39.9 & 6.0 & .3 \\ .2 & 12.6 & 49.8 & 32.8 & 4.7 \\ .1 & 5.5 & 73.3 & 20.4 & .8 \\ .2 & 11.4 & 61.6 & 25.3 & 1.4 \\ .2 & 26.5 & 64.4 & 8.6 & .3 \\ .8 & 32.7 & 61.6 & 4.6 & .2 \\ .8 & 33.5 & 57.8 & 7.7 & .2 \\ 5.1 & 50.9 & 39.3 & 4.7 & .0 \\ 1.9 & 44.0 & 49.7 & 4.1 & .2 \end{bmatrix}$$

(a) What is the dimension of this matrix?
(b) What is a_{35}? a_{53}?
(c) What is the set $\{a_{ij} : i = j, j \leq 5\}$? Compare this with the set $\{a_{ii} : i \leq 5\}$.
(d) What is the greatest entry in the first row?
(e) What is the greatest entry in the first column?
(f) What are the greatest and least numbers in the fifth row?
(g) What are the greatest and least numbers in the fifth column?

2. Three people, A, B, and C, play a game. By the rules, if A beats B, A gets 40 cents from B; if A beats C, A gets 30 cents from C; if B beats A, B gets 35 cents from A; if B beats C, B gets 25 cents from C; if C beats A, C gets 38 cents from A; if C beats B, C gets 32 cents from B. Display these payoffs as a matrix in which the winner is read at the left of each row, and the loser at the top of each column. For a_{ii} write 0. Is $a_{ij} = a_{ji}$ true for any values of i or j?

3. A and B play a game in which each rolls a single die (having six faces showing numerals 1, 2, 3, 4, 5, and 6). If the sum of the numbers appearing on the top faces is even, B pays A that number of dollars. If the sum is odd, then A pays that number of dollars to B. Using positive and negative numbers, display these payoffs in a 6×6 matrix.

4. The diagrams on page 139 represent two-way bus routes connecting towns A, B, C, and D. Describe, in a 4×4 matrix, the number of routes between each pair of towns.

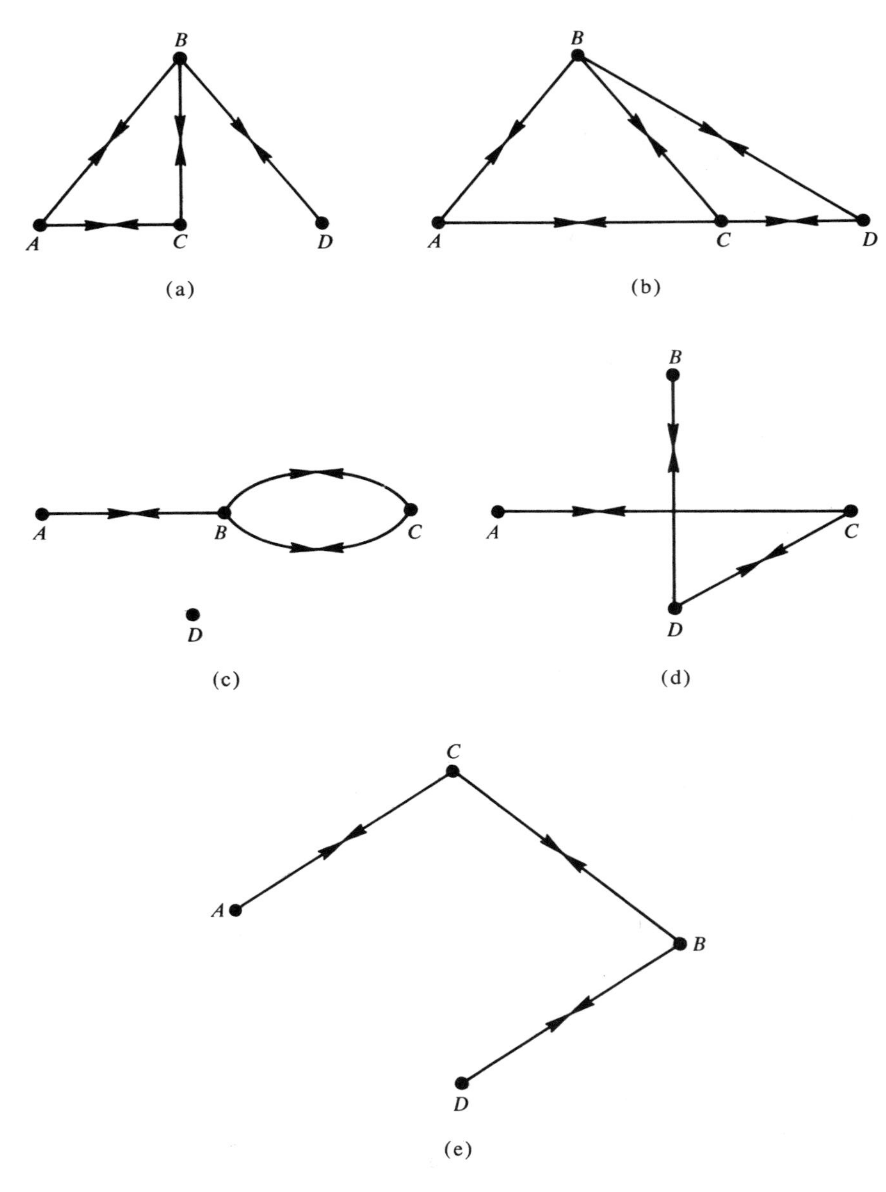

5. For each set of equations listed, write its coefficient matrix and augmented matrix.

(a) $3x + 5y = 8$

$4x - 2y = 0.$

(b) $3x + 2y - z = 3$

$2x - 3y + z = 5.$

(c) $\begin{aligned} 2x + 3y &= 4 \\ x + 2y &= 8 \\ x - 4y &= -4. \end{aligned}$

(d) $\begin{aligned} x + y + z &= 3 \\ x + y \quad &= 2 \\ y + z &= 1. \end{aligned}$

(e) $2x - y = 5.$

(f) $2x - 4y + z = 8.$

6. Draw a graph of the games still to be played in the round robin of Example 3. Write down the adjacency matrix of this graph.

7. The matrix of a magic square is

$$\mathbf{A} = \begin{bmatrix} 12 & 13 & 3 & 6 \\ 7 & 2 & 16 & 9 \\ 14 & 11 & 5 & 4 \\ 1 & 8 & 10 & 15 \end{bmatrix}$$

Find the following sums:

(a) $a_{i1} + a_{i2} + a_{i3} + a_{i4}$ when (i) $i = 1$; (ii) $i = 2$; (iii) $i = 3$; (iv) $i = 4$.

(b) $a_{1j} + a_{2j} + a_{3j} + a_{4j}$ when (i) $j = 1$; (ii) $j = 2$; (iii) $j = 3$; (iv) $j = 4$.

(c) $a_{11} + a_{22} + a_{33} + a_{44}$ (the main diagonal).

(d) The sum of all entries, a_{ij} for which
 (i) $i - j = 1$ or -3.
 (ii) $i - j = 2$ or -2.
 (iii) $i - j = 3$ or -1.

(e) The sum of all entries a_{ij} for which
 (i) $i + j = 5$ (the other diagonal).
 (ii) $i + j = 6$ or 2.
 (iii) $i + j = 7$ or 3.
 (iv) $i + j = 8$ or 4.

3.2 *Matrix Addition and Scalar Multiplication*

Numbers tell how many objects there are in a set or the size of an object. If that were all they did, they would be quite valuable. But their value increases considerably when operations such as addition are available. Operations enable us to make judgments concerning relations between numbers. An illustration of this is the case of two people, one 23 years old and the other 20 years old. To see that one is three years older then the other, we use the operation of subtraction.

Just as numbers become more valuable when operations on them are available, so too for matrices. We present three operations, two on matrices (addition and multiplication), plus a third type of operation, which operates on a pair of dissimilar objects, a number and a matrix.

To motivate addition with matrices, consider the case of a contractor who builds two models of homes, call them I and II. He builds in three towns, Huntington, Smithtown, and Merrick. Table P illustrates how many homes of each model he built in each town in 1969, and Table Q tells the 1970 story.

	I	II
Huntington	8	4
Smithtown	3	2
Merrick	5	1

Table P
(1969)

	I	II
Huntington	3	5
Smithtown	1	2
Merrick	2	4

Table Q
(1970)

One might ask: How many of each model, in each town, did he build for both years? Unhesitatingly we add entries in corresponding locations and write a new table with sums of corresponding entries in the same location. We call the third table, Table R.

	I	II		I	II
Huntington	8	4		3	5
Smithtown	3	2	"+"	1	2
Merrick	5	1		2	4
	Table P (1969)			**Table Q** (1970)	

	I	II		I	II
	8 + 3	4 + 5		11	9
=	3 + 1	2 + 2	=	4	4
	5 + 2	1 + 4		7	5
	Table R (1969 + 1970)			**Table R** (1969 + 1970)	

The corresponding matrices are **P**, **Q**, and **R**:

$$\begin{bmatrix} 8 & 4 \\ 3 & 2 \\ 5 & 1 \end{bmatrix} + \begin{bmatrix} 3 & 5 \\ 1 & 2 \\ 2 & 4 \end{bmatrix}$$

$$\mathbf{P} \quad + \quad \mathbf{Q}$$

$$= \begin{bmatrix} 8+3 & 4+5 \\ 3+1 & 2+2 \\ 5+2 & 1+4 \end{bmatrix} = \begin{bmatrix} 11 & 9 \\ 4 & 4 \\ 7 & 5 \end{bmatrix}$$

$$= \quad \mathbf{R}$$

In this example we have introduced equal matrices. Two matrices are *equal* if they are the same size and their corresponding entries are equal.

It is natural to call **R** the *sum* of **P** and **Q** and the operation itself *addition* with matrices. Observe that addition with matrices makes sense only for two matrices that have the same number of rows and the same number of columns.

EXAMPLE 1

(a) $$\begin{bmatrix} 3 & 2 & 1 \\ -3 & 0 & 2 \end{bmatrix} + \begin{bmatrix} -1 & 2 & -1 \\ 3 & 4 & 3 \end{bmatrix}$$

$$= \begin{bmatrix} 2 & 4 & 0 \\ 0 & 4 & 5 \end{bmatrix}$$

(b) $$\begin{bmatrix} 3 & 2 & -1 \\ 4 & 8 & 7 \\ 6 & 2 & 1 \end{bmatrix} + \begin{bmatrix} 0 & 0 & 0 \\ 0 & 0 & 0 \\ 0 & 0 & 0 \end{bmatrix}$$

$$= \begin{bmatrix} 3 & 2 & -1 \\ 4 & 8 & 7 \\ 6 & 2 & 1 \end{bmatrix}$$

(c) There is no sum

$$\begin{bmatrix} 3 & 2 \\ 1 & 5 \end{bmatrix} + \begin{bmatrix} 1 & 2 \\ 3 & 4 \\ 5 & 6 \end{bmatrix}$$

(d) Solve for all variables if:

$$\begin{bmatrix} x & y \\ 3 & 2 \end{bmatrix} + \begin{bmatrix} 3 & 9 \\ z & w \end{bmatrix} = \begin{bmatrix} 1 & 2 \\ 3 & 4 \end{bmatrix}$$

The sum of the matrices in the left member is

$$\begin{bmatrix} x + 3 & y + 9 \\ 3 + z & 2 + w \end{bmatrix}$$

Therefore, setting corresponding entries of the equal matrices equal yields

$$x + 3 = 1, \quad y + 9 = 2, \quad 3 + z = 3, \quad 2 + w = 4$$

or

$$x = -2, \quad y = -7, \quad z = 0, \quad w = 2$$

Some obvious facts about matrix addition are:

1. The sum of two $m \times n$ matrices is also an $m \times n$ matrix.
2. Addition with matrices is commutative, because addition of their components is commutative. That is $\mathbf{A} + \mathbf{B} = \mathbf{B} + \mathbf{A}$.
3. Addition with matrices is associative, because addition of their components is associative. That is, if $\mathbf{A}$, $\mathbf{B}$, $\mathbf{C}$ are $m \times n$ matrices, then $(\mathbf{A} + \mathbf{B}) + \mathbf{C} = \mathbf{A} + (\mathbf{B} + \mathbf{C})$.
4. In the set of all 2×3 matrices (denoted $M_{2\times 3}$) the *zero* matrix

$$\begin{bmatrix} 0 & 0 & 0 \\ 0 & 0 & 0 \end{bmatrix}$$

has the property that when added to any other 2×3 matrix $\mathbf{A}$ the result is $\mathbf{A}$.

EXAMPLE 2

$$\begin{bmatrix} 2 & 4 & 8 \\ -3 & -4 & 7 \end{bmatrix} + \begin{bmatrix} 0 & 0 & 0 \\ 0 & 0 & 0 \end{bmatrix} = \begin{bmatrix} 2 & 4 & 8 \\ -3 & -4 & 7 \end{bmatrix}$$

The zero matrix for $M_{2\times 3}$ is denoted $\mathbf{0}_{2\times 3}$. For any m and n the zero matrix for the set of $m \times n$ matrices, $M_{m\times n}$ is denoted $\mathbf{0}_{m\times n}$.

A not-so-obvious property of matrix addition is:

5. For any $m \times n$ matrix $\mathbf{A}$ there is another $m \times n$ matrix $\mathbf{B}$ such that $\mathbf{A} + \mathbf{B} = \mathbf{0}_{m\times n}$. $\mathbf{B}$ is obtained by changing the sign of each entry of $\mathbf{A}$. $\mathbf{B}$ and $\mathbf{A}$ are called *additive inverses* of each other.

EXAMPLE 3

$$\begin{bmatrix} 2 & 1 & 5 & 6 \\ -1 & 0 & 4 & -2 \end{bmatrix} + \begin{bmatrix} -2 & -1 & -5 & -6 \\ 1 & 0 & -4 & 2 \end{bmatrix}$$

$$= \begin{bmatrix} 0 & 0 & 0 & 0 \\ 0 & 0 & 0 & 0 \end{bmatrix} = \mathbf{0}_{2\times 4}$$

EXAMPLE 4

Solve for x, y, and w if

$$\begin{bmatrix} x & y \\ z & w \end{bmatrix} + \begin{bmatrix} 3 & 2 \\ 1 & -5 \end{bmatrix} = \begin{bmatrix} 1 & 2 \\ 3 & 4 \end{bmatrix}$$

Add the additive inverse of

$$\begin{bmatrix} 3 & 2 \\ 1 & -5 \end{bmatrix} \quad \text{namely,} \quad \begin{bmatrix} -3 & -2 \\ -1 & 5 \end{bmatrix}$$

to each member of the given matrix equation. This gives

$$\begin{bmatrix} x & y \\ z & w \end{bmatrix} + \begin{bmatrix} 3 & 2 \\ 1 & -5 \end{bmatrix} + \begin{bmatrix} -3 & -2 \\ -1 & 5 \end{bmatrix}$$

$$= \begin{bmatrix} 1 & 2 \\ 3 & 4 \end{bmatrix} + \begin{bmatrix} -3 & -2 \\ -1 & 5 \end{bmatrix}$$

$$\begin{bmatrix} x & y \\ z & w \end{bmatrix} + \begin{bmatrix} 0 & 0 \\ 0 & 0 \end{bmatrix} = \begin{bmatrix} -2 & 0 \\ 2 & 9 \end{bmatrix}$$

$$\begin{bmatrix} x & y \\ z & w \end{bmatrix} = \begin{bmatrix} -2 & 0 \\ 2 & 9 \end{bmatrix}$$

Checking this answer in the original matrix equation gives

$$\begin{bmatrix} -2 & 0 \\ 2 & 9 \end{bmatrix} + \begin{bmatrix} 3 & 2 \\ 1 & -5 \end{bmatrix} = \begin{bmatrix} 1 & 2 \\ 3 & 4 \end{bmatrix} \quad \checkmark$$

Returning to the case of the contractor, one might ask for a production matrix that lists the quantity of each model he is to build if he wishes, in 1970, to triple his production of each model. This is asking: What is 3 times

$$\begin{bmatrix} 3 & 5 \\ 1 & 2 \\ 2 & 4 \end{bmatrix}?$$

The commonsense answer is to triple each entry in this matrix, obtaining

$$\begin{bmatrix} 9 & 15 \\ 3 & 6 \\ 6 & 12 \end{bmatrix}$$

We write

$$3\begin{bmatrix} 3 & 5 \\ 1 & 2 \\ 2 & 4 \end{bmatrix} = \begin{bmatrix} 9 & 15 \\ 3 & 6 \\ 6 & 12 \end{bmatrix}$$

and think of this as multiplication of a number with a matrix. Numbers are sometimes called *scalars*, and this multiplication of a scalar and a matrix is called *scalar multiplication*.

EXAMPLE 5

(a)

$$2\begin{bmatrix} 1 & -1 \\ \frac{1}{2} & \frac{1}{3} \end{bmatrix} = \begin{bmatrix} 2 & -2 \\ 1 & \frac{2}{3} \end{bmatrix}$$

(b)

$$-2\begin{bmatrix}1\\0\\4\end{bmatrix}=\begin{bmatrix}-2\\0\\-8\end{bmatrix}$$

(c) $\frac{1}{2}\begin{bmatrix}2 & 4 & 5\end{bmatrix}=\begin{bmatrix}1 & 2 & \frac{5}{2}\end{bmatrix}$

(d)

$$2\begin{bmatrix}3 & 2 & 1\\0 & 1 & 2\end{bmatrix}+3\begin{bmatrix}-1 & 4 & 2\\2 & 0 & -2\end{bmatrix}$$

$$=\begin{bmatrix}6 & 4 & 2\\0 & 2 & 4\end{bmatrix}+\begin{bmatrix}-3 & 12 & 6\\6 & 0 & -6\end{bmatrix}$$

$$=\begin{bmatrix}3 & 16 & 8\\6 & 2 & -2\end{bmatrix}$$

(e)

$$3\left(\begin{bmatrix}1 & 3\\2 & -1\\4 & 0\end{bmatrix}+\begin{bmatrix}-1 & 2\\1 & 1\\3 & 5\end{bmatrix}\right)=3\begin{bmatrix}0 & 5\\3 & 0\\7 & 5\end{bmatrix}=\begin{bmatrix}0 & 15\\9 & 0\\21 & 15\end{bmatrix}$$

The same result is found if the multiplication by 3 is distributed over each matrix in the sum, as follows:

$$3\begin{bmatrix}1 & 3\\2 & -1\\4 & 0\end{bmatrix}+3\begin{bmatrix}-1 & 2\\1 & 1\\3 & 5\end{bmatrix}=\begin{bmatrix}3 & 9\\6 & -3\\12 & 0\end{bmatrix}$$

$$+\begin{bmatrix}-3 & 6\\3 & 3\\9 & 15\end{bmatrix}=\begin{bmatrix}0 & 15\\9 & 0\\21 & 15\end{bmatrix}$$

(f)

$$(a+b)\begin{bmatrix}3 & 1 & 4\\0 & 2 & -1\end{bmatrix}=\begin{bmatrix}(a+b)3 & (a+b)1 & (a+b)4\\(a+b)0 & (a+b)2 & (a+b)(-1)\end{bmatrix}$$

$$=\begin{bmatrix}3a+3b & a+b & 4a+4b\\0 & 2a+2b & -a-b\end{bmatrix}$$

The same result is obtained if the scalar multiplication is distributed over each addend in the sum $a + b$, as follows:

$$a\begin{bmatrix} 3 & 1 & 4 \\ 0 & 2 & -1 \end{bmatrix} + b\begin{bmatrix} 3 & 1 & 4 \\ 0 & 2 & -1 \end{bmatrix} = \begin{bmatrix} 3a & a & 4a \\ 0 & 2a & -a \end{bmatrix}$$

$$+\begin{bmatrix} 3b & b & 4b \\ 0 & 2b & -b \end{bmatrix} = \begin{bmatrix} 3a + 3b & a + b & 4a + 4b \\ 0 & 2a + 2b & -a \quad -b \end{bmatrix}$$

The following properties of scalar multiplication are easily proved. For all scalars c and d and $m \times n$ matrices $\mathbf{A}$ and $\mathbf{B}$:

1. $1\mathbf{A} = \mathbf{A}$.
2. $(c + d)\mathbf{A} = c\mathbf{A} + d\mathbf{A}$.
3. $c(\mathbf{A} + \mathbf{B}) = c\mathbf{A} + c\mathbf{B}$.
4. $(cd)\mathbf{A} = c(d\mathbf{A})$.

EXAMPLE 6

Solve for x, y, z, and w if

$$2\begin{bmatrix} x & y \\ z & w \end{bmatrix} + 3\begin{bmatrix} 1 & 2 \\ 3 & 4 \end{bmatrix} = 4\begin{bmatrix} 4 & 3 \\ 2 & 1 \end{bmatrix}$$

By scalar multiplication

$$\begin{bmatrix} 2x & 2y \\ 2z & 2w \end{bmatrix} + \begin{bmatrix} 3 & 6 \\ 9 & 12 \end{bmatrix} = \begin{bmatrix} 16 & 12 \\ 8 & 4 \end{bmatrix}$$

Adding,

$$\begin{bmatrix} 2x + 3 & 2y + 6 \\ 2z + 9 & 2w + 12 \end{bmatrix} = \begin{bmatrix} 16 & 12 \\ 8 & 4 \end{bmatrix}$$

Thus

$$2x + 3 = 16, \quad 2y + 6 = 12, \quad 2z + 9 = 8, \quad 2w + 12 = 4$$

or

$$x = 13/2, \quad y = 3, \quad z = -1/2, \quad w = -4$$

Exercises

1. If possible, add. If not, explain why.

(a) $\begin{bmatrix} 1 & 2 \\ 3 & 4 \end{bmatrix} + \begin{bmatrix} -1 & 2 \\ 3 & -2 \end{bmatrix}$

(b) $\begin{bmatrix} 1 & 2 \\ 3 & 4 \end{bmatrix} + \begin{bmatrix} -1 \\ 3 \end{bmatrix}$

(c) $\begin{bmatrix} 3 & 2 & 1 \\ 4 & 6 & 8 \end{bmatrix} + \begin{bmatrix} -3 & -1 & 2 \\ 0 & 0 & 1 \end{bmatrix}$

(d) $\begin{bmatrix} a \\ b \end{bmatrix} + \begin{bmatrix} a & b \end{bmatrix}$

(e) $\begin{bmatrix} 3 & 2 \\ 4 & 5 \\ -2 & -3 \end{bmatrix} + \begin{bmatrix} -3 & -2 \\ -4 & -5 \\ 2 & 3 \end{bmatrix}$

(f) $\begin{bmatrix} a & b \\ c & d \end{bmatrix} + \begin{bmatrix} 0 & 0 \\ 0 & 0 \end{bmatrix}$

2. Express each of the following as a single matrix.

(a) $4\begin{bmatrix} 3 & 2 \\ -1 & 0 \end{bmatrix}$ (b) $\frac{1}{2}\begin{bmatrix} 4 \\ -1 \\ 3 \end{bmatrix}$ (c) $0\begin{bmatrix} 2 & 0 & 4 \\ -2 & 0 & -4 \end{bmatrix}$

3. Express as a single matrix (if possible).

(a) $3\begin{bmatrix} 2 & 3 \\ 1 & 0 \end{bmatrix} - 2\begin{bmatrix} 1 & 4 \\ 2 & -1 \end{bmatrix}$ (b) $a\begin{bmatrix} 1 & 0 \\ 0 & 1 \end{bmatrix} + b\begin{bmatrix} 0 & 1 \\ 1 & 0 \end{bmatrix}$

(c) $a\begin{bmatrix} 1 & 0 \\ 0 & 0 \end{bmatrix} + b\begin{bmatrix} 0 & 1 \\ 0 & 0 \end{bmatrix} + c\begin{bmatrix} 0 & 0 \\ 1 & 0 \end{bmatrix} + d\begin{bmatrix} 0 & 0 \\ 0 & 1 \end{bmatrix}$

(d) $\frac{1}{2}\begin{bmatrix} 2 & 4 & 6 \\ 3 & 1 & 0 \\ 0 & 8 & 2 \end{bmatrix} + \frac{1}{3}\begin{bmatrix} 6 & 9 & 12 \\ 0 & 15 & 2 \\ 0 & -3 & -6 \end{bmatrix}$

4. Solve for a, b, c, d, e, and f (if possible).

(a) $2\begin{bmatrix} a & b & c \\ 4 & 5 & 6 \end{bmatrix} = 3\begin{bmatrix} 1 & 2 & 3 \\ d & e & -f \end{bmatrix}$

(b) $3\begin{bmatrix} a & 1 & c \\ d & -1 & f \end{bmatrix} + 2\begin{bmatrix} 2 & b & 4 \\ -1 & e & 0 \end{bmatrix} = \begin{bmatrix} 0 & 0 & 0 \\ 0 & 0 & 0 \end{bmatrix}$

(c)
$$2\begin{bmatrix} 2a & -d \\ b & -e \\ 3c & -f \end{bmatrix} - 2\begin{bmatrix} 1 & 2 \\ 0 & 1 \\ 0 & 0 \end{bmatrix} = \begin{bmatrix} a & d \\ b & e \\ c & f \end{bmatrix}$$

5. Solve for x and y.

(a) $\begin{bmatrix} x + y \\ x - y \end{bmatrix} = \begin{bmatrix} 7 \\ 3 \end{bmatrix}$ (b) $\begin{bmatrix} x \\ y \end{bmatrix} + 2\begin{bmatrix} x \\ y \end{bmatrix} = \begin{bmatrix} 6 \\ 9 \end{bmatrix}$

(c) $\begin{bmatrix} x \\ y \end{bmatrix} + 2\begin{bmatrix} y \\ x \end{bmatrix} = \begin{bmatrix} 5 \\ 4 \end{bmatrix}$ (d) $\begin{bmatrix} x \\ y \end{bmatrix} = \begin{bmatrix} 3 \\ 4 \end{bmatrix}$

6. Add the adjacency matrices of Example 3 of Section 3.1 and Exercise 6 of Section 3.1. Interpret this sum. Draw the graph that corresponds to this sum. (Such a graph is called a *complete graph.*)

3.3 *Matrix Multiplication*

In the study of mathematics there is a constant effort to obtain new information from old. The procedures for so doing are the various operations. Thus the operation addition gives one number as the sum of two others. The operation multiplication gives one number as the product of two others. The operation pivot exchange applied to a tableau gives a different tableau with a new perspective. The operation matrix addition gives one matrix from two others, and the operation scalar multiplication gives a matrix from a matrix and a number. There is another type of multiplication for matrices in which two matrices are multiplied together, resulting in a third matrix.

Matrix multiplication has its origin in Cayley's investigations of mappings.† It came about because Cayley was investigating the operation composition for which two mappings F and G compose into a new mapping whose domain is that of F and whose codomain is that of G. To see how this composition works it is necessary to use general symbols on the inside of the tableaus of the following mappings. Consider

$F: R^3 \rightarrow R^2$

	x_1	x_2	x_3	
F:	a_1	a_2	a_3	$= y_1$
	b_1	b_2	b_3	$= y_2$

†For those of you whose course of study caused you to omit Section 2.8, read "tableau" for "mapping" wherever it appears.

and

$$G: R^2 \to R^4$$

$$G: \quad \begin{array}{cc} y_1 & y_2 \\ \hline c_1 & c_2 \\ d_1 & d_2 \\ e_1 & e_2 \\ f_1 & f_2 \end{array} \quad \begin{array}{l} \\ = z_1 \\ = z_2 \\ = z_3 \\ = z_4 \end{array}$$

Under F, $(x_1, x_2, x_3) \mapsto (y_1, y_2)$ and under G, $(y_1, y_2) \mapsto (z_1, z_2, z_3, z_4)$. The net effect of the composition F followed by G is $(x_1, x_2, x_3) \mapsto (z_1, z_2, z_3, z_4)$. (See Figure 3.3.)

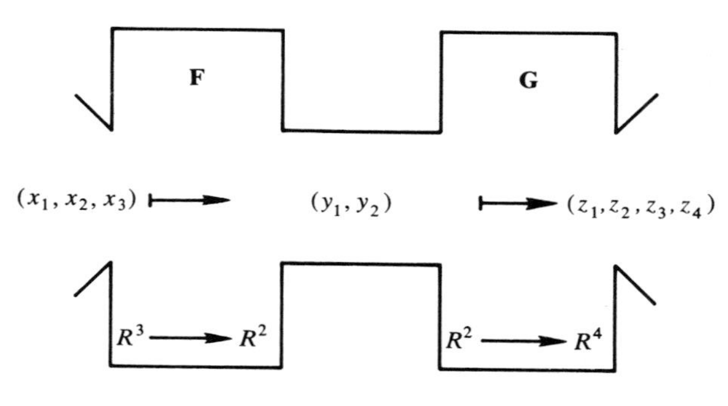

Figure 3.3

Let us examine how this composition is carried out. From row 1 of the mapping F we get an expression for y_1. Replace y_1 in G by this expression so that

$$(1) \quad \begin{array}{cc} a_1x_1 + a_2x_2 + a_3x_3 & y_2 \\ \hline c_1 & c_2 \\ d_1 & d_2 \\ e_1 & e_2 \\ f_1 & f_2 \end{array} \quad \begin{array}{l} \\ = z_1 \\ = z_2 \\ = z_3 \\ = z_4 \end{array}$$

Similarly, from row 2 of F we get $y_2 = b_1x_1 + b_2x_2 + b_3x_3$. Substituting this value for y_2 in (1) we have

(2)

$a_1x_1 + a_2x_2 + a_3x_3$	$b_1x_1 + b_2x_2 + b_3x_3$	
c_1	c_2	$= z_1$
d_1	d_2	$= z_2$
e_1	e_2	$= z_3$
f_1	f_2	$= z_4$

Row 1 of (2) is

$$c_1a_1x_1 + c_1a_2x_2 + c_1a_3x_3 + c_2b_1x_1 + c_2b_2x_2 + c_2b_3x_3 = z_1$$

or, collecting terms,

$$(c_1a_1 + c_2b_1)x_1 + (c_1a_2 + c_2b_2)x_2 + (c_1a_3 + c_2b_3)x_3 = z_1$$

which is written in tableau form as

x_1	x_2	x_3	
$c_1a_1 + c_2b_1$	$c_1a_2 + c_2b_2$	$c_1a_3 + c_2b_3$	$= z_1$

Similarly, from (2) we can get expressions for z_2, z_3, and z_4 (try z_2 yourself), and write a new tableau:

(3)

x_1	x_2	x_3	
$c_1a_1 + c_2b_1$	$c_1a_2 + c_2b_2$	$c_1a_3 + c_2b_3$	$= z_1$
$d_1a_1 + d_2b_1$	$d_1a_2 + d_2b_2$	$d_1a_3 + d_2b_3$	$= z_2$
$e_1a_1 + e_2b_1$	$e_1a_2 + e_2b_2$	$e_1a_3 + e_2b_3$	$= z_3$
$f_1a_1 + f_2b_1$	$f_1a_2 + f_2b_2$	$f_1a_3 + f_2b_3$	$= z_4$

This tableau (3) is a new mapping $R^3 \to R^4$ obtained from the composition of the tableau for F followed by the tableau for G. It is called the *composite* of F and G. To show its origins we denote this mapping by the symbolism

$$G \circ F : R^3 \to R^4$$

where $G \circ F$ is read "F followed by G" or "G operating on F." This new mapping can be denoted by $H : R^3 \to R^4$; however, we use the symbol $G \circ F$ to indicate that the composite map has its origins in F and in G.

It is clear that the composition of two mappings, where the domain of the second (in this case G) contains the codomain of the first, is a new mapping which has as its domain the domain of the first and as its codomain the codomain of the second. What is not yet clear is how to express the relationship that exists between the insides of the tableaus of F, G, and $G \circ F$. It is this relationship that defines multiplication of the matrices **F** and **G**. That there is a pattern is clear from (3). The genius of Cayley is that he was able to express this pattern in the format called matrix multiplication. This matrix multiplication, which sprang from the composition of mappings, then developed "a life of its own," as will be amply illustrated in the rest of this book. To "rediscover" Cayley's work let's look at the matrices of the mappings F, G, and $G \circ F$.

Figure 3.4 shows the multiplication **G** times **F** written **GF**.

$$\begin{bmatrix} c_1 & c_2 \\ d_1 & d_2 \\ e_1 & e_2 \\ f_1 & f_2 \end{bmatrix} \begin{bmatrix} a_1 & a_2 & a_3 \\ b_1 & b_2 & b_3 \end{bmatrix} = \begin{bmatrix} c_1a_1 + c_2b_1 & c_1a_2 + c_2b_2 & c_1a_3 + c_2b_3 \\ d_1a_1 + d_2b_1 & d_1a_2 + d_2b_2 & d_1a_3 + d_2b_3 \\ e_1a_1 + e_2b_1 & e_1a_2 + e_2b_2 & e_1a_3 + e_2b_3 \\ f_1a_1 + f_2b_1 & f_1a_2 + f_2b_2 & f_1a_3 + f_2b_3 \end{bmatrix}$$

Figure 3.4

Let us look at the entry $c_1a_1 + c_2b_1$ in the first row and first column of $G \circ F$. It is obtained by multiplying the entries in the first row of **G** by the corresponding entries in the first column of **F** and then adding the products:

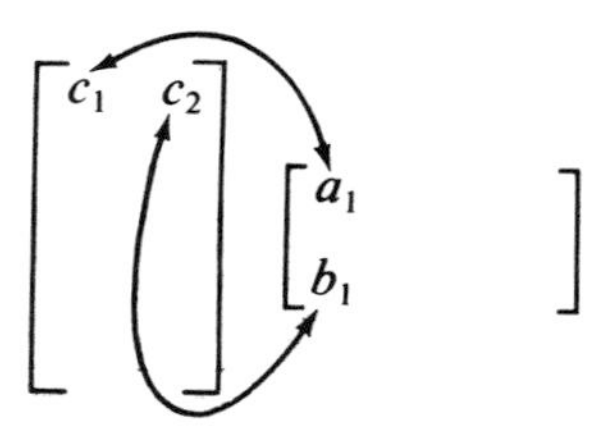

The entry $e_1a_2 + e_2b_2$ is in the third row and second column of $G \circ F$. It is obtained by multiplying the entries in the third row of **G** by the corresponding entries in the second column of **F** and then adding the products:

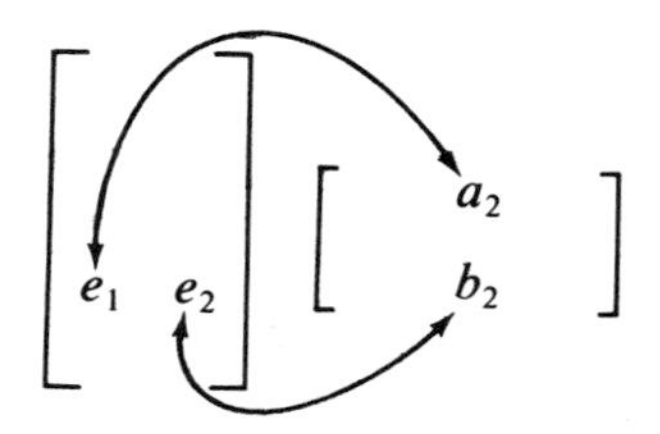

The pattern is clear. To multiply two matrices **A** and **B** together to obtain a third matrix **C** do the following:

1. If the number of entries in a row of **A** is the same as the number of entries in a column of **B**, then we can multiply **AB**. Otherwise, not.
2. If **AB** = **C**, then the entry c_{ij} in **C** (the entry in the ith row and jth column of **C**) is obtained by multiplying corresponding entries of the ith row of **A** and the jth column of **B** and adding these products.

EXAMPLE 1

Let

$$\mathbf{Q} = \begin{bmatrix} 6 & 3 \\ 2 & 7 \\ 4 & -3 \end{bmatrix} \quad \text{and} \quad \mathbf{S} = \begin{bmatrix} 1 & 8 \\ 5 & 9 \end{bmatrix}$$

Then

$$\mathbf{QS} = \begin{bmatrix} 6 & 3 \\ 2 & 7 \\ 4 & -3 \end{bmatrix} \begin{bmatrix} 1 & 8 \\ 5 & 9 \end{bmatrix}$$

$$= \begin{bmatrix} 6 \cdot 1 + 3 \cdot 5 & 6 \cdot 8 + 3 \cdot 9 \\ 2 \cdot 1 + 7 \cdot 5 & 2 \cdot 8 + 7 \cdot 9 \\ 4 \cdot 1 - 3 \cdot 5 & 4 \cdot 8 - 3 \cdot 9 \end{bmatrix} = \begin{bmatrix} 21 & 75 \\ 37 & 79 \\ -11 & 5 \end{bmatrix}$$

EXAMPLE 2

$$\mathbf{SQ} = \begin{bmatrix} 1 & 8 \\ 5 & 9 \end{bmatrix} \begin{bmatrix} 6 & 3 \\ 2 & 7 \\ 4 & -3 \end{bmatrix}$$

is not defined since the number of elements in a row of **S** is not equal to the number of elements in a column of **Q**. This points out that multiplication is not commutative.

EXAMPLE 3

$$\begin{bmatrix} 20 & 15 & 10 \\ 15 & 10 & 20 \end{bmatrix} \begin{bmatrix} 2000 \\ 2500 \\ 3000 \end{bmatrix} = \begin{bmatrix} 107{,}500 \\ 115{,}000 \end{bmatrix}$$

EXAMPLE 4

$$\begin{bmatrix} 1 & 2 & 3 & 4 \end{bmatrix} \begin{bmatrix} -1 \\ -2 \\ 3 \\ -5 \end{bmatrix} = \begin{bmatrix} -16 \end{bmatrix}$$

EXAMPLE 5

$$\begin{bmatrix} -1 \\ -2 \\ 3 \\ -5 \end{bmatrix} \begin{bmatrix} 1 & 2 & 3 & 4 \end{bmatrix} = \begin{bmatrix} -1 & -2 & -3 & -4 \\ -2 & -4 & -6 & -8 \\ 3 & 6 & 9 & 12 \\ -5 & -10 & -15 & -20 \end{bmatrix}$$

Examples 4 and 5 point out that even if **AB** and **BA** both exist, **AB** may not be equal to (or even resemble) **BA**. We do assert that if

$$\mathbf{AB} = \mathbf{C}$$

then **C** has as many rows as **A** and as many columns as **B**.

Also **AB** will exist if and only if the number of columns of **A** is equal to the number of rows of **B**.

EXAMPLE 6

Multiply **AB** and then multiply this result with **C** when

$$\mathbf{A} = \begin{bmatrix} 1 & -1 & 2 \\ -2 & 1 & 3 \end{bmatrix}, \quad \mathbf{B} = \begin{bmatrix} 2 & -1 \\ 1 & 0 \\ 0 & 2 \end{bmatrix}, \quad \mathbf{C} = \begin{bmatrix} 3 & -3 \\ 2 & 4 \end{bmatrix}$$

$$\mathbf{AB} = \begin{bmatrix} 1 & -1 & 2 \\ -2 & 1 & 3 \end{bmatrix} \begin{bmatrix} 2 & -1 \\ 1 & 0 \\ 0 & 9 \end{bmatrix} = \begin{bmatrix} 1 & 3 \\ -3 & 8 \end{bmatrix}$$

$$\begin{bmatrix} 1 & 3 \\ -3 & 8 \end{bmatrix} \begin{bmatrix} 3 & -3 \\ 2 & 4 \end{bmatrix} = \begin{bmatrix} 9 & 9 \\ 7 & 41 \end{bmatrix}$$

Symbolically,

$$(\mathbf{AB})\mathbf{C} = \begin{bmatrix} 9 & 9 \\ 7 & 41 \end{bmatrix}$$

EXAMPLE 7

Using the same **A**, **B**, and **C** as in Example 6, multiply **BC** and then multiply the result with **A** to obtain **A**(**BC**).

$$\mathbf{BC} = \begin{bmatrix} 2 & -1 \\ 1 & 0 \\ 0 & 2 \end{bmatrix} \begin{bmatrix} 3 & -3 \\ 2 & 4 \end{bmatrix} = \begin{bmatrix} 4 & -10 \\ 3 & -3 \\ 4 & 8 \end{bmatrix}$$

$$\begin{bmatrix} 1 & -1 & 2 \\ -2 & 1 & 3 \end{bmatrix} \begin{bmatrix} 4 & -10 \\ 3 & -3 \\ 4 & 8 \end{bmatrix} = \begin{bmatrix} 9 & 9 \\ 7 & 41 \end{bmatrix}$$

Symbolically,

$$\mathbf{A}(\mathbf{BC}) = \begin{bmatrix} 9 & 9 \\ 7 & 41 \end{bmatrix}$$

Examples 6 and 7 hint that whenever multiplication exists, then

$$(\mathbf{AB})\mathbf{C} = \mathbf{A}(\mathbf{BC})$$

That is, matrix multiplication is associative. Indeed, this is true. You can see this for yourself for the 3×3 case by multiplying together

$$\mathbf{A} = \begin{bmatrix} a_{11} & a_{12} & a_{13} \\ a_{21} & a_{22} & a_{23} \\ a_{31} & a_{32} & a_{33} \end{bmatrix}, \quad \mathbf{B} = \begin{bmatrix} b_{11} & b_{12} & b_{13} \\ b_{21} & b_{22} & b_{23} \\ b_{31} & b_{32} & b_{33} \end{bmatrix}, \quad \mathbf{C} = \begin{bmatrix} c_{11} & c_{12} & c_{13} \\ c_{21} & c_{22} & c_{23} \\ c_{31} & c_{32} & c_{33} \end{bmatrix}$$

Since matrix multiplication springs from composition of mappings, we also have that whenever the compositions of mappings make sense (this is the equivalent of saying matrix multiplication is defined), then the composition of the mappings is associative.

The mapping $I_3 : R^3 \rightarrow R^3$,

x_1	x_2	x_3	
1	0	0	$= y_1$
0	1	0	$= y_2$
0	0	1	$= y_3$

is called an *identity mapping.* The reason for this name is obvious. Every domain element (x_1, x_2, x_2) is mapped identically to itself.

$$(\ 1, 1, 1) \mapsto (\ 1, 1, 1)$$
$$(-7, 2, 5) \mapsto (-7, 2, 5) \qquad \text{etc.}$$

For any positive integer n the mapping $I_n : R^n \to R^n$ with a tableau that has a 1 for each diagonal element and zero for all other positions is also called an identity mapping. We will sometimes write I for I_n when the n is clearly understood in the context of our work.

The matrix

$$\mathbf{I}_3 = \begin{bmatrix} 1 & 0 & 0 \\ 0 & 1 & 0 \\ 0 & 0 & 1 \end{bmatrix}$$

is called a 3×3 *identity matrix* or *multiplicative identity matrix:*

$$\mathbf{I}_2 = \begin{bmatrix} 1 & 0 \\ 0 & 1 \end{bmatrix}$$

is a 2×2 identity matrix. The composition of any $n \times n$ mapping M with the identity mapping yields M. This is illustrated for the 2×2 case by

x_1	x_2	
a	b	$= y_1$
c	d	$= y_2$

$\circ$

z_1	z_2	
1	0	$= x_1$
0	1	$= x_2$

$=$

z_1	z_2	
a	b	$= y_1$
c	d	$= y_2$

and

x_1	x_2	
1	0	$= y_1$
0	1	$= y_2$

$\circ$

z_1	z_2	
a	b	$= x_1$
c	d	$= x_2$

$=$

z_1	z_2	
a	b	$= y_1$
c	d	$= y_2$

EXAMPLE 8

Thus for any 2 × 2 matrix

$$\begin{bmatrix} a_{11} & a_{12} \\ a_{21} & a_{22} \end{bmatrix}$$

$\begin{bmatrix} 1 & 0 \\ 0 & 1 \end{bmatrix}$ has the property

$$\begin{bmatrix} 1 & 0 \\ 0 & 1 \end{bmatrix} \begin{bmatrix} a_{11} & a_{12} \\ a_{21} & a_{22} \end{bmatrix} = \begin{bmatrix} a_{11} & a_{12} \\ a_{21} & a_{22} \end{bmatrix}$$

and

$$\begin{bmatrix} a_{11} & a_{12} \\ a_{21} & a_{22} \end{bmatrix} \begin{bmatrix} 1 & 0 \\ 0 & 1 \end{bmatrix} = \begin{bmatrix} a_{11} & a_{12} \\ a_{21} & a_{22} \end{bmatrix}$$

EXAMPLE 9

The multiplicative identity

$$\mathbf{I}_3 = \begin{bmatrix} 1 & 0 & 0 \\ 0 & 1 & 0 \\ 0 & 0 & 1 \end{bmatrix}$$

has the property

$$\begin{bmatrix} 1 & 0 & 0 \\ 0 & 1 & 0 \\ 0 & 0 & 1 \end{bmatrix} \begin{bmatrix} a_{11} & a_{12} & a_{13} \\ a_{21} & a_{22} & a_{23} \\ a_{31} & a_{32} & a_{33} \end{bmatrix} = \begin{bmatrix} a_{11} & a_{12} & a_{13} \\ a_{21} & a_{22} & a_{23} \\ a_{31} & a_{32} & a_{33} \end{bmatrix}$$

and

$$\begin{bmatrix} a_{11} & a_{12} & a_{13} \\ a_{21} & a_{22} & a_{23} \\ a_{31} & a_{32} & a_{33} \end{bmatrix} \begin{bmatrix} 1 & 0 & 0 \\ 0 & 1 & 0 \\ 0 & 0 & 1 \end{bmatrix} = \begin{bmatrix} a_{11} & a_{12} & a_{13} \\ a_{21} & a_{22} & a_{23} \\ a_{31} & a_{32} & a_{33} \end{bmatrix}$$

We have just seen that

$$\mathbf{I}_2 = \begin{bmatrix} 1 & 0 \\ 0 & 1 \end{bmatrix}$$

the multiplicative identity for all 2×2 matrices, has the property that for all 2×2 matrices $\mathbf{A}$, $\mathbf{I}_2\mathbf{A} = \mathbf{A}\mathbf{I}_2 = \mathbf{A}$. Is there another matrix $\mathbf{Y}$ which has this property (that is, $\mathbf{YA} = \mathbf{AY} = \mathbf{A}$)? If so, then

$$\mathbf{YI}_2 = \mathbf{Y} \qquad \text{(proved property of } \mathbf{I}_2\text{)}$$

$$\mathbf{YI}_2 = \mathbf{I}_2 \qquad \text{(because } \mathbf{Y} \text{ is assumed to have this property)}$$

Therefore,

$$\mathbf{Y} = \mathbf{I}_2 \qquad \text{(substitution principle)}$$

Hence there is only one matrix, the multiplicative identity, which has this property for all 2×2 matrices. We sum this up by saying that $\mathbf{I}_2$ is the *unique* multiplicative identity for all 2×2 matrices. Clearly the same proof shows that

$$\mathbf{I}_3 = \begin{bmatrix} 1 & 0 & 0 \\ 0 & 1 & 0 \\ 0 & 0 & 1 \end{bmatrix}$$

is the unique multiplicative identity for all 3×3 matrices. It is easy to show that for any n, $\mathbf{I}_n$ is the unique multiplicative identity for the set of all $n \times n$ matrices. For any n, $n \times n$ matrices are called *square matrices*. The above work shows that each set of square matrices has a multiplicative identity. How about matrices that are not square? Let $\mathbf{A}$ be an $m \times n$ matrix and suppose that matrix $\mathbf{I}$ exists such that

$$\mathbf{AI} = \mathbf{IA} = \mathbf{A}$$

By the multiplicative rules for matrices, $\mathbf{IA} = \mathbf{A}$ implies that $\mathbf{I}$ has as many columns as $\mathbf{A}$ has rows. Therefore, $\mathbf{I}$ has m columns. $\mathbf{AI} = \mathbf{A}$ implies that $\mathbf{A}$ has as many columns as $\mathbf{I}$ has rows. Therefore, $\mathbf{I}$ has n rows. Thus $\mathbf{I}$ is an $n \times m$ matrix. So $\mathbf{IA}$ is an $n \times n$ matrix. But $\mathbf{IA} = \mathbf{A}$. Thus $n = m$, which means that only square matrices have multiplicative identities. Notice that

$$\begin{bmatrix} 1 & 0 \\ 0 & 1 \end{bmatrix} \begin{bmatrix} 2 & 3 & -6 \\ 4 & 7 & 8 \end{bmatrix} = \begin{bmatrix} 2 & 3 & -6 \\ 4 & 7 & 8 \end{bmatrix}$$

However,

$$\begin{bmatrix} 2 & 3 & -6 \\ 4 & 7 & 8 \end{bmatrix} \begin{bmatrix} 1 & 0 \\ 0 & 1 \end{bmatrix}$$

is undefined. Similarly,

$$\begin{bmatrix} 2 & 3 & -6 \\ 4 & 7 & 8 \end{bmatrix} \begin{bmatrix} 1 & 0 & 0 \\ 0 & 1 & 0 \\ 0 & 0 & 1 \end{bmatrix} = \begin{bmatrix} 2 & 3 & -6 \\ 4 & 7 & 8 \end{bmatrix}$$

but

$$\begin{bmatrix} 1 & 0 & 0 \\ 0 & 1 & 0 \\ 0 & 0 & 1 \end{bmatrix} \begin{bmatrix} 2 & 3 & -6 \\ 4 & 7 & 8 \end{bmatrix}$$

is undefined. This points up that for a matrix **Y** to be a multiplicative identity it is not enough that **YA** = **A** for all **A** or that **AY** = **A** for all **A**. It must be true that

$$\mathbf{AY} = \mathbf{YA} = \mathbf{A}$$

In the set of 2 × 2 matrices we ask the question: For each matrix **A**, is there a 2 × 2 matrix **B** such that **AB** = **BA** = **I**? If **B** exists, we call it an *inverse* of **A**. We display a method for finding **B**, if it exists, for a given **A**.

EXAMPLE 10

Find an inverse matrix, if possible, for

$$\mathbf{A} = \begin{bmatrix} 2 & 3 \\ 1 & 2 \end{bmatrix}$$

Let $A : R^2 \rightarrow R^2$ with tableau

x_1	x_2	
2	3	$= y_1$
1	2	$= y_2$

Pivoting as indicated yields

$$(1)\qquad \begin{array}{cc} x_1 & x_2 \\ \hline \multicolumn{1}{|c}{2^*} & \multicolumn{1}{c|}{3} \\ \multicolumn{1}{|c}{1} & \multicolumn{1}{c|}{2} \\ \hline \end{array} \begin{array}{l} \\ = y_1 \\ = y_2 \end{array}$$

$$(2)\qquad \begin{array}{cc} -y_1 & x_2 \\ \hline \multicolumn{1}{|c}{1/2} & \multicolumn{1}{c|}{3/2} \\ \multicolumn{1}{|c}{-1/2} & \multicolumn{1}{c|}{1/2^*} \\ \hline \end{array} \begin{array}{l} \\ = -x_1 \\ = y_2 \end{array}$$

$$(3)\qquad \begin{array}{cc} -y_1 & -y_2 \\ \hline \multicolumn{1}{|c}{2} & \multicolumn{1}{c|}{-3} \\ \multicolumn{1}{|c}{-1} & \multicolumn{1}{c|}{2} \\ \hline \end{array} \begin{array}{l} \\ = -x_1 \\ = -x_2 \end{array}$$

Since all variables have minus signs in (3), the inverse of (1), it can be rewritten as

$$(4)\qquad \begin{array}{cc} y_1 & y_2 \\ \hline \multicolumn{1}{|c}{2} & \multicolumn{1}{c|}{-3} \\ \multicolumn{1}{|c}{-1} & \multicolumn{1}{c|}{2} \\ \hline \end{array} \begin{array}{l} \\ = x_1 \\ = x_2 \end{array}$$

Now we see that

$$\begin{array}{cc} x_1 & x_2 \\ \hline \multicolumn{1}{|c}{2} & \multicolumn{1}{c|}{3} \\ \multicolumn{1}{|c}{1} & \multicolumn{1}{c|}{2} \\ \hline \end{array} \begin{array}{l} \\ = y_1 \\ = y_2 \end{array} \circ \begin{array}{cc} y_1 & y_2 \\ \hline \multicolumn{1}{|c}{2} & \multicolumn{1}{c|}{-3} \\ \multicolumn{1}{|c}{-1} & \multicolumn{1}{c|}{2} \\ \hline \end{array} \begin{array}{l} \\ = x_1 \\ = x_2 \end{array} = \begin{array}{cc} x_1 & x_2 \\ \hline \multicolumn{1}{|c}{1} & \multicolumn{1}{c|}{0} \\ \multicolumn{1}{|c}{0} & \multicolumn{1}{c|}{1} \\ \hline \end{array} \begin{array}{l} \\ = x_1 \\ = x_2 \end{array} = \text{“}I_2\text{”}$$

also

$$\begin{array}{cc} y_1 & y_2 \\ \hline \multicolumn{1}{|c}{2} & \multicolumn{1}{c|}{-3} \\ \multicolumn{1}{|c}{-1} & \multicolumn{1}{c|}{2} \\ \hline \end{array} \begin{array}{l} \\ = x_1 \\ = x_2 \end{array} \circ \begin{array}{cc} x_1 & x \\ \hline \multicolumn{1}{|c}{2} & \multicolumn{1}{c|}{3} \\ \multicolumn{1}{|c}{1} & \multicolumn{1}{c|}{2} \\ \hline \end{array} \begin{array}{l} \\ = y_1 \\ = y_2 \end{array} = \begin{array}{cc} x_1 & x_2 \\ \hline \multicolumn{1}{|c}{1} & \multicolumn{1}{c|}{0} \\ \multicolumn{1}{|c}{0} & \multicolumn{1}{c|}{1} \\ \hline \end{array} \begin{array}{l} \\ = x_1 \\ = x_2 \end{array} = \text{“}I_2\text{”}$$

It is clear that the inverse of the matrix

$$\begin{bmatrix} 2 & 3 \\ 1 & 2 \end{bmatrix} \quad \text{is} \quad \begin{bmatrix} 2 & -3 \\ -1 & 2 \end{bmatrix}$$

EXAMPLE 11

Find the inverse of

$$\mathbf{A} = \begin{bmatrix} 1 & 2 & 3 \\ 2 & -1 & -2 \\ 3 & 1 & 1 \end{bmatrix}$$

Consider the mapping $R^3 \to R^3$ associated with $\mathbf{A}$, and the sequence of pivot exchanges on diagonal entries.

(1)

	x_1	x_2	x_3		
	1*	2	3	=	y_1
	2	−1	−2	=	y_2
	3	1	1	=	y_3

(2)

	$-y_1$	x_2	x_3		
	1	2	3	=	$-x_1$
	−2	−5*	−8	=	y_2
	−3	−5	−8	=	y_3

(3)

	$-y_1$	$-y_2$	x_3		
				=	$-x_1$
				=	$-x_2$
			0	=	y_3

The only entry shown in (3) is 0, in row 3, column 3. This is all we need to know to conclude that the matrix $\mathbf{A}$ has no inverse, since the mapping (1) has no inverse.

The last two examples lead to

THEOREM 3.1. A square matrix of order n has an inverse if and only if the pivot rank of the tableau of the associated mapping is equal to n.

A matrix that has an inverse is called *invertible.* Does an invertible matrix have more than one inverse? If so, let matrix $\mathbf{A}$ have inverses $\mathbf{B}$ and $\mathbf{C}$. Then $\mathbf{AB} = \mathbf{BA} = \mathbf{I}$ and $\mathbf{AC} = \mathbf{CA} = \mathbf{I}$. Therefore, $\mathbf{AB} = \mathbf{AC}$. Multiplying on the left by $\mathbf{B}$ gives $\mathbf{B(AB)} = \mathbf{B(AC)}$, which yields $\mathbf{(BA)B} = \mathbf{(BA)C}$. Therefore, $\mathbf{B} = \mathbf{C}$.

This result leads to

THEOREM 3.2. An invertible matrix has exactly one inverse.

For any invertible matrix $\mathbf{A}$ its inverse is written $\mathbf{A}^{-1}$.

THEOREM 3.3. If two square matrices $\mathbf{A}$ and $\mathbf{B}$ of the same order are invertible, then $\mathbf{AB}$ also has an inverse $(\mathbf{AB})^{-1}$ and $(\mathbf{AB})^{-1} = \mathbf{B}^{-1}\mathbf{A}^{-1}$.

Proof:

$$
\begin{aligned}
(\mathbf{AB})(\mathbf{B}^{-1}\mathbf{A}^{-1}) &= \mathbf{A}(\mathbf{BB}^{-1})\mathbf{A}^{-1} && \text{(associativity)} \\
&= \mathbf{AIA}^{-1} && \text{(definition of inverse)} \\
&= \mathbf{AA}^{-1} && \text{(associativity and property of } \mathbf{I}\text{)} \\
&= \mathbf{I}
\end{aligned}
$$

Therefore, $\mathbf{B}^{-1}\mathbf{A}^{-1} = (\mathbf{AB})^{-1}$

EXAMPLE 12

Let

$$\mathbf{A} = \begin{bmatrix} 2 & 3 \\ 1 & 2 \end{bmatrix} \quad \text{and} \quad \mathbf{B} = \begin{bmatrix} 1 & -4 \\ 3 & 1 \end{bmatrix}$$

We know that

$$\mathbf{A}^{-1} = \begin{bmatrix} 2 & -3 \\ -1 & 2 \end{bmatrix}$$

We obtain $\mathbf{B}^{-1}$ by pivoting.

$$\mathbf{B} = \begin{bmatrix} 1^* & -4 \\ 3 & 1 \end{bmatrix}$$

$$\begin{bmatrix} 1 & -4 \\ -3 & 13^* \end{bmatrix}$$

$$\mathbf{B}^{-1} = \begin{bmatrix} 1/13 & 4/13 \\ -3/13 & 1/13 \end{bmatrix}$$

Note that the pivoting is performed on the diagonal entries of the matrix. It is not necessary to write down the mapping when all we want is the inverse matrix. Now,

$$\mathbf{AB} = \begin{bmatrix} 2 & 3 \\ 1 & 2 \end{bmatrix} \begin{bmatrix} 1 & -4 \\ 3 & 1 \end{bmatrix} = \begin{bmatrix} 11 & -5 \\ 7 & -2 \end{bmatrix}$$

and the theorem states that **AB**'s inverse, $(\mathbf{AB})^{-1}$, is equal to $\mathbf{B}^{-1}\mathbf{A}^{-1}$.

$$(\mathbf{B}^{-1}\mathbf{A}^{-1}) = \begin{bmatrix} 1/13 & 4/13 \\ -3/13 & 1/13 \end{bmatrix} \begin{bmatrix} 2 & -3 \\ -1 & 2 \end{bmatrix} = \begin{bmatrix} -2/13 & 5/13 \\ -7/13 & 11/13 \end{bmatrix}$$

$$(\mathbf{AB})(\mathbf{B}^{-1}\mathbf{A}^{-1}) = \begin{bmatrix} 11 & -5 \\ 7 & -2 \end{bmatrix} \begin{bmatrix} -2/13 & 5/13 \\ -7/13 & 11/13 \end{bmatrix} = \begin{bmatrix} 1 & 0 \\ 0 & 1 \end{bmatrix}$$

confirming that $\mathbf{B}^{-1}\,\mathbf{A}^{-1} = (\mathbf{AB})^{-1}$.

EXAMPLE 13

From Section 2.6 we know that the inverse tableau of

(1)

x_1	x_2	
0	1	$= y_1$
3	7	$= y_2$

is

(2)

$-y_2$	$-y_1$	
0	1	$= -x_2$
$1/3$	$-7/3$	$= -x_1$

However

$$(3)\qquad \begin{bmatrix} 0 & 1 \\ 1/3 & -7/3 \end{bmatrix}$$

is not the inverse of

$$(4)\qquad \begin{bmatrix} 0 & 1 \\ 3 & 7 \end{bmatrix}$$

as you may verify. This is so because the pivoting was not done on the diagonal entries. However, rearranging the rows and columns of (2) and changing all signs of the variables, gives

(5)

y_1	y_2	
$-7/3$	$1/3$	$= x_1$
1	0	$= x_2$

The matrix

$$(6)\qquad \begin{bmatrix} -7/3 & 1/3 \\ 1 & 0 \end{bmatrix}$$

is the inverse of (3) as you may verify.

Exercises

1. If possible, multiply. If not possible, explain why not.

(a) $\begin{bmatrix} 1 & 2 \\ 3 & 6 \end{bmatrix}\begin{bmatrix} 4 \\ -2 \end{bmatrix}$ (b) $\begin{bmatrix} 1 & 2 \\ 3 & 6 \end{bmatrix}\begin{bmatrix} 4 & 2 & 3 \\ 1 & 2 & 3 \end{bmatrix}$

(c) $\begin{bmatrix} 1 & 2 \\ 3 & 6 \end{bmatrix}\begin{bmatrix} 4 & 2 & 3 \end{bmatrix}$

(d) $\begin{bmatrix} 4 & 2 \end{bmatrix}\begin{bmatrix} 1 & 2 \\ 3 & 6 \end{bmatrix}$ (e) $\begin{bmatrix} 4 & 2 \end{bmatrix}\begin{bmatrix} 1 & 2 \\ 3 & 4 \\ 5 & 6 \end{bmatrix}$

(f) $\begin{bmatrix} 1 & 0 \\ 0 & 1 \end{bmatrix} \begin{bmatrix} a & b \\ c & d \end{bmatrix}$ (g) $\begin{bmatrix} 2 & 3 \\ 1 & 2 \end{bmatrix} \begin{bmatrix} 2 & -3 \\ -1 & 2 \end{bmatrix}$

(h) $\begin{bmatrix} a & 0 \\ 0 & b \end{bmatrix} \begin{bmatrix} a & 0 \\ 0 & b \end{bmatrix}$

(i) $\begin{bmatrix} 3 & 2 & 1 \\ 4 & 6 & 8 \end{bmatrix} \begin{bmatrix} -3 & -1 & 2 \\ 0 & 0 & 1 \end{bmatrix}$

(j) $\begin{bmatrix} a & b \\ c & d \end{bmatrix} \begin{bmatrix} a & b \\ c & d \end{bmatrix}$ (k) $\begin{bmatrix} 0 & 0 \\ 0 & 0 \end{bmatrix} \begin{bmatrix} a & b \\ c & d \end{bmatrix}$

(l) $\begin{bmatrix} 0 & 1 \\ 1 & 0 \end{bmatrix} \begin{bmatrix} a & b \\ c & d \end{bmatrix}$

2. Express as a single matrix (if possible):

(a) $a \begin{bmatrix} a & b \\ c & d \end{bmatrix} \begin{bmatrix} x \\ y \end{bmatrix}$

(b) $\begin{bmatrix} 2 & 0 \\ -1 & 1 \end{bmatrix} \begin{bmatrix} 2 & 0 \\ -1 & 1 \end{bmatrix} \begin{bmatrix} 2 & 0 \\ -1 & 1 \end{bmatrix}$

(c) $\begin{bmatrix} a & b \\ c & d \end{bmatrix} \begin{bmatrix} a & b \\ c & d \end{bmatrix} + \begin{bmatrix} a & 0 \\ c & 0 \end{bmatrix} \begin{bmatrix} a & b \\ c & d \end{bmatrix}$

(Do you think multiplication of matrices is distributive over addition?)

(d) $\begin{bmatrix} x & y \end{bmatrix} \begin{bmatrix} 1 & 2 \\ 3 & 4 \end{bmatrix} \begin{bmatrix} x \\ y \end{bmatrix}$

(e) $\begin{bmatrix} x & y & z \end{bmatrix} \begin{bmatrix} 1 & 0 & 2 \\ 1 & 2 & 3 \\ -1 & 2 & 0 \end{bmatrix} \begin{bmatrix} x \\ y \\ z \end{bmatrix}$

(f)

$$\begin{bmatrix} x & y & z \end{bmatrix} \begin{bmatrix} x \\ y \\ z \end{bmatrix} \begin{bmatrix} 1 & 0 & 2 \\ 1 & 2 & 3 \\ -1 & 2 & 0 \end{bmatrix}$$

*3. Verify that

$$\begin{bmatrix} 1/3 & 2/3 \\ 1/3 & 2/3 \end{bmatrix} \begin{bmatrix} 1/5 & 4/5 \\ 2/5 & 3/5 \end{bmatrix} = \begin{bmatrix} 1/3 & 2/3 \\ 1/3 & 2/3 \end{bmatrix}$$

Does this mean that

$$\begin{bmatrix} 1/5 & 4/5 \\ 2/5 & 3/5 \end{bmatrix}$$

is a multiplicative identity? Explain your answer.

4. Find the inverse, if any, of each of the following.

(a) $\begin{bmatrix} 4 & 1 \\ 11 & 3 \end{bmatrix}$ (b) $\begin{bmatrix} 4 & 6 \\ 2 & 3 \end{bmatrix}$

(c) $\begin{bmatrix} 4 & 6 \\ 3 & 2 \end{bmatrix}$ (d) $\begin{bmatrix} 1 & 2 & 3 \\ 3 & 7 & 2 \\ 2 & 3 & 1 \end{bmatrix}$

(e) $\begin{bmatrix} 1 & 0 & 0 \\ 0 & 0 & 1 \\ 0 & 1 & 0 \end{bmatrix}$ (f) $\begin{bmatrix} a & 0 & 0 \\ 0 & b & 0 \\ 0 & 0 & c \end{bmatrix}$ $(abc \neq 0)$

(g) $\begin{bmatrix} 1 & 3 & 0 \\ 0 & 2 & 0 \\ 0 & 5 & 1 \end{bmatrix}$ (h) $\begin{bmatrix} 1 & 2 & 3 \\ 0 & 1 & 4 \\ 0 & 0 & 5 \end{bmatrix}$

*Asterisk indicates difficult exercise.

(i) $\begin{bmatrix} 1 & 2 & 3 & -2 \\ 2 & 5 & 2 & 0 \\ 3 & 7 & 6 & -2 \\ 1 & 3 & -1 & 2 \end{bmatrix}$

5. Show that

$$\mathbf{A} = \begin{bmatrix} 1 & 4 \\ 2 & 3 \end{bmatrix}$$

satisfies $\mathbf{A}^2 - 4\mathbf{A} - 5\mathbf{I} = \mathbf{0}$. ($\mathbf{A}^2$ means $\mathbf{AA}$.)

6. Show that

$$\mathbf{A} = \begin{bmatrix} 0 & 0 \\ 1 & 0 \end{bmatrix}$$

satisfies $\mathbf{A}^2 = \mathbf{0}$. Find matrices other than $\mathbf{A}$ that also satisfy this equation.

7. Let $\mathbf{A}$ be a 3×2 matrix. Show that
(a) There is no 3×2 matrix $\mathbf{X}$ such that either

$$\mathbf{XA} = \mathbf{A} \quad \text{or} \quad \mathbf{AX} = \mathbf{A}$$

(b) No 2×3 matrix $\mathbf{Y}$ exists such that either

$$\mathbf{YA} = \mathbf{A} \quad \text{or} \quad \mathbf{AY} = \mathbf{A}$$

(c) No square matrix $\mathbf{Z}$ exists such that

$$\mathbf{ZA} = \mathbf{A} \quad \text{and} \quad \mathbf{AZ} = \mathbf{A}$$

8.

$$\mathbf{A} = \begin{bmatrix} 0 & 0 & 1 & 0 & 0 \\ 0 & 0 & 1 & 0 & 1 \\ 1 & 1 & 0 & 0 & 1 \\ 0 & 0 & 0 & 0 & 1 \\ 0 & 1 & 1 & 1 & 0 \end{bmatrix}$$

is the adjacency matrix of the graph in Example 3, Section 3.1. Find $\mathbf{AA}$ (written $\mathbf{A}^2$). Find $\mathbf{A}^3 = \mathbf{AAA}$.

A path in a graph is a route along its *edges* (the lines) from vertex to vertex. For instance, there is a path from A to D of length 3 (the length is the number of edges traversed) via C and E and a path from A to D of length 4 via C, B, and E. There are three paths of length 2 from E to E (via D, C, or B). It is a fact that **A** lists the number of paths of length 1 between vertices, $\mathbf{A}^2$ the number of paths of length 2, $\mathbf{A}^3$ those of length 3, etc., including retracings. Thus, for instance, in $\mathbf{A}^3$, the number in the fifth row and fifth column, which represents the number of paths of length 3 from E to E, is 2.

*9. The following operation on matrices is widely used in physics (it does not appear elsewhere in this book). Let **A** and **B** be 2×2 matrices and define their *cross product*, $\mathbf{A} \times \mathbf{B}$, to be given by the rule

$$\mathbf{A} \times \mathbf{B} = \mathbf{AB} - \mathbf{BA}$$

Prove that
(a) $\mathbf{A} \times \mathbf{B} = -(\mathbf{B} \times \mathbf{A})$.
(b) $\mathbf{A} \times \mathbf{A} = \mathbf{O}_2$.
(c) $\mathbf{A} \times \mathbf{I} = \mathbf{I} \times \mathbf{A} = \mathbf{O}_2$.
(d) $\mathbf{A} \times (\mathbf{B} \times \mathbf{C}) = \mathbf{B} \times (\mathbf{A} \times \mathbf{C}) + \mathbf{C} \times (\mathbf{B} \times \mathbf{A})$.
(e) $\mathbf{A} \times (\mathbf{B} + \mathbf{C}) = (\mathbf{A} \times \mathbf{B}) + (\mathbf{A} \times \mathbf{C})$.

3.4 Applications of Matrix Multiplication

The applications are varied and many, including those related to transition matrices, linear programming, systems of equations, mappings, coding, and communication matrices.

APPLICATION 1 TRANSITION MATRICES

As the word "transition" indicates, a set of circumstances is changing from one state to another. As an example, consider the population of a city and the population of its suburbs as people change their residences from one to the other. To keep the example simple, disregard births and deaths and assume that in one year 90% of the city people stay in the city and the remaining 10% move to the suburbs, while 20% of the suburban people move to the city, the remaining 80% staying in the suburbs. These data are displayed in a table as

		To: City	To: Suburbs
From:	City	.9	.1
	Suburbs	.2	.8

Its matrix is

$$\begin{bmatrix} .9 & .1 \\ .2 & .8 \end{bmatrix}$$

Suppose further that at the end of 1971 the city had a population of 5 million and its suburbs had a population of 2 million. This is displayed as

	City	Suburbs
Population	5,000,000	2,000,000

or in matrix form,

$$\begin{bmatrix} 5,000,000 & 2,000,000 \end{bmatrix}$$

To find how many people are in each place at the end of 1972, multiply the two matrices:

$$\begin{bmatrix} 5,000,000 & 2,000,000 \end{bmatrix} \begin{bmatrix} .9 & .1 \\ .2 & .8 \end{bmatrix}$$

To see why this is so, consider the entry of row 1, column 1, of the product:

$$5,000,000(.9) + 2,000,000(.2)$$

This is the number of people who remain in the city plus those who move to the city from the suburbs, at the end of 1972. The entry in row 1, column 2, of the product gives the same breakdown for the suburbs at the end of 1972. The product is

$$\begin{bmatrix} 5,000,000 & 2,000,000 \end{bmatrix} \begin{bmatrix} .9 & .1 \\ .2 & .8 \end{bmatrix} = \begin{bmatrix} 4,900,000 & 2,100,000 \end{bmatrix}$$

To calculate the population in the city and its suburbs for the end of 1973, we multiply the 1972 population by the transition matrix. This is

$$\begin{bmatrix} 4,900,000 & 2,100,000 \end{bmatrix} \begin{bmatrix} .9 & .1 \\ .2 & .8 \end{bmatrix}$$

or

$$\begin{bmatrix} 5{,}000{,}000 & 2{,}000{,}000 \end{bmatrix} \begin{bmatrix} .9 & .1 \\ .2 & .8 \end{bmatrix} \begin{bmatrix} .9 & .1 \\ .2 & .8 \end{bmatrix}$$

Since matrix multiplication is associative, we associate the last two matrices and obtain

$$\begin{bmatrix} 5{,}000{,}000 & 2{,}000{,}000 \end{bmatrix} \begin{bmatrix} .9 & .1 \\ .2 & .8 \end{bmatrix}^2$$

$$\begin{bmatrix} .9 & .1 \\ .2 & .8 \end{bmatrix}^2 = \begin{bmatrix} .83 & .17 \\ .34 & .66 \end{bmatrix}$$

We interpret

$$\begin{bmatrix} .9 & .1 \\ .2 & .8 \end{bmatrix}^2$$

as a transition matrix over a period of two years. This suggests that the transition matrix over a period of n years is

$$\begin{bmatrix} .9 & .1 \\ .2 & .8 \end{bmatrix}^n$$

Such a matrix is of interest to city planners who wish to predict what the population in the city and its suburbs will be in n years. When $n = 3$ we have

$$\begin{bmatrix} .9 & .1 \\ .2 & .8 \end{bmatrix}^3 = \begin{bmatrix} .781 & .199 \\ .438 & .562 \end{bmatrix}$$

As higher powers of the transition matrix are found, the sequence of entries in row 1, column 1, approaches a definite number, as do the sequences for other entries. In fact, as n gets very large, the entries approach those of

$$\begin{bmatrix} 2/3 & 1/3 \\ 2/3 & 1/3 \end{bmatrix}$$

This fact indicates that a stability of population sets in as the years go by. In fact,

$$[5{,}000{,}000 \quad 2{,}000{,}000]\begin{bmatrix} \frac{2}{3} & \frac{1}{3} \\ \frac{2}{3} & \frac{1}{3} \end{bmatrix} = \begin{bmatrix} \frac{14{,}000{,}000}{3} & \frac{7{,}000{,}000}{3} \end{bmatrix}$$

so that the city population stabilizes at about $\frac{14}{3}$ million and the suburb population at about $\frac{7}{3}$ million.

Other situations in which transition matrices are helpful are in the study of

1. Interchange between a liquid and vapor in a vacuum.
2. Heredity.
3. Probability chains (Markov chains).
4. Sociological trends, for instance children following the profession of their parents.

APPLICATION 2 MATRICES FOR SOLVING A SYSTEM OF LINEAR EQUATIONS

Consider the system of two equations, in two variables:

x_1	x_2		
2	3	=	12
1	2	=	7

Let

$$\mathbf{A} = \begin{bmatrix} 2 & 3 \\ 1 & 2 \end{bmatrix}, \quad \mathbf{X} = \begin{bmatrix} x_1 \\ x_2 \end{bmatrix}, \quad \mathbf{B} = \begin{bmatrix} 12 \\ 7 \end{bmatrix}$$

A is the *coefficient matrix* of the system, **X** is the *variables matrix,* and **B** is the *constant matrix.*

Since

$$\mathbf{AX} = \begin{bmatrix} 2 & 3 \\ 1 & 2 \end{bmatrix}\begin{bmatrix} x_1 \\ x_2 \end{bmatrix} = \begin{bmatrix} 2x_1 + 3x_2 \\ x_1 + 2x_2 \end{bmatrix}$$

it is clear that $\mathbf{AX} = \mathbf{B}$ actually says the same thing as the system of equations above. This fact suggests that we can find **X** if we multiply both members of the matrix equation, on the left, by $\mathbf{A}^{-1}$. $\mathbf{A}^{-1}$ has been found, in the preceding section, to be

$$\begin{bmatrix} 2 & -3 \\ -1 & 2 \end{bmatrix}$$

Then

$$\mathbf{A}^{-1}\mathbf{A}\mathbf{X} = \mathbf{A}^{-1}\mathbf{B}$$

or

$$\begin{bmatrix} 2 & -3 \\ -1 & 2 \end{bmatrix} \begin{bmatrix} 2 & 3 \\ 1 & 2 \end{bmatrix} \begin{bmatrix} x_1 \\ x_2 \end{bmatrix} = \begin{bmatrix} 2 & -3 \\ -1 & 2 \end{bmatrix} \begin{bmatrix} 12 \\ 7 \end{bmatrix}$$

$$\begin{bmatrix} 1 & 0 \\ 0 & 1 \end{bmatrix} \begin{bmatrix} x_1 \\ x_2 \end{bmatrix} = \begin{bmatrix} x_1 \\ x_2 \end{bmatrix} = \begin{bmatrix} 3 \\ 2 \end{bmatrix}$$

That is, $x_1 = 3$, $x_2 = 2$, or $(x_1, x_2) = (3, 2)$. There are occasions when the solution to many systems of equations are sought, all having the same coefficient matrix. All of these are solved together by the method of multiply on the left by the inverse of the coefficient matrix, if it exists: for instance, to solve

x_1	x_2	
2	3	$= a$
1	2	$= b$

where $(a, b) = (12, 7)$, $(3, 5)$, $(-3, 7)$, and $(0, 8)$. The set of all these constants is combined in one matrix,

$$\mathbf{B} = \begin{bmatrix} 12 & 3 & -3 & 0 \\ 7 & 5 & 7 & 8 \end{bmatrix}$$

The solutions to $\mathbf{AX} = \mathbf{B}$ are then

$$\mathbf{X} = \mathbf{A}^{-1}\mathbf{B} = \begin{bmatrix} 2 & -3 \\ -1 & 2 \end{bmatrix} \begin{bmatrix} 12 & 3 & -3 & 0 \\ 7 & 5 & 7 & 8 \end{bmatrix}$$

$$= \begin{bmatrix} 3 & -9 & -27 & -24 \\ 2 & 7 & 17 & 16 \end{bmatrix}$$

Thus $(x_1, x_2) = (3, 2), (-9, 7), (-27, 17), (-24, 16)$.

A transformation on a matrix that gives us greater freedom in the order in which matrices can be multiplied and facilitates our work is that of the transpose of a matrix. Let

$$\mathbf{A} = \begin{bmatrix} 3 & 0 & 1 \\ 1 & -1 & 2 \end{bmatrix} \quad \text{and} \quad \mathbf{B} = \begin{bmatrix} 2 \\ 4 \\ -2 \end{bmatrix}$$

Since **A** has three columns and **B** has three rows, it is possible to calculate **AB**. But **B** has one column and **A** has three rows. It is not possible to calculate **BA**. Nevertheless, suppose that for some reason we want **B** or a matrix related to **B** to multiply **A** on the left. This is done by deriving from **A** and **B** two other matrices in which columns appear as rows and rows as columns. These new matrices are called *transposes*. The transpose of **A** is denoted $\mathbf{A}^T$, and that of **B** by $\mathbf{B}^T$.

$$\mathbf{A}^T = \begin{bmatrix} 3 & 1 \\ 0 & -1 \\ 1 & 2 \end{bmatrix} \quad \text{and} \quad \mathbf{B}^T = \begin{bmatrix} 2 & 4 & -2 \end{bmatrix}$$

It is possible to calculate $\mathbf{B}^T\mathbf{A}^T$. (Why?) Indeed,

$$\mathbf{B}^T\mathbf{A}^T = \begin{bmatrix} 2 \cdot 3 + 4 \cdot 0 - 2 \cdot 1 & 1(2) - 1(4) + 2(-2) \end{bmatrix} = \begin{bmatrix} 3 & -6 \end{bmatrix}$$

Notice that

$$\mathbf{AB} = \begin{bmatrix} 3 \cdot 2 + 0 \cdot 4 + 1(-2) \\ 1(2) - 1(4) + 2(-2) \end{bmatrix} = \begin{bmatrix} 3 \\ -6 \end{bmatrix}$$

This suggests the generalization expressed in the following theorem.

THEOREM 3.4. Let **A** have dimension $m \times n$ and **B** have dimension $n \times p$. Then $(\mathbf{AB})^T = \mathbf{B}^T\mathbf{A}^T$ is a matrix of dimension $p \times m$.

Proof: The proof is shown for special cases of matrices **A** and **B** and for a particular entry in $(\mathbf{AB})^T$ and $\mathbf{B}^T\mathbf{A}^T$. This is typical of the proof for any two matrices that can be multiplied and for any pair of corresponding entries.

Let

$$\mathbf{A} = \begin{bmatrix} a_{11} & a_{12} & a_{13} \\ a_{21} & a_{22} & a_{23} \end{bmatrix} \quad \text{and} \quad \mathbf{B} = \begin{bmatrix} b_{11} & b_{12} & b_{13} \\ b_{21} & b_{22} & b_{23} \\ b_{31} & b_{32} & b_{33} \end{bmatrix}$$

Let $\mathbf{AB} = \mathbf{C}$ and consider c_{21}.

$$c_{21} = a_{21}b_{11} + a_{22}b_{21} + a_{23}b_{31}$$

Since c_{21} is in the second row, first column, of $\mathbf{C}$, it is the element in the first row, second column, of $\mathbf{C}^T$. Now since $\mathbf{AB} = \mathbf{C}$ we have that $(\mathbf{AB})^T = \mathbf{C}^T$. Thus c_{21} becomes the entry in row 1, column 2, of $(\mathbf{AB})^T$.

$$\mathbf{B}^T = \begin{bmatrix} b_{11} & b_{21} & b_{31} \\ b_{12} & b_{22} & b_{32} \\ b_{13} & b_{23} & b_{33} \end{bmatrix} \quad \text{and} \quad \mathbf{A}^T = \begin{bmatrix} a_{11} & a_{21} \\ a_{12} & a_{22} \\ a_{13} & a_{23} \end{bmatrix}$$

The entry in row 1, column 2, of $\mathbf{B}^T\mathbf{A}^T$.

$$b_{11}a_{21} + b_{21}a_{22} + b_{31}a_{23}$$

and this is equal to c_{21} above. This comparison can be carried out for each c_{ij}. Hence $(\mathbf{AB})^T = \mathbf{B}^T\mathbf{A}^T$.

It is an interesting fact that the inverse of $\mathbf{A}^T$ is the transpose of $\mathbf{A}^{-1}$. We show this in

THEOREM 3.5. If a matrix $\mathbf{A}$ has an inverse, $\mathbf{A}^{-1}$, then $(\mathbf{A}^{-1})^T = (\mathbf{A}^T)^{-1}$.

Proof: $\mathbf{A}$ has an inverse $\mathbf{A}^{-1}$. Then

$$\begin{aligned} \mathbf{AA}^{-1} &= \mathbf{I} \\ (\mathbf{AA}^{-1})^T &= \mathbf{I}^T \\ (\mathbf{A}^{-1})^T\mathbf{A}^T &= \mathbf{I} \qquad \text{(Theorem 3.4)} \end{aligned}$$

Therefore,

$$(\mathbf{A}^{-1})^T = (\mathbf{A}^T)^{-1} \qquad \text{since } (\mathbf{A}^T)^{-1}\mathbf{A}^T = \mathbf{I}$$

An alternative method to solve $\mathbf{AX} = \mathbf{B}$ is to work with $\mathbf{X}^T\mathbf{A}^T = \mathbf{B}^T$. In this case multiply each member on the right by the inverse of $\mathbf{A}^T$.

Thus to solve

x_1	x_2	
2	3	= 12
1	2	= 7

we begin with

$$\mathbf{A}^{-1} = \begin{bmatrix} 2 & -3 \\ -1 & 2 \end{bmatrix} \quad \text{and} \quad (\mathbf{A}^{-1})^T = \begin{bmatrix} 2 & -1 \\ -3 & 2 \end{bmatrix}$$

Now $\mathbf{X}^T\mathbf{A}^T = \mathbf{B}^T$, and thus $\mathbf{X}^T\mathbf{A}^T(\mathbf{A}^T)^{-1} = \mathbf{B}^T(\mathbf{A}^T)^{-1}$, and since $(\mathbf{A}^T)^{-1} = (\mathbf{A}^{-1})^T$ we have $\mathbf{X}^T\mathbf{A}^T(\mathbf{A}^{-1})^T = \mathbf{B}^T(\mathbf{A}^{-1})^T$. Substituting yields

$$\mathbf{X}^T \begin{bmatrix} 2 & 1 \\ 3 & 2 \end{bmatrix} \begin{bmatrix} 2 & -3 \\ -1 & 2 \end{bmatrix} = \begin{bmatrix} 12 & 7 \end{bmatrix} \begin{bmatrix} 2 & -1 \\ -3 & 2 \end{bmatrix} = \begin{bmatrix} 3 & 2 \end{bmatrix}$$

Thus

$$\begin{bmatrix} x_1 & x_2 \end{bmatrix} = \begin{bmatrix} 3 & 2 \end{bmatrix}$$

APPLICATION 3 FINDING IMAGES UNDER MAPPINGS

Let $M : R^3 \rightarrow R^2$ with tableau

x_1	x_2	x_3	
2	-1	1	$= y_1$
1	2	-1	$= y_2$

and let it be required to find the image of $(2, 1, -3)$ under M. Of course this can be done by a straightforward substitution for x_1, x_2, and x_3 and a simple calculation for y_1 and y_2. It is our purpose here to show this calculation by matrix multiplication.

Let

$$\mathbf{A} = \begin{bmatrix} 2 & -1 & 1 \\ 1 & 2 & -1 \end{bmatrix}, \quad \mathbf{X} = \begin{bmatrix} x_1 \\ x_2 \\ x_3 \end{bmatrix}, \quad \mathbf{Y} = \begin{bmatrix} y_1 \\ y_2 \end{bmatrix}$$

Then

$$\begin{bmatrix} 2 & -1 & 1 \\ 1 & 2 & -1 \end{bmatrix} \begin{bmatrix} x_1 \\ x_2 \\ x_3 \end{bmatrix} = \begin{bmatrix} y_1 \\ y_2 \end{bmatrix} \qquad \text{(verify this)}$$

or

$$\mathbf{AX} = \mathbf{Y}$$

Thus to find the image of $(x_1, x_2, x_3) = (2, -1, 3)$, multiply

$$\begin{bmatrix} 2 & -1 & 1 \\ 1 & 2 & -1 \end{bmatrix} \begin{bmatrix} 2 \\ -1 \\ 3 \end{bmatrix}$$

The product is

$$\begin{bmatrix} 8 \\ -3 \end{bmatrix}$$

Therefore, $(y_1, y_2) = (8, -3)$.

Observe, in passing, that this operation is available for finding the images of several values of (x_1, x_2, x_3) simultaneously as in Application 2.

An alternative method uses transposes of **A**, **X**, and **Y**. In this case Theorem 3.4 is invoked and

$$\mathbf{X}^T\mathbf{A}^T = \mathbf{Y}^T$$

or

$$\begin{bmatrix} x_1 & x_2 & x_3 \end{bmatrix} \begin{bmatrix} 2 & 1 \\ -1 & 2 \\ 1 & -1 \end{bmatrix} = \begin{bmatrix} y_1 & y_2 \end{bmatrix}$$

We have already seen how the product of matrices can be used to find the matrix of the composite of two mappings.

Let **F** be the matrix associated with $F: R^m \to R^n$ and **G** the matrix associated with $G: R^n \to R^p$; then **GF** is the matrix associated with $G \circ F$. The image of $(x_1, x_2, \ldots, x_m)$, under $G \circ F$ is

$$\mathbf{GF} \begin{bmatrix} x_1 \\ x_2 \\ \vdots \\ x_m \end{bmatrix} = \begin{bmatrix} z_1 \\ z_2 \\ \vdots \\ z_p \end{bmatrix}$$

or

$$\begin{bmatrix} x_1 & x_2 & \cdots & x_m \end{bmatrix} \mathbf{F}^T \mathbf{G}^T = \begin{bmatrix} z_1 & z_2 & \cdots & z_p \end{bmatrix}$$

APPLICATION 4 LINEAR PROGRAMS

Given the dual linear programs

	x_1	x_2	x_3	-1	
v_1	a_{11}	a_{12}	a_{13}	b_1	$= -y_1$
v_2	a_{21}	a_{22}	a_{23}	b_2	$= -y_2$
-1	c_1	c_2	c_3	0	$= M$
	$= u_1$	$= u_2$	$= u_3$	$= m$	

let

$$\mathbf{A} = \begin{bmatrix} a_{11} & a_{12} & a_{13} \\ a_{21} & a_{22} & a_{23} \end{bmatrix}$$

$$\mathbf{X} = \begin{bmatrix} x_1 \\ x_2 \\ x_3 \end{bmatrix}, \qquad \mathbf{Y} = \begin{bmatrix} y_1 \\ y_2 \end{bmatrix}$$

$$\mathbf{B} = \begin{bmatrix} b_1 \\ b_2 \end{bmatrix}, \qquad \mathbf{C} = \begin{bmatrix} c_1 & c_2 & c_3 \end{bmatrix}$$

$$\mathbf{V} = \begin{bmatrix} v_1 & v_2 \end{bmatrix}, \qquad \mathbf{U} = \begin{bmatrix} u_1 & u_2 & u_3 \end{bmatrix}$$

$$\mathbf{M} = \begin{bmatrix} M \end{bmatrix}, \qquad \mathbf{m} = \begin{bmatrix} m \end{bmatrix}$$

Then the programs can be expressed as: find nonnegative **X**, **Y**, **U**, and **V** to maximize

$$\mathbf{CX} = \mathbf{M}$$

subject to

$$\mathbf{AX} - \mathbf{B} = -\mathbf{Y} \qquad (\text{or } \mathbf{AX} + \mathbf{Y} = \mathbf{B})$$

and to minimize

$$\mathbf{VB} = \mathbf{m}$$

subject to

$$\mathbf{VA} - \mathbf{C} = \mathbf{U}$$

(To say that $\mathbf{X}$ is nonnegative means that none of its components are negative.)

APPLICATION 5 CODING WITH A MATRIX

The title refers to the coding of messages. It is offered in the spirit of recreation. Let us code the message GOOD LUCK. To begin with, each letter is replaced by a numeral.

A	*B*	*C*	*D*	*E*	*F*	*G*	*H*	*I*	*J*	*K*	*L*	*M*	*N*
1	2	3	4	5	6	7	8	9	10	11	12	13	14

O	*P*	*Q*	*R*	*S*	*T*	*U*	*V*	*W*	*X*	*Y*	*Z*
15	16	17	18	19	20	21	22	23	24	25	26

The message then becomes 7, 15, 15, 4, 12, 21, 3, 11.

To disguise the message we first write it in square matrices. The message is now

$$\mathbf{A} = \begin{bmatrix} 7 & 15 \\ 15 & 4 \end{bmatrix} \text{ and } \mathbf{B} = \begin{bmatrix} 12 & 21 \\ 3 & 11 \end{bmatrix}$$

Then each matrix is multiplied on the right by a coding matrix, say

$$\mathbf{C} = \begin{bmatrix} 2 & 3 \\ 1 & 2 \end{bmatrix}$$

The message is now hidden in the products:

$$\mathbf{AC} = \begin{bmatrix} 29 & 51 \\ 34 & 53 \end{bmatrix} \quad \text{and} \quad \mathbf{BC} = \begin{bmatrix} 45 & 78 \\ 17 & 31 \end{bmatrix}$$

It is dispatched in the form 29, 51, 34, 53, 45, 78, 17, 31. The recipient restructures $\mathbf{AC}$ and $\mathbf{BC}$ from the dispatch and then is able to recover $\mathbf{A}$ and $\mathbf{B}$ by multiplying on the right by a decoding matrix. The decoding matrix is $\mathbf{C}^{-1}$, the inverse of the coding matrix $\mathbf{C}$ since $\mathbf{CC}^{-1} = \mathbf{I}$ and therefore $\mathbf{ACC}^{-1} = \mathbf{A}$ and $\mathbf{BCC}^{-1} = \mathbf{B}$.

$$\mathbf{C}^{-1} = \begin{bmatrix} 2 & -3 \\ -1 & 2 \end{bmatrix}$$

$$(\mathbf{AC})\mathbf{C}^{-1} = \begin{bmatrix} 29 & 51 \\ 34 & 53 \end{bmatrix} \begin{bmatrix} 2 & -3 \\ -1 & 2 \end{bmatrix} = \begin{bmatrix} 7 & 15 \\ 15 & 4 \end{bmatrix} = \mathbf{A}$$

$$(\mathbf{BC})\mathbf{C}^{-1} = \begin{bmatrix} 45 & 78 \\ 17 & 31 \end{bmatrix} \begin{bmatrix} 2 & -3 \\ -1 & 2 \end{bmatrix} = \begin{bmatrix} 12 & 21 \\ 3 & 11 \end{bmatrix} = \mathbf{B}$$

The message is now translatable back into English. To form a new code select a different coding matrix $\mathbf{D}$ and provide the recipient with $\mathbf{D}^{-1}$ as the decoding matrix.

APPLICATION 6 COMMUNICATION MATRICES

Suppose that A, B, C, and D are nations such that A and B have diplomatic relations (and therefore communicate directly with each other), as do B and C as well as A and D. These relations are diagrammed as

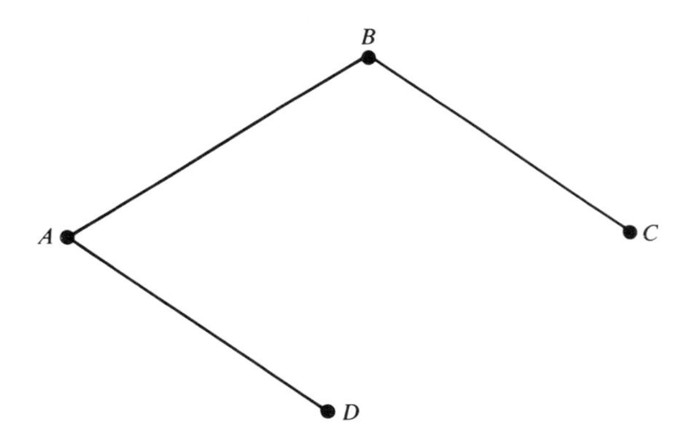

The communicative possibilities are displayed in a *communication table:*

	A	*B*	*C*	*D*
A	0	1	0	1
B	1	0	1	0
C	0	1	0	0
D	1	0	0	0

whose matrix is

$$\mathbf{M} = \begin{bmatrix} 0 & 1 & 0 & 1 \\ 1 & 0 & 1 & 0 \\ 0 & 1 & 0 & 0 \\ 1 & 0 & 0 & 0 \end{bmatrix}$$

where $m_{ij} = 1$ means there is direct communication and $m_{ij} = 0$ means there is none. (This assumes that no country has a direct communication route with itself.) Since an avenue of communication is two ways, $m_{ij} = m_{ji}$. Now consider **M M**, or

$$\mathbf{M}^2 = \begin{bmatrix} 2 & 0 & 1 & 0 \\ 0 & 2 & 0 & 1 \\ 1 & 0 & 1 & 0 \\ 0 & 1 & 0 & 1 \end{bmatrix}$$

The entries in $\mathbf{M}^2$ represent the number of two-stage communication routes between countries. The 2 in row 1, column 1, means there are 2 two-stage communication routes between A and itself. (These are from A to B to A, and A to D to A.) The 0 in row 1, column 2, means there are no two-stage communication routes from A to B. The 1 in row 3 column 1, represents the only two-stage communication route from C to A (via B).

$$\mathbf{MMM} = \mathbf{M}^3$$

represents the number of three-stage communication routes between the countries,

$$\mathbf{M}^3 = \begin{bmatrix} 0 & 3 & 0 & 2 \\ 3 & 0 & 2 & 0 \\ 0 & 2 & 0 & 1 \\ 2 & 0 & 1 & 0 \end{bmatrix}$$

To verify one of these, consider the three-stage communication routes from C. They are C to B to C to B, C to B to A to B, and C to B to A to D. Verify these in row 3 of $\mathbf{M}^3$.

In general $\mathbf{M}^n$ gives the number of n-stage communication routes between countries. (One four-stage communication route from A to A is given by A to B to A to B to A. There are four other four-stage communication routes from A to A. What are they?) The communication matrix **M** is clearly the adjacency matrix of the graph of the communication network.

Thus the meaning of $\mathbf{M}^2$ and $\mathbf{M}^3$ is to be expected based on our previous work with graphs.

Exercises

1. Using the coding method described in Application 5 with the coding matrix

$$\mathbf{C} = \begin{bmatrix} 2 & 3 \\ 1 & 2 \end{bmatrix}$$

code each of the following messages:

(a) COME HOME.

(b) WHERE ARE YOU. (Group as follows: WHER/EARE/YOUX. The X fills the empty space in the last 2×2 matrix.)

2. Using the decoding matrix

$$\mathbf{D} = \begin{bmatrix} 2 & -3 \\ -1 & 2 \end{bmatrix}$$

decode the following messages:

(a) 58, 97, 27, 53, 25, 49, 27, 53.

(b) 30, 51, 52, 89, 35, 65, 51, 87.

3. If a coding matrix is used to multiply on the right, must the decoding matrix also be used to multiply on the right? Explain

4. Solve the following systems of equations with one matrix multiplication:

(a) $$\begin{aligned} 2x + 3y &= a, \\ x + 2y &= b \end{aligned} \qquad (a, b) = (0, 0), (1, -1), (3, 2), (8, 4).$$

(b) $$\begin{aligned} x + y \phantom{{}+ 3z} &= a \\ x + \phantom{2y +{}} 2z &= b, \\ x + 2y + 3z &= c \end{aligned} \qquad (a, b, c) = (2, 3, 6), (0, 0, 0), (1, 1, 1).$$

5. Using matrix multiplication, find the images of (3, 2, 1), (1, 2, −1), and (0, 1, 1) under the mapping $R^3 \to R^2$ with tableau

x_1	x_2	x_3	
2	0	3	$= y_1$
1	−1	4	$= y_2$

6. Let

$$\mathbf{X} = \begin{bmatrix} x_1 \\ x_2 \\ x_3 \end{bmatrix}, \quad \mathbf{Y} = \begin{bmatrix} y_1 \\ y_2 \end{bmatrix}, \quad \mathbf{A} = \begin{bmatrix} 1 & 2 & 3 \\ 4 & 5 & 6 \end{bmatrix}$$

Describe the tableau of the mapping (if any) defined by each of the following:
(a) $\mathbf{AX} = \mathbf{Y}$. (b) $\mathbf{A}^T\mathbf{X}^T = \mathbf{Y}^T$. (c) $\mathbf{XA} = \mathbf{Y}$.
(d) $\mathbf{X}^T\mathbf{A}^T = \mathbf{Y}^T$. (e) $\mathbf{A}^{-1}\mathbf{Y} = \mathbf{X}$. (f) $\mathbf{A}^T\mathbf{Y} = \mathbf{X}$.

7. Water and water vapor are confined in a closed tank. Assume that 20% of the water evaporates in 1 hour while 10% of the vapor condenses to water. Write a transition matrix to describe the hourly change. Find the transition matrix for a 2-hour period.

8. Using the data in Exercise 7 and starting with 100 units of water and 0 units of vapor, calculate the number of units of water and vapor at the end of (a) 1 hour; (b) 2 hours.

9. Among four countries A, B, C, and D, there is direct communication only between A and C, B and D, and C and D.
(a) Write a communication matrix for these data.
(b) Write a two-stage communication matrix.
(c) Write a three-stage communication matrix.

3.5 Some Special Square Matrices

EXAMPLE 1 THE UNIT OR IDENTITY MATRIX AND THE ZERO MATRIX

In each set of $n \times n$ matrices (denoted M_n) there is one zero matrix $\mathbf{0}$ (or $\mathbf{0}_n$) and one $\mathbf{I}$ (or $\mathbf{I}_n$) such that for all $\mathbf{A}$ in M_n, $\mathbf{I}_n\mathbf{A} = \mathbf{AI}_n = \mathbf{A}$.

EXAMPLE 2 DIAGONAL MATRICES

Let each entry $a_{ij} = 0$ if $i \neq j$. For $i = j$, $a_{ij} = d_i$, where d_i is a number. This matrix is called a *diagonal matrix*.

$$\mathbf{A} = \begin{bmatrix} 3 & 0 & 0 \\ 0 & 2 & 0 \\ 0 & 0 & -1 \end{bmatrix}$$

is a diagonal matrix. In this case $d_1 = 3$, $d_2 = 2$, and $d_3 = -1$.

The product or sum of two diagonal matrices of the same order is a diagonal matrix. For instance,

$$\begin{bmatrix} 3 & 0 & 0 \\ 0 & 2 & 0 \\ 0 & 0 & -1 \end{bmatrix} + \begin{bmatrix} 4 & 0 & 0 \\ 0 & -1 & 0 \\ 0 & 0 & 3 \end{bmatrix} = \begin{bmatrix} 7 & 0 & 0 \\ 0 & 1 & 0 \\ 0 & 0 & 2 \end{bmatrix}$$

and

$$\begin{bmatrix} 3 & 0 & 0 \\ 0 & 2 & 0 \\ 0 & 0 & -1 \end{bmatrix} \begin{bmatrix} 4 & 0 & 0 \\ 0 & -1 & 0 \\ 0 & 0 & 3 \end{bmatrix} = \begin{bmatrix} 12 & 0 & 0 \\ 0 & -2 & 0 \\ 0 & 0 & -3 \end{bmatrix}$$

From this example it is clear that, for any two diagonal matrices **D** and **E**, **DE** has diagonal entries $d_1e_1, d_2e_2, \ldots, d_ne_n$ and that **ED** has the same entries. Hence multiplication of diagonal matrices is commutative. If **D** has no zero diagonal entries, then $\mathbf{D}^{-1}$ has diagonal entries

$$\frac{1}{d_1}, \frac{1}{d_2}, \ldots, \frac{1}{d_n}$$

(See Exercises 12 and 13 of Section 2.6.)

EXAMPLE 3 TRIANGULAR MATRICES

(See Exercises 14 and 15 of Section 2.6.) A *triangular matrix* is a square matrix that has all zeros above the diagonal or all zeros below the diagonal. If it has all zeros above the diagonal, it is called *lower triangular*, and in the other case, it is called *upper triangular*.

$$\mathbf{A} = \begin{bmatrix} 1 & 0 & 0 \\ 2 & -1 & 0 \\ 1 & 2 & 2 \end{bmatrix}$$

is a lower triangular matrix of dimension 3×3. Now take three identity matrices of dimension 3×3. In one of these replace the first column by the first column of **A** giving

$$\mathbf{A}_1 = \begin{bmatrix} 1 & 0 & 0 \\ 2 & 1 & 0 \\ 1 & 0 & 1 \end{bmatrix}$$

In another of the 3×3 identity matrices replace the second column by the second column of $\mathbf{A}$.

$$\mathbf{A}_2 = \begin{bmatrix} 1 & 0 & 0 \\ 0 & -1 & 0 \\ 0 & 2 & 1 \end{bmatrix}$$

In the third such, replace the third column by the third column of $\mathbf{A}$.

$$\mathbf{A}_3 = \begin{bmatrix} 1 & 0 & 0 \\ 0 & 1 & 0 \\ 0 & 0 & 2 \end{bmatrix}$$

The reader should show that

$$\mathbf{A} = \mathbf{A}_1\mathbf{A}_2\mathbf{A}_3$$

This procedure generalizes so that if $\mathbf{A}$ is $n \times n$ lower triangular, then

$$\mathbf{A} = \mathbf{A}_1\mathbf{A}_2\mathbf{A}_3 \cdots \mathbf{A}_n$$

where $\mathbf{A}_1$ is obtained by replacing the first column of $\mathbf{I}_n$ by the first column of $\mathbf{A}$, $\mathbf{A}_2$ is obtained by replacing the second column of $\mathbf{I}_n$ by the second column of $\mathbf{A}$, etc.

$$\mathbf{A} = \begin{bmatrix} 2 & 1 & 3 \\ 0 & -2 & 5 \\ 0 & 0 & -1 \end{bmatrix}$$

is an upper triangle matrix. As above, form

$$\mathbf{A}_1 = \begin{bmatrix} 2 & 0 & 0 \\ 0 & 1 & 0 \\ 0 & 0 & 1 \end{bmatrix}, \qquad \mathbf{A}_2 = \begin{bmatrix} 1 & 1 & 0 \\ 0 & -2 & 0 \\ 0 & 0 & 1 \end{bmatrix},$$

$$\mathbf{A}_3 = \begin{bmatrix} 1 & 0 & 3 \\ 0 & 1 & 5 \\ 0 & 0 & -1 \end{bmatrix}$$

The reader is invited to show that

$$\mathbf{A} = \mathbf{A}_3\mathbf{A}_2\mathbf{A}_1$$

For an $n \times n$ upper triangular matrix $\mathbf{A}$ the generalization suggested is true. Form $\mathbf{A}_1, \mathbf{A}_2, \ldots, \mathbf{A}_n$ as above and then

$$\mathbf{A} = \mathbf{A}_n\mathbf{A}_{n-1} \cdots \mathbf{A}_1$$

It is apparent that the sum of two triangular matrices of the same order is a triangular matrix. It may be a surprise to find that the product of two triangular matrices is also a triangular matrix. (You are asked to prove this in an exercise.)

EXAMPLE 4 SYMMETRIC AND SKEW SYMMETRIC MATRICES

A matrix $\mathbf{A}$ is symmetric if $a_{ij} = a_{ji}$ for all i and j. This is synonymous with saying that $\mathbf{A}$ is symmetric if $\mathbf{A} = \mathbf{A}^T$. All diagonal matrices are symmetric. Since, in any graph if vertex B is connected to vertex D (for instance), then D is connected to B; it follows that all adjacency matrices are symmetric. The sum of two symmetric matrices is a symmetric matrix, and the inverse of a symmetric matrix, if it exists, is also symmetric. A matrix is called *skew symmetric* if $\mathbf{A} = -\mathbf{A}^T$.

$$\begin{bmatrix} 0 & -1 & -2 \\ 1 & 0 & -1 \\ 2 & 1 & 0 \end{bmatrix}$$

is skew symmetric. Clearly all diagonal elements must be zero for a skew symmetric matrix.

EXAMPLE 5 ELEMENTARY MATRICES

In many texts dealing with matrices, elementary matrices are the building blocks upon which all other invertible matrices are constructed. For us, the pivot matrices of Section 3.6 perform the role of building blocks. We include the elementary matrices in this discussion because of the role they play for others. There are three types of elementary matrices. Each is obtained from an identity matrix $\mathbf{I}$ by performing exactly one of three operations on the rows of $\mathbf{I}$.

Type 1: Type 1 elementary matrices are obtained from an identity matrix by multiplying a single row of the identity matrix by a nonzero scalar. Thus

$$\begin{bmatrix} 2 & 0 & 0 \\ 0 & 1 & 0 \\ 0 & 0 & 1 \end{bmatrix}, \quad \begin{bmatrix} 1 & 0 & 0 \\ 0 & -\frac{3}{4} & 0 \\ 0 & 0 & 1 \end{bmatrix}, \quad \text{and} \quad \begin{bmatrix} 1 & 0 & 0 \\ 0 & 1 & 0 \\ 0 & 0 & -4 \end{bmatrix}$$

are elementary matrices of type 1.

Type 2: Type 2 elementary matrices are obtained from an identity matrix by adding a scalar multiple of one row of $\mathbf{I}$ to another row of $\mathbf{I}$. Thus

$$\begin{bmatrix} 1 & 1 & 0 & 0 \\ 0 & 1 & 0 & 0 \\ 0 & 0 & 1 & 0 \\ 0 & 0 & 0 & 1 \end{bmatrix}$$

is of type 2 and obtained by adding row 2 to row 1, and

$$\begin{bmatrix} 1 & 0 & 0 & 0 \\ 0 & 1 & 0 & 3 \\ 0 & 0 & 1 & 0 \\ 0 & 0 & 0 & 1 \end{bmatrix}$$

is of type 2, obtained by adding 3 times row 4 to row 2.

Type 3: Type 3 elementary matrices are obtained from an identity matrix by permuting (interchanging) any two rows.

$$\begin{bmatrix} 0 & 1 \\ 1 & 0 \end{bmatrix}, \quad \begin{bmatrix} 1 & 0 & 0 \\ 0 & 0 & 1 \\ 0 & 1 & 0 \end{bmatrix}, \quad \text{and} \quad \begin{bmatrix} 1 & 0 & 0 & 0 & 0 \\ 0 & 1 & 0 & 0 & 0 \\ 0 & 0 & 0 & 0 & 1 \\ 0 & 0 & 0 & 1 & 0 \\ 0 & 0 & 1 & 0 & 0 \end{bmatrix}$$

are type 3 elementary matrices.

Elementary matrices are all invertible. The inverse of a type 1 elementary matrix $\mathbf{A}$ is obtained by multiplying the distinguished row of the identity matrix **I** by the reciprocal of the scalar used to obtain **A**. Thus the inverse of

$$\begin{bmatrix} 3 & 0 \\ 0 & 1 \end{bmatrix} \quad \text{is} \quad \begin{bmatrix} \frac{1}{3} & 0 \\ 0 & 1 \end{bmatrix}$$

The inverse of a type 2 matrix **A** is obtained by adding the negative of the added row. Thus the inverse of

$$\begin{bmatrix} 1 & 0 & 0 \\ 0 & 1 & 2 \\ 0 & 0 & 1 \end{bmatrix} \quad \text{is} \quad \begin{bmatrix} 1 & 0 & 0 \\ 0 & 1 & -2 \\ 0 & 0 & 1 \end{bmatrix}$$

The inverse of a type 3 matrix **A** is itself.

Type 1 and type 2 elementary matrices generate *pivot matrices,* which are discussed in the next section. A product of type 3 elementary matrices is called a *permutation matrix,* and these appear in Chapter 8.

Exercises

1. Given two diagonal matrices, both of order 3, **A** and **B**. Prove that
   (a) $\mathbf{A} + \mathbf{B} = \mathbf{B} + \mathbf{A}$, and each sum is a diagonal matrix.
   (b) $\mathbf{AB} = \mathbf{BA}$ and each product is a diagonal matrix.
   (c) $\mathbf{A}^{-1}$ (if it exists) is a diagonal matrix. (See Exercise 13 of Section 2.6.)

2. Prove that
   (a) If **A** is a square matrix, then $\mathbf{A} + \mathbf{A}^T$ is symmetric.
   *(b) If **A** and **B** are symmetric and $\mathbf{AB} = \mathbf{BA}$, then **AB** is symmetric.
   *(c) The product of two symmetric matrices of the same order need not be symmetric. (Give an example.)
   (d) If **A** is square, then $\mathbf{A} - \mathbf{A}^T$ is skew symmetric.

3. Given two upper triangular matrices, **A** and **B**, both of order 3. Prove that
   (a) $\mathbf{A} + \mathbf{B}$ is an upper triangular matrix.
   (b) **AB** is an upper triangular matrix.
   (c) **BA** is an upper triangular matrix.
   (d) $\mathbf{AB} \neq \mathbf{BA}$ (in general).
   (e) $\mathbf{A}^{-1}$, if it exists, is upper triangular. (See Exercises 14 and 15 of Section 2.6.)

4. Repeat Exercise 3 for lower triangular matrices.

5. Why must a skew symmetric matrix have all zeros on the diagonal?

6. Three elementary matrice of type 2 with a $\neq 0$, b $\neq 0$, c $\neq 0$ are

$$\mathbf{A} = \begin{bmatrix} 1 & 0 & 0 \\ a & 1 & 0 \\ 0 & 0 & 1 \end{bmatrix}, \quad \mathbf{B} = \begin{bmatrix} 1 & 0 & 0 \\ 0 & 1 & 0 \\ b & 0 & 1 \end{bmatrix}, \quad \mathbf{C} = \begin{bmatrix} 1 & 0 & 0 \\ 0 & 1 & 0 \\ 0 & c & 1 \end{bmatrix}$$

Is **AB** = **BA**? **AC** = **CA**? **BC** = **CB**? For type 2, when do products commute?

7. Find the inverse, if any, of the permutation matrix:

$$\begin{bmatrix} 0 & 1 & 0 \\ 0 & 0 & 1 \\ 1 & 0 & 0 \end{bmatrix}$$

8. Find the inverse, if any, of each of the following:

$$\text{(a)} \begin{bmatrix} 1 & 0 & 0 \\ 0 & 0 & 1 \\ 0 & 1 & 0 \end{bmatrix} \quad \text{(b)} \begin{bmatrix} 3 & 0 & 0 \\ 0 & 4 & 0 \\ 0 & 0 & -5 \end{bmatrix} \quad \text{(c)} \begin{bmatrix} 1 & 2 & 3 \\ 0 & 1 & 4 \\ 0 & 0 & 1 \end{bmatrix}$$

9. Prove that
(a) The inverse of an elementary matrix, of order 3, of type 3 is itself.
(b) The inverse of

$$\begin{bmatrix} a & 0 & 0 \\ 0 & 1 & 0 \\ 0 & 0 & 1 \end{bmatrix} \quad (a \neq 0) \quad \text{is} \quad \begin{bmatrix} \frac{1}{a} & 0 & 0 \\ 0 & 1 & 0 \\ 0 & 0 & 1 \end{bmatrix}$$

(c) The inverse of

$$\begin{bmatrix} a & 0 & 0 \\ 0 & 1 & 0 \\ b & 0 & 1 \end{bmatrix} \quad (a \neq 0) \quad \text{is} \quad \begin{bmatrix} \frac{1}{a} & 0 & 0 \\ 0 & 1 & 0 \\ -\frac{b}{a} & 0 & 1 \end{bmatrix}$$

10. Let

$$\mathbf{A} = \begin{bmatrix} a_{11} & a_{12} & a_{13} \\ a_{21} & a_{22} & a_{23} \\ a_{31} & a_{32} & a_{33} \end{bmatrix}$$

Multiply **A** on the left by

(a) $\begin{bmatrix} 0 & 1 & 0 \\ 1 & 0 & 0 \\ 0 & 0 & 1 \end{bmatrix}$ (b) $\begin{bmatrix} b & 0 & 0 \\ 0 & 1 & 0 \\ 0 & 0 & 1 \end{bmatrix}$ (c) $\begin{bmatrix} 1 & 0 & b \\ 0 & 1 & c \\ 0 & 0 & 1 \end{bmatrix}$

11. Multiply **A** of Exercise 10 on the right by the elementary matrices listed in that exercise. Describe the effect on **A** of each multiplication.

***12.** Prove: $\mathbf{P}^{-1} = \mathbf{P}^T$ if **P** is a product of type 3 elementary matrices.

3.6 Pivot Matrices

Pivot matrices are building blocks of other matrices. To show this we need the concept of pivot step, which essentially is pivot-exchange without the exchange.

EXAMPLE 1

Let

$$\mathbf{A} = \begin{bmatrix} 1 & 5 \\ 3 & 4 \\ 2 & 6 \end{bmatrix}$$

be the matrix corresponding to the mapping

	x_1	x_2		
	1	5	=	y_1
F:	3	4	=	y_2
	2*	6	=	y_3

If x_1 and y_3 are exchanged, the tableau becomes

$-y_3$	x_2		
$-\frac{1}{2}$	2	=	y_1
$-\frac{3}{2}$	-5	=	y_2
$\frac{1}{2}$	3	=	$-x_1$

and its matrix is

$$\mathbf{B} = \begin{bmatrix} -\frac{1}{2} & 2 \\ -\frac{3}{2} & -5 \\ \frac{1}{2} & 3 \end{bmatrix}$$

We go directly (without considering the mappings) from matrix **A** to matrix **B** by pivoting (without exchange) on the pivot entry 2 in **A**. This operation on matrices is called a *pivot step*. A pivot step is performed on a matrix. Since there are no variables in evidence, there is clearly no exchange.

EXAMPLE 2

Pivot on -2 in **A** to obtain **B**, where

$$\mathbf{A} = \begin{bmatrix} 1 & 2 & 3 \\ 0 & -4 & 5 \\ 4 & -2^* & 2 \end{bmatrix}$$

Then

$$\mathbf{B} = \begin{bmatrix} 5 & 1 & 5 \\ -8 & -2 & 1 \\ -2 & -\frac{1}{2} & -1 \end{bmatrix}$$

There is another way to obtain **B** from **A**. Let

$$\mathbf{I}_3 = \begin{bmatrix} 1 & 0 & 0 \\ 0 & 1 & 0 \\ 0 & 0 & 1 \end{bmatrix}$$

Replace the third column of $\mathbf{I}_3$ by the second column of **B**, the post pivot column. This yields

$$\mathbf{I}_B = \begin{bmatrix} 1 & 0 & 1 \\ 0 & 1 & -2 \\ 0 & 0 & -\frac{1}{2} \end{bmatrix}$$

Now multiply $\mathbf{I}_B$ (called a *pivot matrix* of **A** on -2) with **A**

$$\begin{bmatrix} 1 & 0 & 1 \\ 0 & 1 & -2 \\ 0 & 0 & -\frac{1}{2} \end{bmatrix}\begin{bmatrix} 1 & 2 & 3 \\ 0 & -4 & 5 \\ 4 & -2 & 2 \end{bmatrix} = \begin{bmatrix} 5 & 0 & 5 \\ -8 & 0 & 1 \\ -2 & 1 & -1 \end{bmatrix} = \mathbf{B}_R$$

Replace the second column of $\mathbf{B}_R$ by the post pivot column to get $\mathbf{B}$:

$$\begin{bmatrix} 5 & 1 & 5 \\ -8 & -2 & 1 \\ -2 & -\frac{1}{2} & 1 \end{bmatrix}$$

EXAMPLE 3

Using the method of Example 2, obtain the pivot matrix $\mathbf{I}_B$, matrix $\mathbf{B}_R$, and matrix $\mathbf{B}$ from matrix $\mathbf{A}$ using 2 as the pivot element.

$$\mathbf{A} = \begin{bmatrix} 4 & -4 & 5 & 2 \\ 1 & 5 & 2^* & -4 \\ -6 & -10 & -9 & 8 \\ 6 & -2 & 8 & 0 \\ -3 & -9 & -5 & 7 \end{bmatrix}$$

We formalize the procedure of Example 2. The matrix $\mathbf{A}$ is 5×4. Since it has 5 rows we select

$$\mathbf{I}_5 = \begin{bmatrix} 1 & 0 & 0 & 0 & 0 \\ 0 & 1 & 0 & 0 & 0 \\ 0 & 0 & 1 & 0 & 0 \\ 0 & 0 & 0 & 1 & 0 \\ 0 & 0 & 0 & 0 & 1 \end{bmatrix}$$

The post pivot column obtained by pivoting on 2 is

$$\begin{bmatrix} -\frac{5}{2} \\ \frac{1}{2} \\ \frac{9}{2} \\ -4 \\ \frac{5}{2} \end{bmatrix}$$

The pivot 2 is in the second row of **A**. Therefore, we replace the second column of $\mathbf{I}_5$ with the post pivot column, yielding

$$\mathbf{I}_B = \begin{bmatrix} 1 & -5/2 & 0 & 0 & 0 \\ 0 & 1/2 & 0 & 0 & 0 \\ 0 & 9/2 & 1 & 0 & 0 \\ 0 & -4 & 0 & 1 & 0 \\ 0 & 5/2 & 0 & 0 & 1 \end{bmatrix}$$

Now $\mathbf{I}_B \mathbf{A} = \mathbf{B}_R$:

$$\begin{bmatrix} 1 & -5/2 & 0 & 0 & 0 \\ 0 & 1/2 & 0 & 0 & 0 \\ 0 & 9/2 & 1 & 0 & 0 \\ 0 & -4 & 0 & 1 & 0 \\ 0 & 5/2 & 0 & 0 & 1 \end{bmatrix} \begin{bmatrix} 4 & -4 & 5 & 2 \\ 1 & 5 & 2 & -4 \\ -6 & -10 & -9 & 8 \\ 6 & -2 & 8 & 0 \\ -3 & -9 & -5 & 7 \end{bmatrix}$$

$$= \begin{bmatrix} 3/2 & -33/2 & 0 & 12 \\ 1/2 & 5/2 & 1 & -2 \\ -3/2 & 25/2 & 0 & -10 \\ 2 & -22 & 0 & 16 \\ -1/2 & 7/2 & 0 & -3 \end{bmatrix} = \mathbf{B}_R$$

Now in $\mathbf{B}_R$ replace the third column by the post pivot column. This yields

$$\mathbf{B} = \begin{bmatrix} 3/2 & -33/2 & -5/2 & 12 \\ 1/2 & 5/2 & 1/2 & -2 \\ -3/2 & 25/2 & 9/2 & -10 \\ 2 & -22 & -4 & 16 \\ -1/2 & 7/2 & 5/2 & -3 \end{bmatrix}$$

That **B** is obtained from **A** by a pivot step can be verified directly (or see Example 5 of Section 2.2).

There are three key selections to make in using a pivot matrix in the process of Examples 2 and 3. It can be proved that this process works for all matrices.

1. If **A** is $m \times n$, select $\mathbf{I}_m$.
2. If the pivot element is a_{ij}, replace the ith column of $\mathbf{I}_m$ by the post pivot column obtained by pivoting on a_{ij}. (This is done so that the reciprocal of the pivot element replaces the 1 in the ith column of $\mathbf{I}_m$.) The matrix so obtained is designated $\mathbf{I}_B$ and called a pivot matrix of **A** on a_{ij}.
3. Replace the jth column of $\mathbf{B}_R$ by the post pivot column of (2).

   This resultant matrix is **B**, which is the matrix obtainable from **A** by performing a pivot step on **A** using a_{ij} as the pivot element.

EXAMPLE 4

Find the inverse of the pivot matrix:

$$\mathbf{P} = \begin{bmatrix} 1 & 0 & a \\ 0 & 1 & b \\ 0 & 0 & c \end{bmatrix}, c \neq 0$$

Notice that even though we do not know the origin of **P**, we can see it is a pivot matrix since all columns except the third are from an identity matrix. Now since c replaces 1 in $\mathbf{I}_3$, it is the reciprocal of a pivot element and thus not equal to zero. Therefore, it can be a pivot element itself. Pivoting on c yields

$$\mathbf{P}^{-1} = \begin{bmatrix} 1 & 0 & -\frac{a}{c} \\ 0 & 1 & -\frac{b}{c} \\ 0 & 0 & \frac{1}{c} \end{bmatrix}$$

as can be verified by multiplying

$$\begin{bmatrix} 1 & 0 & a \\ 0 & 1 & b \\ 0 & 0 & c \end{bmatrix} \begin{bmatrix} 1 & 0 & -\frac{a}{c} \\ 0 & 1 & -\frac{b}{c} \\ 0 & 0 & \frac{1}{c} \end{bmatrix} = \begin{bmatrix} 1 & 0 & 0 \\ 0 & 1 & 0 \\ 0 & 0 & 1 \end{bmatrix}$$

This result is extended to

THEOREM 3.6. All pivot matrices have inverses which themselves are pivot matrices. The inverse of a pivot matrix is obtained from the pivot matrix by pivoting on that element which replaced the 1 in the identity matrix.

EXAMPLE 5

Find $\mathbf{I}_B^{-1}$ when

$$\mathbf{I}_B = \begin{bmatrix} 1 & 0 & 2 \\ 0 & 1 & 5 \\ 0 & 0 & 3 \end{bmatrix}$$

The 3 must be the pivot element. Thus

$$\mathbf{I}_B^{-1} = \begin{bmatrix} 1 & 0 & -2/3 \\ 0 & 1 & -5/3 \\ 0 & 0 & 1/3 \end{bmatrix}$$

The reader should verify this result by showing that

$$\mathbf{I}_B \mathbf{I}_B^{-1} = \mathbf{I}_3$$

Theorem 3.6 enables us to express $\mathbf{A}$ as

$$\mathbf{A} = \mathbf{I}_B^{-1}\mathbf{B}_R \tag{1}$$

This follows since

$$\mathbf{I}_B \mathbf{A} = \mathbf{B}_R$$

and multiplying both sides of this equation on the left by $\mathbf{I}_B^{-1}$ yields

$$\mathbf{I}_B^{-1}\mathbf{I}_B \mathbf{A} = \mathbf{I}_B^{-1}\mathbf{B}_R$$

or

$$\mathbf{IA} = \mathbf{I}_B^{-1}\mathbf{B}_R$$

from which (1) follows. Equation (1) suggests that the mapping associated with $\mathbf{A}$ is expressible as a composition of two mappings, at least one of which is invertible ($\mathbf{I}_B^{-1}$).

In Section 3.5 we saw that any triangular matrix without zeros on the diagonal is expressible as a product of what we now know are pivot matrices. For instance, the lower triangular matrix

$$\mathbf{A} = \begin{bmatrix} 1 & 0 & 0 \\ 2 & -1 & 0 \\ 1 & 2 & 2 \end{bmatrix}$$

is expressible as

$$\mathbf{A} = \mathbf{A}_1 \mathbf{A}_2 \mathbf{A}_3 \tag{1}$$

where

$$\mathbf{A}_1 = \begin{bmatrix} 1 & 0 & 0 \\ 2 & 1 & 0 \\ 1 & 0 & 1 \end{bmatrix}, \qquad \mathbf{A}_2 = \begin{bmatrix} 1 & 0 & 0 \\ 0 & -1 & 0 \\ 0 & 2 & 1 \end{bmatrix},$$

$$\mathbf{A}_3 = \begin{bmatrix} 1 & 0 & 0 \\ 0 & 1 & 0 \\ 0 & 0 & 2 \end{bmatrix}$$

Each of $\mathbf{A}_1$, $\mathbf{A}_2$, and $\mathbf{A}_3$ has an inverse since each is a pivot matrix. Thus

$$(\mathbf{A}_1 \mathbf{A}_2 \mathbf{A}_3)^{-1} = \mathbf{A}_3^{-1} \mathbf{A}_2^{-1} \mathbf{A}_1^{-1}$$

[from $(\mathbf{AB})^{-1} = \mathbf{B}^{-1}\mathbf{A}^{-1}$].

From (1),

$$\mathbf{A}^{-1} = (\mathbf{A}_1 \mathbf{A}_2 \mathbf{A}_3)^{-1}$$

and thus

$$\mathbf{A}^{-1} = \mathbf{A}_3^{-1} \mathbf{A}_2^{-1} \mathbf{A}_1^{-1} \tag{2}$$

From Theorem 3.6,

$$\mathbf{A}_3^{-1} = \begin{bmatrix} 1 & 0 & 0 \\ 0 & 1 & 0 \\ 0 & 0 & \tfrac{1}{2} \end{bmatrix}, \qquad \mathbf{A}_2^{-1} = \begin{bmatrix} 1 & 0 & 0 \\ 0 & -1 & 0 \\ 0 & 2 & 1 \end{bmatrix},$$

$$\mathbf{A}_1^{-1} = \begin{bmatrix} 1 & 0 & 0 \\ -2 & 1 & 0 \\ -1 & 0 & 1 \end{bmatrix}$$

From (2),

$$\mathbf{A}^{-1} = \begin{bmatrix} 1 & 0 & 0 \\ 0 & 1 & 0 \\ 0 & 0 & \tfrac{1}{2} \end{bmatrix} \begin{bmatrix} 1 & 0 & 0 \\ 0 & -1 & 0 \\ 0 & 2 & 1 \end{bmatrix} \begin{bmatrix} 1 & 0 & 0 \\ -2 & 1 & 0 \\ -1 & 0 & 1 \end{bmatrix}$$

$$= \begin{bmatrix} 1 & 0 & 0 \\ 0 & -1 & 0 \\ 0 & 1 & \tfrac{1}{2} \end{bmatrix} \begin{bmatrix} 1 & 0 & 0 \\ -2 & 1 & 0 \\ -1 & 0 & 1 \end{bmatrix} = \begin{bmatrix} 1 & 0 & 0 \\ 2 & -1 & 0 \\ -\tfrac{5}{2} & 1 & \tfrac{1}{2} \end{bmatrix}$$

We check that $\mathbf{AA}^{-1} = \mathbf{I}$:

$$\begin{bmatrix} 1 & 0 & 0 \\ 2 & -1 & 0 \\ 1 & 2 & 2 \end{bmatrix} \begin{bmatrix} 1 & 0 & 0 \\ 2 & -1 & 0 \\ -\tfrac{5}{2} & 1 & \tfrac{1}{2} \end{bmatrix} = \begin{bmatrix} 1 & 0 & 0 \\ 0 & 1 & 0 \\ 0 & 0 & 1 \end{bmatrix}$$

Notice that $\mathbf{A}^{-1}$ is also lower triangular. A similar result holds for upper triangular matrices.

We sum up in

THEOREM 3.7. If an $n \times n$ matrix $\mathbf{A}$ has no zeros on the diagonal and if $\mathbf{A}$ is lower triangular, then $\mathbf{A}^{-1}$ is $n \times n$ lower triangular and $\mathbf{A}^{-1} = \mathbf{A}_n^{-1}\mathbf{A}_{n-1}^{-1} \ldots \mathbf{A}_1^{-1}$, where $\mathbf{A}_1, \ldots, \mathbf{A}_n$ are the pivot matrices defined in Section 3.5.

If $\mathbf{A}$ is upper triangular, then $\mathbf{A}^{-1}$ is $n \times n$ upper triangular and

$$\mathbf{A}^{-1} = \mathbf{A}_1^{-1}\mathbf{A}_2^{-1} \ldots \mathbf{A}_n^{-1}.$$

Notice that the pivot matrix

$$\begin{bmatrix} a & 0 & 0 \\ b & 1 & 0 \\ c & 0 & 1 \end{bmatrix}$$

is obtainable from the type 1 elementary matrix

$$\begin{bmatrix} a & 0 & 0 \\ 0 & 1 & 0 \\ 0 & 0 & 1 \end{bmatrix}$$

and the type 2 elementary matrices

$$\begin{bmatrix} 1 & 0 & 0 \\ b & 1 & 0 \\ 0 & 0 & 1 \end{bmatrix}, \quad \begin{bmatrix} 1 & 0 & 0 \\ 0 & 1 & 0 \\ c & 0 & 1 \end{bmatrix}$$

by

$$\begin{bmatrix} a & 0 & 0 \\ b & 1 & 0 \\ c & 0 & 1 \end{bmatrix} =$$

$$\begin{bmatrix} a & 0 & 0 \\ 0 & 1 & 0 \\ 0 & 0 & 1 \end{bmatrix} \begin{bmatrix} 1 & 0 & 0 \\ b & 1 & 0 \\ 0 & 0 & 1 \end{bmatrix} \begin{bmatrix} 1 & 0 & 0 \\ 0 & 1 & 0 \\ c & 0 & 1 \end{bmatrix}$$

Exercises

1. Given

$$\mathbf{A} = \begin{bmatrix} 1 & 2 & 3 \\ 4 & 5 & 6 \\ -3 & -2 & -1 \end{bmatrix}$$

find the pivot matrix $\mathbf{I}_B$ of $\mathbf{A}$ on: (a) 1; (b) 2; (c) 3.

2. For each pivot matrix $\mathbf{I}_B$ in Exercise 1, find $\mathbf{B}$, the matrix that would have resulted from a pivot step.

3. Explain why

$$\begin{bmatrix} 2 & 0 & 0 \\ 3 & 1 & 0 \\ 4 & 0 & 0 \end{bmatrix}$$

is not a pivot matrix.

4. Given

$$\mathbf{A} = \begin{bmatrix} 1 & 2 \\ 4 & 5 \\ -3 & 6 \end{bmatrix}$$

find the pivot matrix $\mathbf{I}_B$ of $\mathbf{A}$ on: (a) 1; (b) 2; (c) 6.

5. For each pivot matrix $\mathbf{I}_B$ in Exercise 4, find $\mathbf{B}$.

6. Given

$$\mathbf{A} = \begin{bmatrix} 1 & 2 & 3 \\ 4 & 5 & 6 \end{bmatrix}$$

carry out the instructions in Exercises 4 and 5.

7. Find the inverse of each of the following pivot matrices:

(a) $\begin{bmatrix} 1 & 2 \\ 0 & 3 \end{bmatrix}$ (b) $\begin{bmatrix} 4 & 0 \\ 5 & 1 \end{bmatrix}$

(c) $\begin{bmatrix} 1 & 0 & 3 \\ 0 & 1 & 4 \\ 0 & 0 & 2 \end{bmatrix}$ (d) $\begin{bmatrix} 1 & 1 & 0 & 0 \\ 0 & -1 & 0 & 0 \\ 0 & 2 & 1 & 0 \\ 0 & 3 & 0 & 1 \end{bmatrix}$

8. Show that the product of the elementary matrices ($b \neq 0$)

$$\begin{bmatrix} 1 & a & 0 \\ 0 & 1 & 0 \\ 0 & 0 & 1 \end{bmatrix} \begin{bmatrix} 1 & 0 & 0 \\ 0 & b & 0 \\ 0 & 0 & 1 \end{bmatrix} \begin{bmatrix} 1 & 0 & 0 \\ 0 & 1 & 0 \\ 0 & c & 1 \end{bmatrix}$$

is a pivot matrix.

9. Express each matrix in Exercise 7 as the product of elementary matrices.

10. Find the inverse of the upper triangular matrix

$$\begin{bmatrix} 2 & 1 & 3 \\ 0 & -2 & 5 \\ 0 & 0 & -1 \end{bmatrix}$$

3.7 Chapter Review

Matrices and operations on matrices are defined. The operations are

1. Addition.
2. Scalar multiplication.
3. Matrix multiplication.
4. Pivot step.

Matrix multiplication is defined in terms of composition of mappings. Special matrices include $\mathbf{0}_n$, $\mathbf{I}_n$, symmetric matrices, skew symmetric matrices, diagonal matrices, triangular matrices, elementary matrices, and pivot matrices. The concept of the inverse of a matrix is studied. All diagonal and triangular matrices with no zeros on the diagonal are shown to have inverses. All pivot matrices have inverses. All triangular matrices $\mathbf{T}$ that have inverses are expressible as a product of certain pivot matrices and thus $\mathbf{T}^{-1}$ is expressible as the product (in reverse order) of the inverses of the pivot matrices.

Applications of matrices are studied, including

1. Transition matrices.
2. Systems of linear equations.
3. Mappings.
4. Linear programs.
5. Coding.
6. Communications.

The theorems of this chapter are:

Theorem 3.1. A square matrix of order n has an inverse if and only if the pivot rank of the tableau of the associated mapping is equal to n.

Theorem 3.2. An invertible matrix has exactly one inverse.

Theorem 3.3. If two square matrices $\mathbf{A}$ and $\mathbf{B}$ of the same order are invertible, then $\mathbf{AB}$ also has an inverse $(\mathbf{AB})^{-1}$ and $(\mathbf{AB})^{-1} = \mathbf{B}^{-1}\mathbf{A}^{-1}$.

Theorem 3.4. Let $\mathbf{A}$ have dimension $m \times n$ and $\mathbf{B}$ have dimension $n \times p$. Then $(\mathbf{AB})^T = \mathbf{B}^T\mathbf{A}^T$ is a matrix of dimension $p \times m$.

Theorem 3.5. If a matrix $\mathbf{A}$ has an inverse, $\mathbf{A}^{-1}$, then $(\mathbf{A}^{-1})^T = (\mathbf{A}^T)^{-1}$.

Theorem 3.6. All pivot matrices have inverses which themselves are pivot matrices. The inverse of a pivot matrix is obtained from the pivot matrix by pivoting on that element which replaced the 1 in the identity matrix.

Theorem 3.7. If an $n \times n$ matrix $\mathbf{A}$ has no zeros on the diagonal and if $\mathbf{A}$ is lower triangular given by

$$\mathbf{A} = \mathbf{A}_1\mathbf{A}_2\cdots\mathbf{A}_n$$

where $\mathbf{A}_1, \mathbf{A}_2, \ldots, \mathbf{A}_n$ are pivot matrices, then

$$\mathbf{A}^{-1} = \mathbf{A}_n^{-1}\mathbf{A}_{n-1}^{-1}\cdots\mathbf{A}_1^{-1}.$$

Chapter Review Exercises

1. Solve for x, y, z, w:

$$2\begin{bmatrix} x & 3 \\ 2 & y \end{bmatrix} + 3\begin{bmatrix} 1 & 2 \\ 3 & 4 \end{bmatrix} = 4\begin{bmatrix} 1 & z \\ w & 1 \end{bmatrix}$$

2. Let

$$\mathbf{A} = \begin{bmatrix} 3 & 1 \\ 4 & 2 \end{bmatrix}, \quad \mathbf{B} = \begin{bmatrix} a & b \\ 0 & c \end{bmatrix}$$

For what values of a, b, and c is it true that $\mathbf{AB} = \mathbf{BA}$?

3. (a) Using

$$\mathbf{C} = \begin{bmatrix} 3 & 1 \\ 5 & 2 \end{bmatrix}$$

as a coding matrix, code the following message:

W I L L C O M E S O O N

(b) What matrix is needed to decode the coded message?

4. Describe each of the following two-way bus routes between towns by a matrix

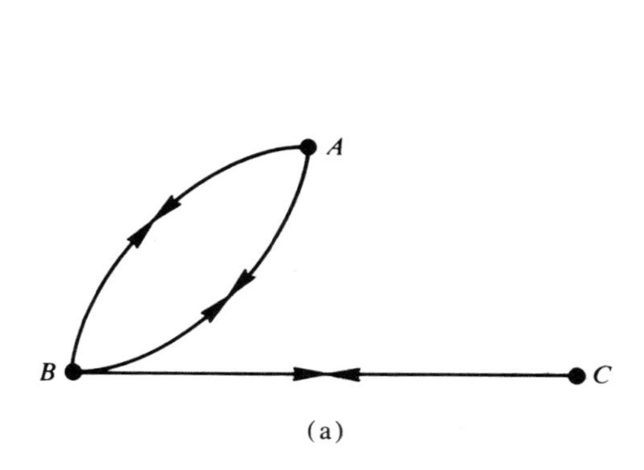

(a)

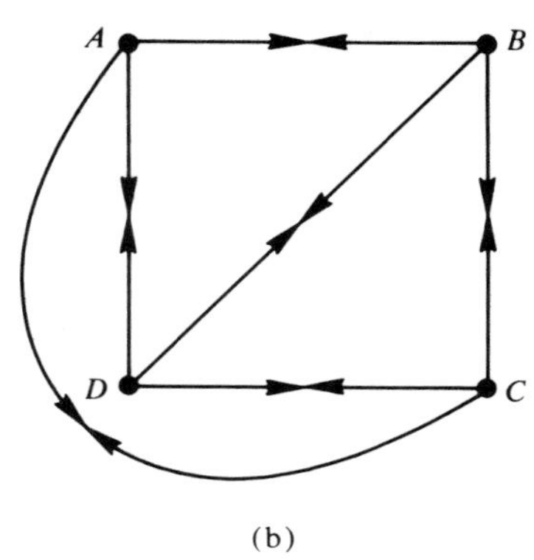

(b)

5. Find the inverse, if any, of each of the following:

(a) $\begin{bmatrix} 0 & a \\ -a & 0 \end{bmatrix}, a \neq 0$

(b) $\begin{bmatrix} -1 & 2 & -3 \\ 1 & 0 & 2 \\ 2 & 2 & 3 \end{bmatrix}$

(c) $\begin{bmatrix} 3 & 0 & 0 \\ 0 & 2 & 0 \\ 0 & 0 & 1 \end{bmatrix}$

(d) $\begin{bmatrix} 2 & 3 & -1 \\ 0 & 1 & -2 \\ 0 & 0 & 4 \end{bmatrix}$

(e) $\begin{bmatrix} 3 & 0 & 0 \\ 2 & 1 & 0 \\ 1 & 0 & 1 \end{bmatrix}$

(f) $\begin{bmatrix} 1 & 2 & 0 & 0 \\ 0 & 3 & 0 & 0 \\ 0 & 4 & 1 & 0 \\ 0 & 5 & 0 & 1 \end{bmatrix}$

6. Given the inverse of

$$\begin{bmatrix} 5 & -7 & -1 \\ -8 & 11 & 2 \\ 4 & -5 & -1 \end{bmatrix}$$

is

$$\begin{bmatrix} 1 & 2 & 3 \\ 0 & 1 & 2 \\ 4 & 3 & 1 \end{bmatrix}$$

Solve the following systems of equations, where

$$(a, b, c) = (0, 0, 0), (1, 2, 3), \text{and } (3, 2, 1)$$

x	y	z	
5	−7	−1	$= a$
−8	11	2	$= b$
4	−5	−1	$= c$

7. Given mapping $F: R^3 \to R^2$ with tableau

x_1	x_2	x_3	
3	1	2	$= y_1$
4	−1	−3	$= y_2$

Let **A** be the matrix of the inside of the tableau,

$$\mathbf{X} = [x_1 \quad x_2 \quad x_3], \qquad \mathbf{Y} = [y_1 \quad y_2]$$

(a) Express the equations in the tableau in terms of $\mathbf{X}^T$ and $\mathbf{Y}^T$.
(b) Express the equations in the tableau in terms of $\mathbf{A}^T$.
(c) With one matrix multiplication find the images of (1, 2, −1), (2, −1, 1), and (−2, 1, 0).

8. Given the dual linear program with tableau

	x_1	x_2	x_3	-1	
v_1	3	1	2	4	$= -y_1$
v_2	4	−1	−3	6	$= -y_2$
-1	2	5	3	0	$= M$
	$= u_1$	$= u_2$	$= u_3$	$= m$	

write the dual linear program in matrix format.

9. Given the graph G:

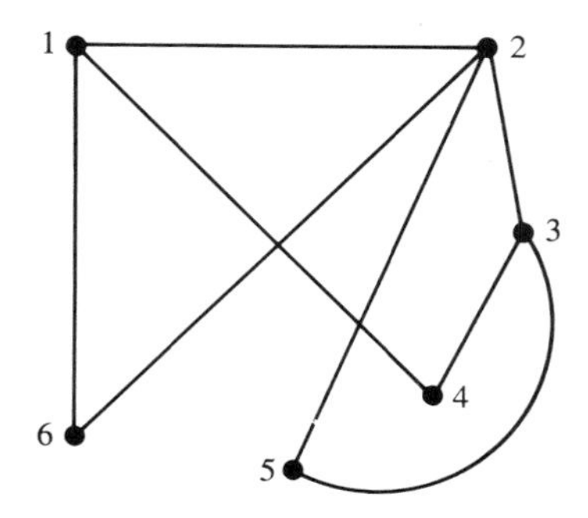

(a) Write its adjacency matrix $\mathbf{A}$.
(b) By counting the number of paths of length 2, try to obtain $\mathbf{A}^2$. Check by multiplying $\mathbf{AA}$.
(c) Draw the complementary graph H of G (the graph consisting of the same vertices as G and all edges not in G).
(d) Write the adjacency matrix $\mathbf{B}$ of H.
(e) Add $\mathbf{A} + \mathbf{B}$. (Why do you think you get this answer?)

10. Factor into pivot matrices

$$\mathbf{A} = \begin{bmatrix} 1 & 0 & 0 & 0 \\ -2 & 4 & 0 & 0 \\ -3 & 2 & 3 & 0 \\ 2 & -5 & -1 & 6 \end{bmatrix}$$

11. Find the inverse of $\mathbf{A}$ above using pivot matrices.

12. Find $\mathbf{A}^T$. Factor it into pivot matrices.

13. Find $(\mathbf{A}^T)^{-1}$ as in Exercise 11.

14. Draw tentative conclusions from Exercises 10–13.

CLA

CHAPTER 4

Systems of Linear Equations

In this chapter, we are interested in finding the set of all solutions to a given system of linear equations. (The word system will be interpreted to mean system of linear equations.) As might be expected, the pivot exchange procedure will once again play a prominent role. We use pivot exchange to develop the pivot reduction algorithm.

4.1 The Pivot Reduction Algorithm

EXAMPLE 1

Solve for (x_1, x_2, x_3) in the system

(1)

x_1	x_2	x_3	-1	
2*	-1	0	20	$= 0$
-1	2	-1	0	$= 0$
0	-1	2	100	$= 0$

This system can also be written as

$$\begin{cases} 2x_1 - x_2 \qquad\qquad -20 = 0 \\ -x_1 + 2x_2 - x_3 \qquad\qquad = 0 \\ \qquad -x_2 + 2x_3 - 100 = 0 \end{cases}$$

or as

$$\begin{cases} 2x_1 - x_2 \qquad = 20 \\ -x_1 + 2x_2 - x_3 = 0 \\ \qquad -x_2 + 2x_3 = 100 \end{cases}$$

Recall from Chapter 2 that to solve for the variables at the top of a tableau, exchange them to the right. Pivot exchanging as indicated in (1) yields

(2)

0	x_2	x_3	-1	
$\frac{1}{2}$	$-\frac{1}{2}$	0	10	$= -x_1$
$\frac{1}{2}$	$\frac{3}{2}$	-1	10	$= 0$
0	-1	2	100	$= 0$

An obvious question is: "Since we cannot pivot on a 0, can we exchange a 0?" The answer is, "yes." Consider the 0 as a symbol to be exchanged. As long as we do not divide by this symbol, it can be exchanged like any other symbol.

The first column of (2) contributes nothing to the system since all numbers in that column are multiplied by 0. Therefore, we eliminate this column and obtain

(3)

x_2	x_3	-1	
$-\frac{1}{2}$	0	10	$= -x_1$
$\frac{3}{2}$*	-1	10	$= 0$
-1	2	100	$= 0$

Pivoting in (3) gives

(4)

0	x_3	-1	
$\frac{1}{3}$	$-\frac{1}{3}$	$\frac{40}{3}$	$= -x_1$
$\frac{2}{3}$	$-\frac{2}{3}$	$\frac{20}{3}$	$= -x_2$
$\frac{2}{3}$	$\frac{4}{3}$	$\frac{320}{3}$	$= 0$

As with (2), the first column can be eliminated without affecting the solution set of the system. Eliminate it.

(5)

x_3	-1	
$-\frac{1}{3}$	$\frac{40}{3}$	$= -x_1$
$-\frac{2}{3}$	$\frac{20}{3}$	$= -x_2$
$\frac{4}{3}$*	$\frac{320}{3}$	$= 0$

Pivoting gives

(6)

0	-1	
$\frac{1}{4}$	40	$= -x_1$
$\frac{1}{2}$	60	$= -x_2$
$\frac{3}{4}$	80	$= -x_3$

Once again eliminating the 0 column gives us a reduced form:

(7)

-1	
40	$= -x_1$
60	$= -x_2$
80	$= -x_3$

In (7) we read that

$$(x_1, x_2, x_3) = (40, 60, 80)$$

is a solution of the system (7) and the only solution of (7). But (1) and (7) have the same solutions. Thus, (40, 60, 80) is the only solution to (1). Substituting (40, 60, 80) in (1) gives

(8)

40	60	80	-1		
2	-1	0	20	$= 0$	✓
-1	2	-1	0	$= 0$	✓
0	-1	2	100	$= 0$	✓

A quick check verifies the consistency of (8) and confirms that $(x_1, x_2, x_3) = (40, 60, 80)$ is a solution.

EXAMPLE 2

Let $M:R^3 \to R^2$ be a mapping given by

(1)

x_1	x_2	x_3	
3	1	−3	$= y_1$
3	−2	4	$= y_2$

Find the set of preimages of $(y_1, y_2) = (3, 7)$.

Substituting $y_1 = 3$, $y_2 = 7$ in (1) gives

(2)

x_1	x_2	x_3	
3	1	−3	= 3
3	−2	4	= 7

Row 1 of (2) is

$$3x_1 + x_2 - 3x_3 = 3$$

which can be written as

$$3x_1 + x_2 - 3x_3 - 3 = 0$$

Similar manipulation of row 2 shows that (2) is equivalent to

(3)

x_1	x_2	x_3	−1	
3	1*	−3	3	= 0
3	−2	4	7	= 0

The set of solutions to the system (3) is the same as the set of preimages of $(y_1, y_2) = (3, 7)$ for the mapping (1). Using the pivot as indicated in (3) we obtain (4):

(4)

x_1	0	x_3	−1	
3	1	−3	3	$= -x_2$
9	2	−2	13	= 0

Deleting the 0 column gives

$$(5)\qquad \begin{array}{ccc} x_1 & x_3 & -1 \\ \hline 3 & -3 & 3 \\ 9^* & -2 & 13 \end{array} \begin{array}{l} \\ = -x_2 \\ = 0 \end{array}$$

Continuing we get

$$(6)\qquad \begin{array}{ccc} 0 & x_3 & -1 \\ \hline -\tfrac{1}{3} & -\tfrac{7}{3} & -\tfrac{4}{3} \\ \tfrac{1}{9} & -\tfrac{2}{9} & \tfrac{13}{9} \end{array} \begin{array}{l} \\ = -x_2 \\ = -x_1 \end{array}$$

$$(7)\qquad \begin{array}{cc} x_3 & -1 \\ \hline -\tfrac{7}{3} & -\tfrac{4}{3} \\ -\tfrac{2}{9} & \tfrac{13}{9} \end{array} \begin{array}{l} \\ = -x_2 \\ = -x_1 \end{array}$$

From (7) we see that for any given number x_3,

$$x_1 = \tfrac{2}{9}x_3 + \tfrac{13}{9}$$
$$x_2 = \tfrac{7}{3}x_3 - \tfrac{4}{3}$$

If $x_3 = 4$, $x_1 = \tfrac{7}{3}$, and $x_2 = 8$. So one solution to (7) is

$$(x_1, x_2, x_3) = (\tfrac{7}{3}, 8, 4)$$

The triple $(\tfrac{7}{3}, 8, 4)$ is also one of the preimages of (3, 7).

If $x_3 = 7$, $(x_1, x_2, x_3) = (3, 15, 7)$, and this too is a solution to (7) and a preimage of (3, 7).

Steps (4) and (6) are for explanatory purposes only. The actual work you show in solving the system in Example 2 is

$$(3)\qquad \begin{array}{cccc} x_1 & x_2 & x_3 & -1 \\ \hline 3 & 1^* & -3 & 3 \\ 3 & -2 & 4 & 7 \end{array} \begin{array}{l} \\ = 0 \\ = 0 \end{array}$$

$$(5)\qquad \begin{array}{ccc} x_1 & x_3 & -1 \\ \hline 3 & -3 & 3 \\ 9^* & -2 & 13 \end{array} \begin{array}{l} \\ = -x_2 \\ = 0 \end{array}$$

(7)

x_3	-1	
$-7/3$	$-4/3$	$= -x_2$
$-2/9$	$-13/9$	$= -x_1$

We need some new terms to describe what is going on. The system of linear equations in (1) of Example 1 and (3) of Example 2 are tableaus in which variables and -1 appear at the top and zeros at the right. Such a tableau is called the *standard form* of a system. With each exchange of a variable and a zero, a column is deleted and the variable is exchanged (with change of sign) to the right, where it becomes a basic variable. This procedure is called *pivot reduction*. The number of possible deletions is the same as the number of exchanges—the pivot rank of the initial tableau. When all possible deletions are made, the last tableau in the sequence is called a *fully reduced form* or *explicit form* of the initial standard form tableau. The complete procedure is called the *pivot reduction algorithm*.

An immediate conclusion is that the number of basic variables in the fully reduced form is the pivot rank of the initial tableau, in standard form.

EXAMPLE 3

Find a fully reduced form of the system in

(1)

w	x	y	z	-1	
3	2	8	1	-4	$= 0$
-1*	2	-4	5	0	$= 0$

Using pivot reduction as indicated,

(2)

x	y	z	-1	
8	-4*	16	-4	$= 0$
-2	4	-5	0	$= -w$

(3)

x	z	-1	
-2	-4	1	$= -y$
6	11	-4	$= -w$

To check, recall that pivot exchange, and therefore pivot reduction preserves solution sets. Use (3) to find one solution. For instance, let $(x, z) = (1, 1)$. Then $y = 7$ and $w = -21$. Now check in (1):

−21	1	7	1	−1		
3	2	8	1	−4	= 0	✓
−1	2	−4	5	0	= 0	✓

We know that some members of the codomain of a mapping may have no preimages, some exactly one, and some many preimages. It is therefore no surprise that some systems of linear equations have no solution, some have exactly one, and some have solution sets with many elements. It is a fact that all three of these types of solution sets are obtained by pivot reduction.

The pivot reduction algorithm consists of the following steps:

1. Put a system of equations in standard form.
2. Use the fundamental pivot-exchange algorithm on the columns headed by variables, eliminating each column as its variable is exchanged with 0.
3. Continue until the fully reduced form is reached.

Exercises

Convert each of the systems of linear equations in Exercises 1–10 to fully reduced forms.

1.

x	y	-1	
2	3	4	= 0
5	8	8	= 0

2.

x	y	-1	
2	3	4	= 0
6	9	12	= 0

3.

x	y	-1	
2	3	4	= 0
6	9	14	= 0

4.

x_1	x_2	-1	
1	2	3	= 0
−1	3	2	= 0
3	1	2	= 0

5.

x_1	x_2	-1	
1	2	3	= 0

6.

x	-1	
1	2	$= 0$
3	4	$= 0$
5	6	$= 0$

7.

x_1	x_2	x_3	-1	
2	4	-1	3	$= 0$
1	2	3	-1	$= 0$

8.

x_1	x_2	x_3	
3	2	1	$= 0$
1	2	3	$= 0$
5	2	-1	$= 0$

9.

x_1	x_2	x_3	-1	
2	1	-1	0	$= 0$
0	-3	-3	-4	$= 0$
1	2	1	2	$= 0$
4	5	1	4	$= 0$

10.

x_1	x_2	x_3	x_4	-1	
2	-1	3	4	5	$= 0$
1	3	-4	1	-1	$= 0$
3	2	-1	5	4	$= 0$

11. Given a mapping $M:R^2 \rightarrow R^3$ with tableau

x_1	x_2	1	
1	2	3	$= y_1$
2	-1	-4	$= y_2$
-1	3	5	$= y_3$

Set up the system of linear equations in standard form that can be used to find the preimage(s), if any, of

(a) $(y_1, y_2, y_3) = (0, 0, 0)$.
(b) $(y_1, y_2, y_3) = (1, 5, 7)$.
(c) $(y_1, y_2, y_3) = (a, b, c)$.

12. A manufacturer produces two brands of fertilizer, *A* and *B*. Each pound of brand *A* contains 15 units of chemical *X* and 8 units of chemical *Y*. Each pound of brand B contains 25 units of chemical *X* and 40 units of chemical *Y*. Set up the equations, in standard form, ready for solution, which are used to find how many pounds are needed to provide 100 units of chemical X and 120 units of chemical Y.

13. A system of four linear equations contains five variables and the pivot rank of its tableau is 3. How many columns are there in its fully reduced tableau? How many basic variables does it have? How many nonbasic variables?

4.2 Inconsistent Systems

We know that a member of the codomain of a mapping that is not in the image set of the mapping has no preimage. How does this manifest itself when we try to find an image for such a member of the codomain? In other words, how can we tell that a system of linear equations has no solution? The first example answers this question.

EXAMPLE 1

Solve the system

x_1	x_2	-1		
3	2	5	=	0
1	-1	2	=	0
2	3	6	=	0

Using the pivot reduction algorithm gives

x_1	x_2	-1		
3	2	5	=	0
1*	-1	2	=	0
2	3	6	=	0

x_2	-1		
5*	-1	=	0
-1	2	=	$-x_1$
5	2	=	0

-1		
$-\frac{1}{5}$	=	$-x_2$
$\frac{9}{5}$	=	$-x_1$
3	=	0

The last tableau is fully reduced since all x's appear at the right. Its equations are

$$-\tfrac{1}{5} = x_2, \qquad \tfrac{9}{5} = x_1, \qquad -3 = 0$$

But there are no values of x_1 and x_2 for which $-3 = 0$ is true! Since we are looking for values of x_1 and x_2 for which all equations are true, we conclude that the last tableau has no solutions. Since the last tableau has the same solution set as the initial tableau, it follows that the initial tableau also has no solutions.

EXAMPLE 2

Solve

x_1	x_2	x_3	-1		
1	1	2	2	=	0
1	-1	1	4	=	0
3	1	5	6	=	0

Pivot reduction yields

x_1	x_2	x_3	-1		
1*	1	2	2	=	0
1	-1	1	4	=	0
3	1	5	6	=	0

x_2	x_3	-1		
1	2	2	=	$-x_1$
-2	-1*	2	=	0
-2	-1	0	=	0

x_2	-1		
-3	6	$=$	$-x_1$
2	-2	$=$	$-x_3$
0	-2	$=$	0

No additional pivoting is possible. Why? The last tableau is therefore fully reduced. Since no values of (x_1, x_2, x_3) exist for which $0x_2 + 2 = 0$ (the last equation in the last tableau), there are no solutions to the initial system.

EXAMPLE 3

Solve

$$\begin{cases} 3x = 8 - 2y \\ 6x - 2 = 7 - 4y \end{cases}$$

Before using pivot reduction these equations have to be put in standard form, which is

x	y	-1		
3	2	8	$=$	0
6	4	9	$=$	0

Using pivot reduction as indicated gives

x	y	-1		
3*	2	8	$=$	0
6	4	9	$=$	0

y	-1		
$2/3$	$8/3$	$=$	$-x$
0	-7	$=$	0

No further reduction is possible. The last equation, $0y + 7 = 0$, has no solution. Thus the initial system has no solution.

Systems such as those in Examples 1, 2, and 3 that have no solutions are called *inconsistent.* This is to say that the system harbors a contradiction. The presence of this contradiction becomes visible in the fully reduced tableau.

There is sufficient evidence in the three examples for the theorem that governs the "no solution" systems.

THEOREM 4.1. If in a fully reduced form of a system of linear equations there appears an equation of the form $b = 0$, when $b \neq 0$, then the system is inconsistent, having no solutions.

COROLLARY. A system of linear equations is consistent if and only if its coefficient matrix and its augmented matrix have the same pivot rank.

EXAMPLE 4

Let M be the mapping with tableau

x_1	x_2	
$3/4$	$1/4$	$= y_1$
$1/2$	$1/2$	$= y_2$
$1/4$	$3/4$	$= y_3$

Find the preimages, if any, of $(y_1, y_2, y_3) = (1, -2, -3)$.

The standard form of

x_1	x_2	
$3/4$	$1/4$	$= 1$
$1/2$	$1/2$	$= -2$
$1/4$	$3/4$	$= -3$

is

x_1	x_2	-1	
$3/4$	$1/4$	1	$= 0$
$1/2$	$1/2$	-2	$= 0$
$1/4$	$3/4$	-3	$= 0$

Pivot as indicated:

x_1	x_2	-1	
$3/4$*	$1/4$	1	$= 0$
$1/2$	$1/2$	-2	$= 0$
$1/4$	$3/4$	-3	$= 0$

x_2	-1	
$1/3$	$4/3$	$= -x_1$
$1/3$*	$-8/3$	$= 0$
$2/3$	$-10/3$	$= 0$

-1	
4	$= -x_1$
-8	$= -x_2$
2	$= 0$

The last equation, $2 = 0$, announces that $(1, -2, -3)$ has no preimage.

Exercises

1. Show that each of the following systems is inconsistent.

(a)

x_1	x_2	-1	
3	2	8	$= 0$
6	4	9	$= 0$

(b)

x	y	-1	
2	3	5	$= 0$
3	-2	1	$= 0$

(c)

x_1	x_2	x_3	-1	
1	1	2	2	$= 0$
2	0	3	0	$= 0$
3	1	5	6	$= 0$

(d)

x_1	-1	
3	2	$= 0$
6	4	$= 0$
9	7	$= 0$

(e)

x	y	-1	
1	2	5	$= 0$
2	-1	5	$= 0$
3	4	11	$= 0$

(f)

x	y	z	-1	
1	2	-3	4	$= 0$
2	-3	5	5	$= 0$
3	-1	2	10	$= 0$

2. For what values of a does each of the following systems have no solution?

(a)

x_1	x_2	-1	
1	2	5	$= 0$
2	4	a	$= 0$

(b)

x_1	x_2	-1	
1	2	5	$= 0$
2	3	9	$= 0$
3	5	a	$= 0$

(c)

x_1	x_2	x_3	-1	
1	3	-1	4	$= 0$
0	-1	8	6	$= 0$
2	5	6	a	$= 0$

3. For each mapping listed, show that the zero n-tuple of the codomain has no preimage.

(a) $R^2 \to R^3$

x_1	x_2	1	
3	-1	1	$= y_1$
2	1	4	$= y_2$
1	-2	5	$= y_3$

(b) $R^3 \to R^4$

x_1	x_2	x_3	1	
2	1	-3	4	$= y_1$
1	2	5	0	$= y_2$
-1	3	-2	1	$= y_3$
2	6	0	6	$= y_4$

4. For each mapping listed, show that the given member of its codomain has no preimage.

(a) $R^3 \to R^3$

x_1	x_2	x_3	
1	0	1	$= y_1$
0	1	-1	$= y_2$
2	2	0	$= y_3$

$$(y_1, y_2, y_3) = (-1, -1, -5)$$

(b) $R^2 \to R^4$

x_1	x_2	
3	1	$= y_1$
1	-1	$= y_2$
2	0	$= y_3$
0	3	$= y_4$

$$(y_1, y_2, y_3, y_4) = (-1, -2, -6, -3)$$

4.3 Systems with Unique Solutions

These systems correspond to mappings in which each image has exactly one preimage. They, too, are solved by pivot reduction.

EXAMPLE 1

Solve

x	y	-1	
8	5	35	$= 0$
4	7	31	$= 0$

Pivot reduction yields

x	y	-1		
8^*	5	35	=	0
4	7	31	=	0

y	-1		
$5/8$	$35/8$	=	$-x$
$9/2^*$	$27/2$	=	0

-1		
$5/2$	=	$-x$
3	=	$-y$

Therefore, $(x, y) = (5/2, 3)$.

Check:

$5/2$	3	-1			
8	5	35	=	0	✓
4	7	31	=	0	✓

Both equations are true.

In this example the pivot rank of the initial tableau is 2, as is the number of variables in the system.

EXAMPLE 2

Solve

x_1	x_2	x_3	-1		
2	-3	1	4	=	0
1	1	-1	1	=	0
1	-2	2	7	=	0

Pivoting as indicated gives

x_1	x_2	x_3	-1		
2	-3	1	4	=	0
1*	1	-1	1	=	0
1	-2	2	7	=	0

x_2	x_3	-1		
-5	3	2	=	0
1	-1	1	=	$-x_1$
-3*	3	6	=	0

x_3	-1		
-2*	-8	=	0
0	3	=	$-x_1$
-1	-2	=	$-x_2$

-1		
4	=	$-x_3$
3	=	$-x_1$
2	=	$-x_2$

Thus $(x_1, x_2, x_3) = (3, 2, 4)$ is the solution.

Check:

3	2	4	-1		
2	-3	1	4	= 0	✓
1	1	-1	1	= 0	✓
1	-2	2	7	= 0	✓

All three equations are true. Observe that the pivot rank of the initial tableau and the number of variables are 3.

EXAMPLE 3

Solve

x_1	x_2	-1		
2	1	5	=	0
1	-2	0	=	0
5	-2	8	=	0

Proceeding as usual we get

x_1	x_2	-1		
2	1	5	=	0
1*	-2	0	=	0
5	-2	8	=	0

x_2	-1		
5*	5	=	0
-2	0	=	$-x_1$
8	8	=	0

-1		
1	=	$-x_2$
2	=	$-x_1$
0	=	0

The last equation in the fully reduced tableau ($0 = 0$) is true for all values of (x_1, x_2) and therefore is disregarded. We read the solution in the remaining rows: $(x_1, x_2) = (2, 1)$.

Check:

2	1	-1		
2	1	5	= 0	✓
1	-2	0	= 0	✓
5	-2	8	= 0	✓

Here, too, the pivot rank of the initial tableau and the number of variables (both 2) are equal. We have enough evidence to formulate the theorem that governs consistent systems with unique solutions.

THEOREM 4.2. A consistent system of linear equations has exactly one solution if the pivot rank of its initial tableau is equal to the number of variables in the system.

Proof: Since the pivot rank is equal to the number of variables, all variables can be exchanged to the right of the tableau. This proves the theorem.

THEOREM 4.3. A system of m linear equations in n variables does not have a unique solution if $m < n$.

Proof: Since $m < n$, not all the variables can be moved to the right in the fully reduced form. This leaves some to remain at the top (serving as nonbasic variables) in addition to the -1 column. Therefore, the solution is either nonexistent or multiple—in either case it is not unique.

EXAMPLE 4

Solve

x_1	x_2	x_3	x_4	-1		
1*	0	0	1	8	=	0
0	1	1	0	9	=	0
0	0	1	-1	10	=	0
-1	-1	1	0	11	=	0

x_2	x_3	x_4	-1		
0	0	1	8	=	$-x_1$
1*	1	0	9	=	0
0	1	-1	10	=	0
-1	1	1	19	=	0

x_3	x_4	-1	
0	1	8	$= -x_1$
1	0	9	$= -x_2$
1*	-1	10	$= 0$
2	1	28	$= 0$

x_4	-1	
1	8	$= -x_1$
1	-1	$= -x_2$
-1	10	$= -x_3$
3*	8	$= 0$

-1	
$16/3$	$= -x_1$
$-11/3$	$= -x_2$
$38/3$	$= -x_3$
$8/3$	$= -x_4$

Thus $(x_1, x_2, x_3, x_4) = (16/3, -11/3, 38/3, 8/3)$. The check is left for the reader.

EXAMPLE 5

Given $M: R^3 \to R^3$ with tableau

x_1	x_2	x_3	1	
1	0	1	2	$= y_1$
0	1	-1	3	$= y_2$
2	2	1	5	$= y_3$

find the preimage, if any, of

$$(y_1, y_2, y_3) = (3, 2, 7)$$

The system of equations that needs to be solved is

$$\begin{aligned} x_1 \qquad\quad + x_3 + 2 &= 3 \\ x_2 - x_3 + 3 &= 2 \\ 2x_1 + 2x_2 + x_3 + 5 &= 7 \end{aligned}$$

or, in standard form,

x_1	x_2	x_3	-1		
1	0	1	1	=	0
0	1	−1	−1	=	0
2	2	1	2	=	0

By the algorithm,

x_1	x_2	x_3	-1		
1*	0	1	1	=	0
0	1	−1	−1	=	0
2	2	1	2	=	0

x_2	x_3	-1		
0	1	1	=	$-x_1$
1*	−1	−1	=	0
2	−1	0	=	0

x_3	-1		
1	1	=	$-x_1$
−1	−1	=	$-x_2$
1*	2	=	0

-1	
-1	$= -x_1$
1	$= -x_2$
2	$= -x_3$

Thus $(x_1, x_2, x_3) = (-1, 1, 2)$. Check in the tableau of the original mapping:

-1	1	2	1		
1	0	1	2	$= 3$	✓
0	1	-1	3	$= 2$	✓
2	2	1	5	$= 7$	✓

This solution is presentable in the following manner:

x_1	x_2	x_3	-1	
1*	0	1	1	$= 0$
0	1	-1	-1	$= 0$
2	2	1	2	$= 0$
	0	1	1	$= -x_1$
	1*	-1	-1	$= 0$
	2	-1	0	$= 0$
		1	1	$= -x_1$
		-1	-1	$= -x_2$
		1*	2	$= 0$
			-1	$= -x_1$
			1	$= -x_2$
			2	$= -x_3$

The reversed-staircase arrangement of tableaus occurs when pivots are chosen in consecutive columns.

Exercises

Solve the following.

1.

x	y	-1	
2	3	1	$= 0$
3	-2	8	$= 0$

2.

x	y	-1	
1	3	10	$= 0$
2	5	16	$= 0$

3.

x_1	x_2	-1	
5	-3	12	$= 0$
2	-1	5	$= 0$

4.

x_1	x_2	x_3	-1	
1	1	-1	-2	$= 0$
1	-2	-2	1	$= 0$
2	3	1	1	$= 0$

5.

x_1	x_2	x_3	-1	
1	0	4	1	$= 0$
2	1	1	3	$= 0$
-1	1	1	1	$= 0$

6.

x_1	x_2	x_3	-1	
1	1	1	5	$= -1$
0	2	1	20	$= 0$
-5	-1	-2	0	$= 3$

7.

x_1	x_2	x_3	-1	
1	-3	2	-1	$= 0$
0	2	-3	3	$= 0$
3	5	0	-2	$= 0$
3	3	3	-5	$= 0$

8.

x_1	x_2	x_3	-1	
3	0	-4	0	$= 0$
6	4	0	-1	$= 0$
0	8	2	5	$= 0$
3	4	4	-1	$= 0$

9.

x_1	x_2	x_3	x_4	-1	
1	1	1	1	0	$= 0$
2	1	-1	1	-3	$= 0$
1	1	0	-1	2	$= 0$
3	2	1	0	3	$= 0$

10.

x_1	x_2	x_3	-1	
1	4	2	19	$= 0$
2	1	2	19	$= 0$
2	3	1	18	$= 0$
1	5	2	19	$= 0$

11.

x_1	x_2	x_3	x_4	x_5	-1	
1	1	1	1	1	3	$= 0$
1	0	1	0	1	6	$= 0$
0	1	0	1	0	-3	$= 0$
0	1	1	0	1	3	$= 0$
1	-1	1	-1	1	9	$= 0$

12.

x_1	x_2	-1	
3	2	5	$= 0$
2	3	0	$= 0$
1	1	1	$= 0$
2	1	4	$= 0$

13. The ends A and B of an insulated wire are kept at constant temperatures of 50° and 80°, respectively. Let X, Y, and Z be interior points of the wire as shown.

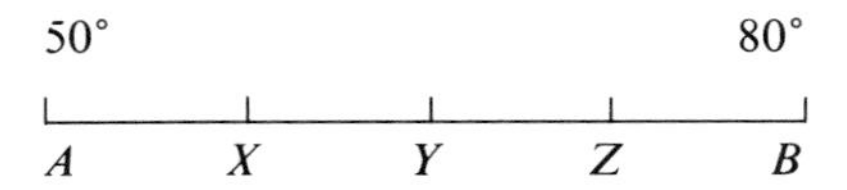

Assume that the temperatures at X, Y, and Z are averages of the two named points on each side. Find the temperatures at X, Y, and Z.

4.4 Systems with Many Solutions

The systems of this section correspond to mappings in which an image has many preimages.

EXAMPLE 1

Solve

x	y	z	-1		
1	-1	2	7	=	0
2	3	-1	-11	=	0

Use pivot reduction as indicated.

(1)

x	y	z	-1		
1*	-1	2	7	=	0
2	3	-1	-11	=	0

(2)

y	z	-1		
-1	2	7	=	$-x$
5*	-5	-25	=	0

(3)

z	-1		
1	2	=	$-x$
-1	-5	=	$-y$

Tableau (3) is fully reduced. It has two basic variables, x and y, and one nonbasic variable, z. For each value assigned to z, x and y can be determined from (3) such that (x, y, z) satisfies the initial system. For instance, if $z = 0$, then $x = 2$ and $y = -5$. Substituting these values in (1) yields

2	−5	0	−1	
1	−1	2	7	= 0
2	3	−1	−11	= 0

which is true.

If $z = 10$, then $x = -8$ and $y = 5$,

−8	5	10	−1	
1	−1	2	7	= 0
2	3	−1	−11	= 0

is also true.

In general, from (3), for any value of z, $x = -z + 2$ and $y = z - 5$. Substituting these x and y values in (1) gives

$-z + 2$	$z - 5$	z	−1	
1	−1	2	7	= 0
2	3	−1	−11	= 0

This tableau is true for any number z as we now verify. For row 1,

$$1(-z + 2) - 1(z - 5) + 2(z) - 1(7) = 0$$

is true for all z, and for row 2,

$$2(-z + 2) + 3(z - 5) - 1(z) - 11(-1) = 0$$

is true for all z.

Thus for each value of z, $(x, y, z) = (-z + 2, z - 5, z)$ is a solution. Since z can be any real number, the original system has many solutions. The solution set is designated

$$\{(x, y, z) : x = -z + 2, y = z - 5; z \text{ any real number}\}$$

The tableau

x	y	z	-1	
1	-1	2	7	$= 0$
2	3	-1	-11	$= 0$

is in standard form. A fully reduced or explicit form is

z	-1	
1	2	$= -x$
-1	-5	$= -y$

Whereas the solution set of a system is implicit in the standard form, it becomes eminently visible in the explicit form. In fact, we may call an explicit form the *solution* of the system. Notice that we said "an" explicit form, not "the," for there can be more than one explicit form for a given system. This will be illustrated in a subsequent example.

It is clear that each pivot exchange introduces an additional basic variable. Hence, as was said earlier, the number of basic variables in an explicit form is the pivot rank of the original system in standard form. This is the theorem of this section, stated below.

THEOREM 4.4. Let a consistent system of linear equations in standard form have n variables and pivot rank r. Then the number of basic variables in an explicit form is r and the number of nonbasic variables is $n - r$.

EXAMPLE 2

Solve

x_1	x_2	x_3	-1	
1	1	3	8	$= 0$
0	1	3	6	$= 0$
1	2	6	14	$= 0$

Pivot reduction yields

x_1	x_2	x_3	-1		
1*	1	3	8	=	0
0	1	3	6	=	0
1	2	6	14	=	0

x_2	x_3	-1		
1	3	8	=	$-x_1$
1*	3	6	=	0
1	3	6	=	0

x_3	-1		
0	2	=	$-x_1$
3	6	=	$-x_2$
0	0	=	0

The last tableau is a fully reduced form of the initial one. The solution set is

$$\{(x_1, x_2, x_3): x_1 = 2, x_2 = -3x_3 + 6; x_3 \text{ any real number}\}$$

Check:

2	$-3x_3 + 6$	x_3	-1		
1	1	3	8	=	0
0	1	3	6	=	0
1	2	6	14	=	0

The reader is invited to verify each row. In this solution the nonbasic variable is x_3. An alternative solution is shown below, in which the nonbasic variable is x_2.

x_1	x_2	x_3	-1		
1*	1	3	8	=	0
0	1	3	6	=	0
1	2	6	14	=	0

x_2	x_3	-1		
1	3	8	$=$	$-x_1$
1	3*	6	$=$	0
1	3	6	$=$	0

x_2	-1		
0	2	$=$	$-x_1$
$\frac{1}{3}$	2	$=$	$-x_3$
0	0	$=$	0

The solution set, according to this solution, is

$$\{(x_1, x_2, x_3): x_1 = 2, x_3 = -\tfrac{1}{3}x_2 + 2; x_2 \text{ any real number}\}$$

A comparison of the two designations of the solution set shows that they agree in $x_1 = 2$ but differ in $x_2 = -3x_3 + 6$ and $x_3 = -\frac{1}{3}x_2 + 2$. An examination will show that they are indeed equivalent. Hence the two designations actually define the same set of solutions.

It is an interesting exercise for the reader to show that in this example, although x_2 and x_3 can act as the nonbasic variables in the different designations of the same solution set, it is not possible to have x_1 serve in the same manner. In any case it is quite clear that there are a finite number of explicit forms for a given system in standard form.

EXAMPLE 3

Solve

x_1	x_2	x_3	-1	
1	2	-3	7	$= 0$
2	4	-6	14	$= 0$

Pivoting as usual yields

x_1	x_2	x_3	-1	
1*	2	-3	7	$= 0$
2	4	-6	14	$= 0$

x_2	x_3	-1	
2	-3	7	$= -x_1$
0	0	0	$= 0$

The pivot rank is 1. The fully reduced tableau has two nonbasic variables, x_2 and x_3. The solution set is

$$\{(x_1, x_2, x_3): x_1 = -2x_1 + 3x_3 + 7;\ x_2, x_3 \text{ any real numbers}\}$$

The check is left for the interested reader. This system has three explicit forms.

EXAMPLE 4

If

x_1	x_2	x_3	x_4	-1	
1*	2	-3	1	10	$= 0$
1	-2	3	2	4	$= 0$
1	1	1	1	0	$= 0$

then

x_2	x_3	x_4	-1	
2	-3	1	10	$= -x_1$
-4	6	1*	-6	$= 0$
-1	4	0	-10	$= 0$

x_2	x_3	-1	
6	-9	16	$= -x_1$
-4	6	-6	$= -x_4$
-1*	4	-10	$= 0$

x_3	-1	
15	-44	$= -x_1$
-10	34	$= -x_4$
-4	10	$= -x_2$

The solution set is

$$\{(x_1, x_2, x_3, x_4): x_1 = -15x_3 - 44, x_2 = 4x_3 + 10, x_4 = 10x_3 + 34; \quad x_3 \text{ any real number}\}$$

Check:

$-15x_3 - 44$	$4x_3 + 10$	x_3	$10x_3 + 34$	-1	
1	2	-3	1	10	$= 0$
1	-2	3	2	4	$= 0$
1	1	1	1	0	$= 0$

The reader is invited to verify each row.

Exercises

Solve the systems in Exercises 1–12.

1.

x	y	-1	
2	5	3	$= 0$
4	10	6	$= 0$

2.

x	y	-1	
3	-2	3	$= 0$
9	-6	9	$= 0$

3.

x	y	z	-1	
1	2	1	1	$= 0$
2	1	0	3	$= 0$
3	0	-1	5	$= 0$

4.

x_1	x_2	x_3	-1	
2	1	3	-3	$= 0$
3	1	0	24	$= 0$

5.

x_1	x_2	x_3	-1	
3	2	-1	5	$= 0$

6.

x_1	x_2	x_3	-1	
2	-3	4	4	$= 0$
4	0	8	15	$= 0$
2	-9	4	16	$= 0$

7.

x_1	x_2	-1	
1	1	5	$= 0$
2	-3	15	$= 0$
5	2	28	$= 0$

8.

x_1	x_2	x_3	-1	
1	2	3	5	$= 0$
1	3	5	3	$= 0$

9.

x_1	x_2	x_3	-1	
1	2	3	5	$= 0$
1	3	5	3	$= 0$
1	5	9	-1	$= 0$

10.

x_1	x_2	x_3	x_4	
3	2	5	1	$= 0$
6	4	10	2	$= 0$

11.

x_1	x_2	x_3	x_4	-1	
2	1	3	5	9	$= 0$
3	0	1	7	14	$= 0$
1	2	5	3	4	$= 0$
5	4	11	13	22	$= 0$

12.

x_1	x_2	x_3	x_4	x_5	-1	
1	0	1	0	1	9	$= 0$
0	1	0	1	0	6	$= 0$
1	0	0	1	0	5	$= 0$

13. Given the mapping $R^2 \to R^3$ with tableau

x_1	x_2	-1	
1	2	3	$= y_1$
2	5	7	$= y_2$
3	6	2	$= y_3$

Find the set of preimages of
(a) $(y_1, y_2, y_3) = (0, 0, -7)$.
(b) $(y_1, y_2, y_3) = (0, 0, 2)$.
(c) $(y_1, y_2, y_3) = (3, 1, 2)$.

14. Prove that

x	y	-1	
a	b	e	$= 0$
c	d	f	$= 0$

has a unique solution if $ad \neq bc$. If $ad = bc$, what then?

4.5 Systems of Homogeneous Linear Equations

In a homogeneous equation the constant term is zero. We give systems of homogeneous linear equations special attention because they occupy an important place in mathematical theory and occur frequently. The thing to notice about these systems is that the "-1" column contains only zeros. Pivotal operations leave any column of only zeros unaltered. Hence the "-1" column can be omitted when solving such systems. However, if at any time it is found advantageous to restore this column of zeros, there is nothing to stop us from doing so.

EXAMPLE 1

Solve

x	y	z	
1	2	1	$= 0$
2	-1	-3	$= 0$
3	4	1	$= 0$

Pivoting gives

x	y	z		
1*	2	1	=	0
2	−1	−3	=	0
3	4	1	=	0

y	z		
2	1	=	$-x$
−5*	−5	=	0
−2	−2	=	0

z		
−1	=	$-x$
1	=	$-y$
0	=	0

The solution set is

$$\{(x, y, z) : x = z, y = -z; z \text{ any real number}\}$$

Perhaps the reader anticipated the solution $(x, y, z) = (0, 0, 0)$. Indeed, it is the solution for $z = 0$. But it is not the only one, for z is a nonbasic variable in the fully reduced tableau. Thus the system has many solutions.

Notice that every system of homogeneous linear equations has the zero n-tuple as a solution. Thus a system of homogeneous linear equations cannot be an inconsistent system.

EXAMPLE 2

Solve

x	y	z		
1	2	1	=	0
2	−1	−3	=	0
3	4	2	=	0

This system has the same entries as those of Example 1, except for the entry in row 3, column 3.

(1)

x	y	z	
1*	2	1	$= 0$
2	-1	-3	$= 0$
3	4	2	$= 0$

(2)

y	z	
2	1	$= -x$
-5*	-5	$= 0$
-2	-1	$= 0$

(3)

z	
-1	$= -x$
1	$= -y$
1*	$= 0$

(4)

0	
1	$= -x$
-1	$= -y$
1	$= -z$

In this case we do not remove the 0 column and read $x = 0$, $y = 0$, $z = 0$. $(x, y, z) = (0, 0, 0)$ is the only solution. Another approach is to reinstate the (-1) column, converting (3) into (5). Pivoting as indicated yields (6), from which the same result is read.

(5)

z	-1	
-1	0	$= -x$
1	0	$= -y$
1*	0	$= 0$

(6)

-1		
0	$=$	$-x$
0	$=$	$-y$
0	$=$	$-z$

The system in Example 1 has three variables and pivot rank 2. The system in Example 2 has three variables and pivot rank 3. The contrast in solution sets is reflected in the theorem of this section.

THEOREM 4.5. If a system of homogeneous linear equations in n variables has pivot rank n, then its solution set contains only the zero n-tuple.

The zero n-tuple is called a *trivial solution* of a system of homogeneous linear equations.

THEOREM 4.6. If the pivot rank of a system of homogeneous linear equations in n variables is less than n, the system has a nontrivial solution.

Proof: This follows from the fact that a fully reduced tableau of the system contains nonbasic variables which can be assigned any values.

EXAMPLE 3

Solve

x	y		
3*	2	$=$	0
6	4	$=$	0
-3	-2	$=$	0
30	20	$=$	0

A reduced tableau of this system is

y		
$2/3$	$=$	$-x$
0	$=$	0
0	$=$	0
0	$=$	0

Its solution set is

$$\{(x, y) : x = -\tfrac{2}{3}y;\ y \text{ any real number}\}$$

The purpose of this example is to show that the number of equations in a system is not a factor that determines the nature of its solution set.

As Theorem 4.6 indicates, the determining factors are the number of variables in the system and its pivot rank. These determine whether or not a fully reduced tableau of the system has nonbasic variabes.

EXAMPLE 4

Solve

x_1	x_2	x_3	x_4	x_5		
1*	4	0	0	5	=	0
0	0	1	0	−3	=	0
0	0	0	1	7	=	0
	−4	0	0	5	=	$-x_1$
	0	1*	0	−3	=	0
	0	0	1	7	=	0
	−4		0	5	=	$-x_1$
	0		0	−3	=	$-x_3$
	0		1*	7	=	0
	−4			5	=	$-x_1$
	0			−3	=	$-x_3$
	0			7	=	$-x_4$

(The reader may find this arrangement of pivot reduction convenient.)

The solution set is

$$\{(x_1, x_2, x_3, x_4, x_5) : x_1 = 4x_2 - 5x_5,\ x_3 = 3x_5,\ x_4 = -7x_5;\quad x_2, x_5 \text{ any real numbers}\}$$

This set may also be designated

$$\{(x_1, x_2, x_3, x_4, x_5) = (4x_2 - 5x_5, x_2, 3x_5, -7x_5, x_5);$$
$$x_2, x_5 \text{ any real numbers}\}$$

For every system of nonhomogeneous linear equations there is a system of homogeneous linear equations with the same coefficient matrix. The question we consider now is: What is the relation between their solution sets?

EXAMPLE 5

Let A be the system

x_1	x_2	x_3	-1	
1	1	1	1	$= 0$
2	1	-1	6	$= 0$

Let B be the homogeneous system obtained from A by eliminating the constants.

x_1	x_2	x_3	
1	1	1	$= 0$
2	1	-1	$= 0$

Find the relationship between the solution sets of A and B.

We get the solution set of B in the usual way:

x_1	x_2	x_3	
1*	1	1	$= 0$
2	1	-1	$= 0$
	1	1	$= -x_1$
	-1*	-3	$= 0$
		-2	$= -x_1$
		3	$= -x_2$

We have $x_1 = 2x_3$, $x_2 = -3x_3$. Thus any solution (x_1, x_2, x_3) is given by

$$(1) \qquad (x_1, x_2, x_3) = (2x_3, -3x_3, x_3)$$

Or in set notation, the set of all solutions of B, a homogeneous system is given by

$$(2) \qquad \{(x_1, x_2, x_3) : x_1 = 2x_3, x_2 = -3x_3; x_3 \text{ any real number}\}$$

A solution of A is $(x_1, x_2, x_3) = (7, -7, 1)$. That is,

7	−7	1	−1	
1	1	1	1	= 0
2	1	−1	6	= 0

is true, as you can verify.

We assert that the general solution to A is obtained by adding the specific solution of A, $(7, -7, 1)$, to the general solution of B, $(2x_3, -3x_3, x_3)$. That is, $(2x_3 + 7, -3x_3 - 7, x_3 + 1)$ is the general solution of A. This is verified from

$2x_3 + 7$	$-3x_3 - 7$	$x_3 + 1$	-1	
1	1	1	1	= 0
2	1	−1	6	= 0

From row 1

$$(1)(2x_3 + 7) + (1)(-3x_3 - 7) + (1)(x_3 + 1) + (1)(-1) = 0$$
$$2x_3 + 7 - 3x_3 - 7 + x_3 + 1 - 1 = 0$$

which is true. Similarly, row 2 is true. Thus the solution set for A is

$$(3) \qquad \{(2x_3 + 7, -3x_3 - 7, x_3 + 1); x_3 \text{ any real number}\}$$

A natural question is: Why did we select $(7, -7, 1)$ as the solution of A? The answer is any solution to A can be used to get its general solution —provided the general solution to B is known. Thus you can verify that $(5, -4, 0)$ is a solution of A and adding to $(2x_3, -3x_3, x_3)$ gives $(2x_3 + 5, -3x_3 - 4, x_3)$ as a form for the general solution of A. Thus the solution set for A is

$$(4) \qquad \{(2x_3 + 5, -3x_3 - 4, x_3); x_3 \text{ is any real number}\}$$

The reader should verify that (3) and (4) are the same set. In general, if (a_1, a_2, a_3) is any solution to A, then the general solution to A is

$$\{(2x_3 + a_1, -3x_3 + a_2, x_3 + a_3); x_3 \text{ any real number}\}$$

This is verified by substitution in the tableau for A, keeping in mind that since (a_1, a_2, a_3) is a solution, then

a_1	a_2	a_3	-1	
1	1	1	1	$= 0$
2	1	-1	6	$= 0$

is true. Thus

$$a_1 + a_2 + a_3 = 1 \quad \text{and} \quad 2a_1 + a_2 - a_3 = 6$$

Example 5 illustrates how to prove

THEOREM 4.7. Let A be a system of nonhomogeneous linear equations in n variables and let B be the corresponding system of homogeneous linear equations having the same coefficient matrix. Let $(x_1, x_2, \ldots, x_n) = (a_1, a_2, \ldots, a_n)$ be a particular solution to A. Then the general solution to A is the set of n-tuples in which the ith component is a_i plus the ith component of the general solution of B.

Theorem 4.7 is useful in reducing the number of calculations when solving a system of linear equations, if a particular solution is known. This is illustrated in the next example.

EXAMPLE 6

Given that

x_1	x_2	x_3	x_4	-1	
1	2	-1	1	3	$= 0$
2	3	4	1	4	$= 0$

has a solution $(x_1, x_2, x_3, x_4) = (2, -1, 0, 3)$ (verify this), find its general solution.

Using Theorem 4.7, solve the homogeneous system

x_1	x_2	x_3	x_4	
1*	2	-1	1	$= 0$
2	3	4	1	$= 0$

x_2	x_3	x_4	
2	-1	1	$= -x_1$
-1*	6	-1	$= 0$

x_3	x_4	
11	-1	$= -x_1$
-6	1	$= -x_2$

The solution set for the homogeneous system is

$$\{(x_1, x_2, x_3, x_4) = (-11x_3 + x_4, 6x_3 - x_4, x_3, x_4);\ x_3, x_4 \text{ any real numbers}\}$$

Therefore, the desired solution set is

$$\{(x_1, x_2, x_3, x_4) = (-11x_3 + x_4 + 2, 6x_3 - x_4 - 1, x_3, x_4 + 3);\ x_3, x_4 \text{ any real numbers}\}$$

Exercises

Solve the systems in Exercises 1–12.

1.

x	y	
2	3	$= 0$
3	2	$= 0$

2.

x	y	
2	3	$= 0$
3	-2	$= 0$

3.

x	y	
2	3	$= 0$
4	6	$= 0$
6	9	$= 0$

4.

x_1	x_2	x_3	
1	-3	2	$= 0$
1	-2	-1	$= 0$
2	-1	3	$= 0$

5.

a	b	c	
2	3	-5	$= 0$
1	-2	1	$= 0$
4	13	-17	$= 0$

6.

x_1	x_2	x_3	
1	0	1	$= 0$
2	3	-1	$= 0$

7.

x_1	x_2	x_3	
1	1	1	$= 0$
2	1	-1	$= 0$
1	-1	3	$= 0$

8.

x_1	x_2	x_3	
2	7	4	$= 0$
1	1	1	$= 0$
3	-2	1	$= 0$

9.

x_1	x_2	x_3	x_4	
1	1	1	1	$= 0$
1	4	3	2	$= 0$
1	-1	0	-1	$= 0$
4	1	2	3	$= 0$

10.

x_1	x_2	x_3	x_4	
1	2	7	4	$= 0$
-1	2	-3	-8	$= 0$
1	-1	4	7	$= 0$

11.

x_1	x_2	x_3	x_4	x_5	
1	4	0	0	3	$= 0$
0	0	1	0	-2	$= 0$
0	0	0	1	5	$= 0$

12.

x_1	x_2	x_3	x_4	x_5	
1	-1	0	0	0	$= 0$
0	1	-1	0	0	$= 0$
0	0	0	1	-1	$= 0$
0	0	0	0	1	$= 0$

13. Prove that a system of m homogeneous linear equations in n variables has a nontrivial solution if $m < n$.

4.6 Gaussian Elimination

For computational purposes, in astronomy and geodesy, the great K. F. Gauss (1777–1855) invented a systematic procedure, now known as *Gaussian elimination*. His purpose was to have an algorithm to solve a large system of linear equations having a unique solution. Gaussian elimination is regarded by modern numerical computation experts as the most efficient direct method for solving such systems. From our viewpoint Gaussian elimination is the forerunner of pivot exchange and as such is the origin of elementary linear algebra.

EXAMPLE 1

(See Section 4.3, Example 4.) Find the unique solution to

(1)

x_1	x_2	x_3	x_4	-1	
1	0	0	1	8	= 0
0	1	1	0	9	= 0
0	0	1	−1	10	= 0
−1	−1	1	0	11	= 0

Before proceeding we point out the difference between the pivot reduction algorithm of Section 4.1 and Gaussian elimination. [For geodetic reasons, Wilhelm Jordan (1819–1904) devised the pivot reduction algorithm, also known as Gauss-Jordan elimination, as a variant of Gaussian elimination.] In pivot reduction each equation is transformed each time we pivot exchange. In Gaussian elimination we *store* the equation of the pivot after each pivot exchange and transform the remaining equations in the next pivot exchange. In tableau language, by pivot reduction, we eliminate one column at each pivot exchange, while by Gaussian elimination we eliminate one column at each pivot exchange and store one row after each pivot exchange. Gaussian elimination is illustrated in Figure 4.1.

Let us analyze Figure 4.1. Tableau (1) represents the original system. Pivoting in (1) as indicated yields (2). The first row of (2) is the only equation that contains x_1. This equation is put in storage and is not involved in subsequent pivot exchanges. Thus (3) has one less row than does (2). In (3) the equation that contains x_2 is put in storage and pivoting is applied to the

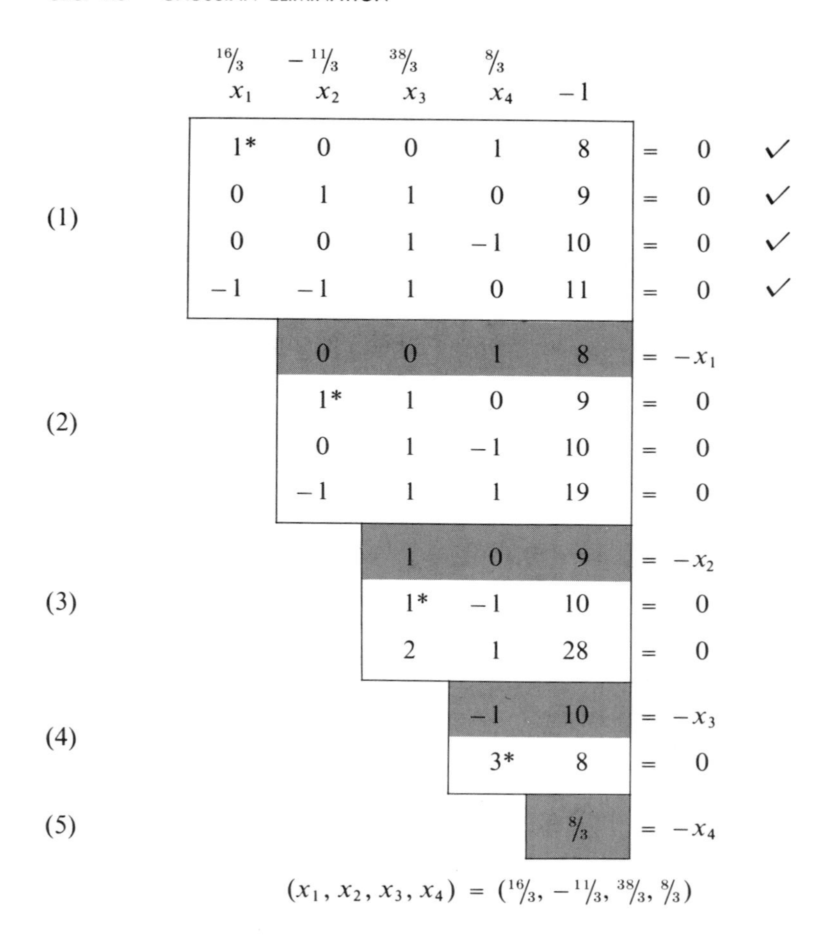

Figure 4.1

remaining two rows of (3), yielding (4). In (4) the x_3 equation is put in storage and pivoting yields (5). To obtain the values of the variables we use a procedure called *back substitution* which is carried out as follows. From (5) we have $x_4 = 8/3$. Back-substitute this value for x_4 to the top of Figure 4.1. From row 1 of tableau (4), and from $x_4 = 8/3$, $x_3 = 38/3$. Back-substituting this value for x_3 to the top we read, from row 1 of tableau (3), that $x_2 = -11/3$. Back-substituting once again we read, from row 1 of (2), that $x_1 = 16/3$. Back-substituting this value for x_1 we use tableau (1) to check the result.

In this example we choose as pivot the largest number (in absolute value) in the left-hand column. This procedure, favored by numerical analysts, minimizes the roundoff error when calculating with a computer. This method is called *partial pivoting*. Another method used by numerical analysts is *complete pivoting*. This method selects as pivot the largest (in absolute value) number in the tableau, that is not in the -1 column.

EXAMPLE 2

(See Section 4.3, Example 2.) Solve by Gaussian elimination.

3	2	4				
x_1	x_2	x_3	-1			
2*	−3	1	4	=	0	✓
1	−2	2	7	=	0	✓
1	1	−1	1	=	0	✓
	−3/2	1/2	2	=	$-x_1$	
	−1/2	3/2	5	=	0	
	5/2*	−3/2	−1	=	0	
		6/5*	24/5	=	0	
		−3/5	−2/5	=	$-x_2$	
			4	=	$-x_3$	

$$(x_1, x_2, x_3) = (3, 2, 4)$$

Exercises

Solve using Gaussian elimination. (See exercises in Section 4.3.) If a computer is available use partial pivoting.

1.

x	y	-1		
2	3	1	=	0
3	−2	8	=	0

2.

x	y	-1		
1	3	10	=	0
2	5	16	=	0

3.

x_1	x_2	x_3	-1		
1	1	−1	−2	=	0
1	−2	−2	1	=	0
2	3	1	1	=	0

4.

x_1	x_2	x_3	-1	
1	0	4	1	= 0
2	1	1	3	= 0
−1	1	1	1	= 0

5.

x_1	x_2	x_3	x_4	-1	
1	1	1	1	0	= 0
2	1	−1	1	−3	= 0
1	1	0	−1	2	= 0
3	2	1	0	3	= 0

6.

x_1	x_2	x_3	x_4	x_5	-1	
1	1	1	1	1	3	= 0
1	0	1	0	1	6	= 0
0	1	0	1	0	−3	= 0
0	1	1	0	1	3	= 0
1	−1	1	−1	1	9	= 0

4.7 Chapter Review

1. Systems of m linear equations in n variables in standard form are written as a tableau as follows:

x_1	x_2	$\cdots$	x_n	-1	
a_{11}	a_{12}	$\cdots$	a_{1n}	b_1	= 0
a_{21}	a_{22}	$\cdots$	a_{2n}	b_2	= 0
$\vdots$	$\vdots$		$\vdots$	$\vdots$	
a_{m1}	a_{m2}	$\cdots$	a_{mn}	b_m	= 0

The system is solved by pivot exchange, exchanging a variable and a zero. Each pivot exchange has the effect of deleting the column of the pivot and introducing the variable of the column as a basic variable at the right. This algorithm is called the *pivot reduction algorithm.*

2. A system of linear equations is solved when its tableau is fully reduced. If a row appears such as $b = 0$, where $b \neq 0$, the system has no solution and is called an *inconsistent system.*
3. If a consistent system of linear equations in n variables has pivot rank n, the system has a unique solution. If the pivot rank r is less than n, then a fully reduced tableau contains r basic variables and $n - r$ nonbasic variables. Such systems have many solutions. A solution is obtained by assigning values to each nonbasic variable.
4. A system of homogeneous linear equations is also solvable by pivot reduction. The basic theorem is

   Theorem 4.6. If the pivot rank of a system of homogeneous linear equations in n variables is less than n, the system has a nontrivial solution.

   The solution to a system of nonhomogeneous linear equations is derived from the solution to a system of homogeneous linear equations having the same coefficient matrix. This is stated in

   Theorem 4.7. Let A be a system of nonhomogeneous linear equations in n variables and let B be the corresponding system of homogeneous linear equations having the same coefficient matrix. Let $(x_1, x_2, \ldots, x_n) = (a_1, a_2, \ldots, a_n)$ be a particular solution to A. Then the general solution to A is the set of n-tuples in which the ith component is a_i plus the ith component of the general solution of B.

5. Gaussian elimination is used to solve a large system of linear equations having a unique solution. It is the forerunner of pivot exchange, and as such is the origin of elementary linear algebra.

Chapter Review Exercises

In these exercises use the pivot reduction algorithm.

1. Solve

x	y	z	-1	
1	3	2	1	$= 0$
1	2	2	3	$= 0$
3	7	6	5	$= 0$

2. Solve

x_1	x_2	x_3	-1	
-1	-1	2	10	$= 0$
-1	1	0	11	$= 0$
1	2	-1	9	$= 0$

3. Solve

$$3x + 2y = 8$$

4. Solve

x	y	z	w	-1	
9	-1	0	0	1	$= 0$
0	8	-2	0	-4	$= 0$
0	0	7	-3	-17	$= 0$

5. Given mapping $M : R^3 \to R^3$ with tableau

x_1	x_2	x_3	1	
1	-4	3	1	$= y_1$
-2	8	-6	2	$= y_2$
2	-8	6	3	$= y_3$

find the set of preimages of
(a) $(y_1, y_2, y_3) = (0, 0, 0)$.
(b) $(y_1, y_2, y_3) = (2, 0, 4)$.
(c) $(y_1, y_2, y_3) = (2, 0, 5)$.

6. State a condition that is both necessary and sufficient for a system of homogeneous linear equations to have a nontrivial solution.

7. Solve

$$\begin{cases} x_1 + x_2 + x_3 + x_4 + x_5 + x_6 = 6 \\ x_1 - x_2 + x_3 - x_4 + x_5 - x_6 = 0 \end{cases}$$

CLA

CHAPTER 5

Vector Spaces and Subspaces

The word *vector* has its origins in physics, where it is used to represent quantities that have both magnitude and direction. Among such quantities are force, velocity, and acceleration. Geometrically such quantities are pictured as directed line segments, where the length of the line is a measure of the magnitude of the vector and the direction is indicated by an arrow.

In some sense such vectors can be thought of as equal if they have the same magnitude and direction. Thus all the vectors drawn in the plane in Figure 5.1 are thought of as equal because they have the same magnitude and direction.

Since we regard all vectors that have the same magnitude and direction as being equal, we may regard any one of them as being representative of all.

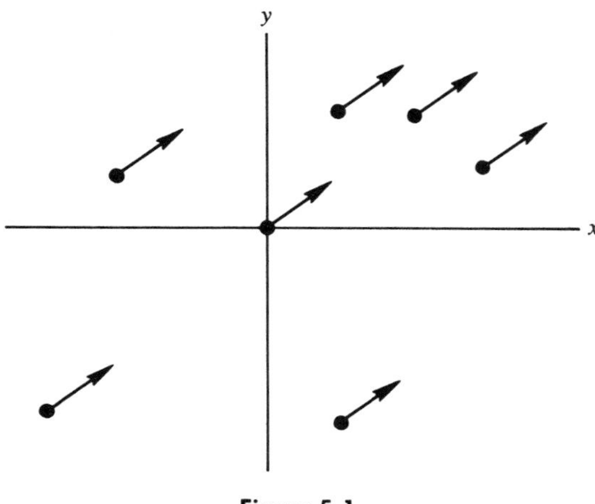

Figure 5.1

In Figure 5.1 the vector originating at the origin (vectors terminate at the arrowhead) is selected as representative. Two things are intuitively clear.

1. Any vector in the plane is representable by a vector originating at the origin. (See Figure 5.1.)
2. No two distinct vectors at the origin are equal.

This means that we can study vectors by studying only vectors that originate at the origin. From the viewpoint of linear algebra vectors become interesting because they can be described in terms of ordered n-tuples and addition and scalar multiplication can be defined on the n-tuples.

To see how vectors are expressible as ordered n-tuples examine Figure 5.2. Since we consider all vectors as originating at the origin we distinguish them by their terminal points. In two dimensions each such terminal point is located by its (x,y) coordinates. Some vectors are identified in Figure 5.2(a). In Figure 5.2(b) a three-dimensional vector is identified as an ordered 3-tuple (triple) $(x, y, z) = (1, 2, 5)$. Clearly there is no diagram for an ordered 4-tuple since we cannot draw pictures in four dimensions.

Once again a concept is defined which then develops a life of its own. While we cannot draw pictures, we can talk about ordered 4-tuples of numbers, ordered 5-tuples of numbers, and in fact ordered n-tuples of numbers (where n is any positive integer). From the viewpoint of mathematics it is the ordered n-tuple that is called the *vector*. When defining addition of vectors, scalar multiplication of vectors, and so on, we will examine two-dimensional pictures so that the reason for using certain words will be clear. These pictures are to be thought of as just that, pictures, hopefully clarifying ideas.

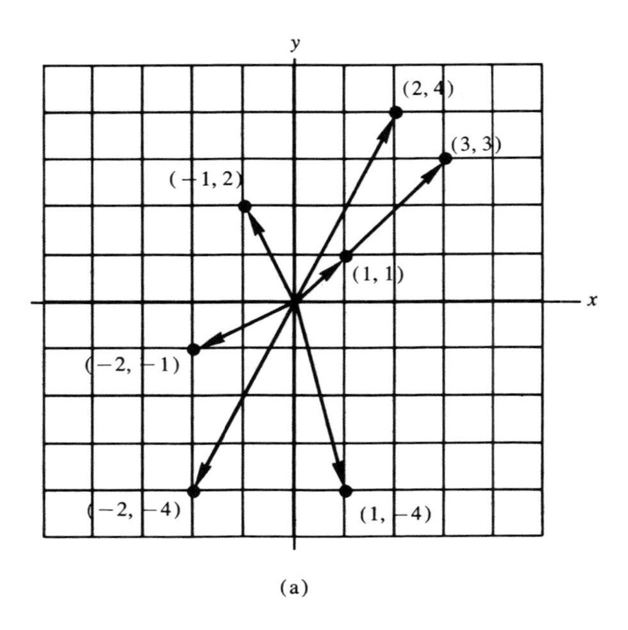

(a)

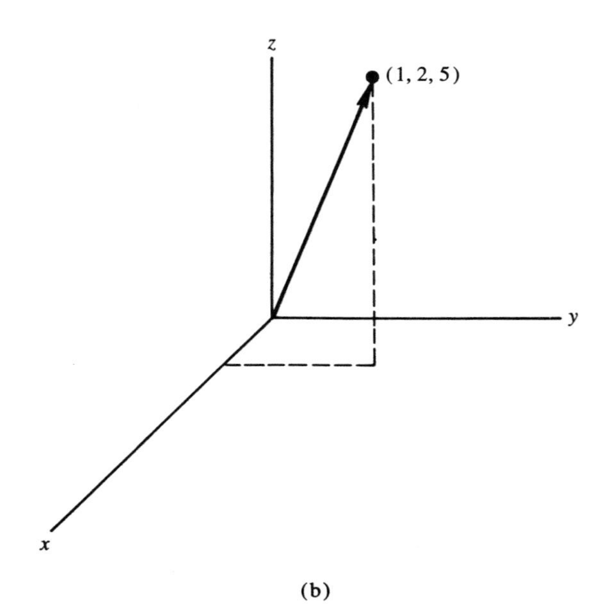

(b)

Figure 5.2

5.1 Scalar Multiplication and Vector Lines

In Figure 5.3 we see that certain vectors lie on a line. Selecting any one of these vectors, say (3, 2), we see that there is a relationship between this

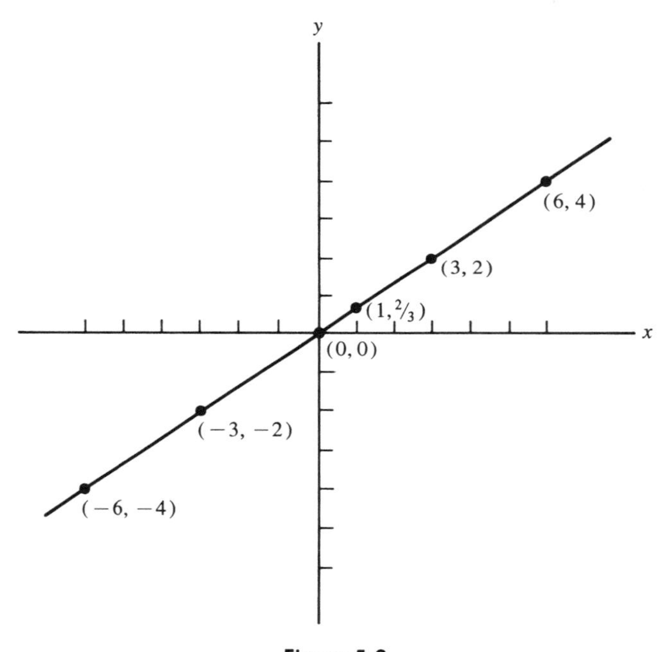

Figure 5.3

vector and the others on the line. For instance, the vector (6, 4) is expressed as 2 (3, 2). We write

$$(6,\ 4) = 2\,(3,\ 2)$$

We call (6, 4) a *scalar multiple* of (3, 2), where 2 is the *scalar*. All the other indicated vectors are also scalar multiples of (3, 2).

$$\begin{aligned}
(1,\ \tfrac{2}{3}) &= \tfrac{1}{3}(3,\ 2)\\
(-3,\ -2) &= -1\,(3,\ 2)\\
(-6,\ -4) &= -2\,(3,\ 2)\\
(0,\ 0) &= 0\,(3,\ 2)\\
(6,\ 4) &= 2\,(3,\ 2)\\
(3,\ 2) &= 1\,(3,\ 2)
\end{aligned}$$

(0, 0) identifies the origin, but it also is a 2-tuple lying on the line. It may be thought of as a special vector with no magnitude and no direction.

By geometric means it is easy to establish that any vector on the line in Figure 5.3 is a scalar multiple of (3, 2). This is why we say that the set of all scalar multiples of (3, 2) is a *vector line*. This set is written in many ways, among which are

(1) $\{(x, y): (x, y) = t(3, 2); t \text{ any real number}\}$

(2) Letting $\mathbf{X} = (x, y)$ (capital letters are used for vectors and small letters for scalars):

$$\{\mathbf{X}: \mathbf{X} = t(3, 2); t \text{ any real number}\}$$

(3)
$$\left\{ (x, y):\ \begin{array}{c|cc|} t & 3 & 2 \\ \hline & = x & = y \end{array}\ ; t \text{ any real number} \right\}$$

(4) $\{(x, y): x = 3t; y = 2t; t \text{ any real number}\}$

(5)

t	3	2
	$= x$	$= y$

The tableau (5) for the vector line is the one we use most often. It is convenient when displaying vector lines (or other vector sets) to use the columnar tableau form. The variable t, called a *parameter*, is on the left and the other variables are at the bottom. From (5), a *parametric* representation of the vector line, we obtain by pivot exchange

(5)

t	3*	2
	$= x$	$= y$

(6)

x	$1/3$	$2/3$
	$= t$	$= y$

The second column of (6) reads

$$y = \tfrac{2}{3}x$$

This equation expresses the graph of the line on which all multiples of (3, 2) lie. It is a *nonparametric* representation of the line, expressing directly the relationship between the components x and y of the vectors. We can in (6), if we are not interested in the parameter, drop the first column, and then the nonparametric form is

(7)
$$\begin{array}{r|c|}
\cline{2-2}
x & 2/3 \\
\cline{2-2}
\multicolumn{1}{r}{} & \multicolumn{1}{c}{= y}
\end{array}$$

Since each vector of the vector line is a multiple of (3, 2), it is natural to say that (3, 2) *generates* (*spans*) the vector line. Since other vectors may generate the same line, the term "generate" must be treated with caution. It is intuitively clear that any one of the vectors except (0, 0) that are in the vector line generated by (3, 2) can itself generate the vector line. Since (0, 0) lies on all lines, it cannot generate any lines. The vector (6, 4) generates the same line as the vector (3, 2).

$$\begin{aligned}
(\ 1,\ 2/3) &= \ \ 1/6\,(6, 4) \\
(-3, -2) &= -1/2\,(6, 4) \\
(-6, -4) &= -1\ (6, 4) \\
(\ 0,\ \ 0) &= \ \ 0\ (6, 4) \\
(\ 6,\ \ 4) &= \ \ 1\ (6, 4) \\
(\ 3,\ \ 2) &= \ \ 1/2\,(6, 4)
\end{aligned}$$

The vector line generated by (6, 4) is, in tableau form,

(8)
$$\begin{array}{r|cc|}
\cline{2-3}
t & 6^* & 4 \\
\cline{2-3}
\multicolumn{1}{r}{} & = x & \multicolumn{1}{c}{= y}
\end{array}$$

Pivoting yields

(9)
$$\begin{array}{r|cc|}
\cline{2-3}
x & 1/6 & 2/3 \\
\cline{2-3}
\multicolumn{1}{r}{} & = t & \multicolumn{1}{c}{= y}
\end{array}$$

which shows that y and x are in the same relationship as in (7). This example contains the germ of

THEOREM 5.1. Let

(1)
$$\begin{array}{c|cc|} \cline{2-3} t & a^* & b \\ \cline{2-3} & = x & = y \end{array}$$

be the vector line α generated by vector $(a, b) \neq (0, 0)$. Then the vector line

(2)
$$\begin{array}{c|cc|} \cline{2-3} t & ca^* & cb \\ \cline{2-3} & = x & = y \end{array}$$

generated by $c(a, b) = (ca, cb)$ with $c \neq 0$ is also α.

Proof: Either a or b is not zero or else all we have is the zero vector. Assume that $a \neq 0$. Pivoting as indicated in (1) yields

(3)
$$\begin{array}{c|cc|} \cline{2-3} x & \frac{1}{a} & \frac{b}{a} \\ \cline{2-3} & = t & = y \end{array}$$

Pivoting as indicated in (2) yields

(4)
$$\begin{array}{c|cc|} \cline{2-3} x & \frac{1}{ca} & \frac{b}{a} \\ \cline{2-3} & = t & = y \end{array}$$

In both (3) and (4) $y = (b/a)x$ and the conclusion follows. Note that we have no interest in how t is expressed in terms of x. In a sense we have "eliminated" the parameter, the last column

$$\begin{array}{c|c|} \cline{2-2} x & \frac{b}{a} \\ \cline{2-2} & = y \end{array}$$

giving the nonparametric form of the line.

The procedure by which t is eliminated is called a *parametric elimination step,* another use for pivot exchange.

Small Greek letters α (alpha), β (beta), and so on, will be used to represent vector lines. We write

α: t	a	b
	$= x$	$= y$

which is read "the vector line α generated by the vector (a, b)."

EXAMPLE 1

Obtain a nonparametric representation of the vector line

β: t	1	4	2
	$= x$	$= y$	$= z$

This is a vector line in three dimensions (an ordered set of 3-tuples, which means it is in R^3). Since we are extending the notion of vector line to dimensions greater than three (where pictures are not available) we do not graph β. To obtain a nonparametric representation of β, we pivot on any element. Pivoting on 1 gives

(1)

t	1*	4	2
	$= x$	$= y$	$= z$

(2)

x	1	4	2
	$= t$	$= y$	$= z$

Eliminating the column of the parameter t yields

(3)

x	4	2
	$= y$	$= z$

Tableau (3) is a nonparametric representation of β. It is an *explicit form* (*reduced form*) of β with x the nonbasic variable and y and z the basic variables. We leave it to the reader to show that the two other explicit forms (pivot on 4 or on 2 of (2)) are equivalent to (3). In an explicit form it is easy to answer such questions as: Is (20, 80, 40) a vector in β? (Is it?) The reader is advised to generate the line with some multiple of (1, 4, 2) and show that this multiple yields the same explicit form for β.

EXAMPLE 2

Obtain a nonparametric form of the vector line β in R^5 whose tableau is

(1)

t	2	-1^*	3	-2	5
	$= x_1$	$= x_2$	$= x_3$	$= x_4$	$= x_5$

Pivoting yields

x_2	-2	-1	-3	2	-5
	$= x_1$	$= t$	$= x_3$	$= x_4$	$= x_5$

Dropping the t column gives the explicit form

(2)

x_2	-2	-3	2	-5
	$= x_1$	$= x_3$	$= x_4$	$= x_5$

In the parametric form all the components of each vector are written as basic variables all in terms of the parameter t.

In the nonparametric form, four of the components of each vector are written as basic variables in terms of the fifth component, which serves as the nonbasic variable.

Three common presentations for the parametric form are

1. $$\begin{cases} x_1 = 2t \\ x_2 = -t \\ x_3 = 3t \\ x_4 = -2t \\ x_5 = 5t. \end{cases}$$

2. $(x_1, x_2, x_3, x_4, x_5) = t(2, -1, 3, -2, 5)$.
3. $\mathbf{X} = t(2, -1, 3, -2, 5)$.

Two common presentations for the nonparametric form are

1. $$\begin{cases} x_1 = -2x_2 \\ x_3 = -3x_2 \\ x_4 = 2x_2 \\ x_5 = -5x_2. \end{cases}$$

2. $(x_1, x_2, x_3, x_4, x_5) = (-2x_2, x_2, -3x_2, 2x_2, -5x_2)$.

EXAMPLE 3

$$\begin{aligned} 3[2(1, 3, -2)] &= 3(2, 6, -4) = (6, 18, -12) = 2(3, 9, -6) \\ &= 2[3(1, 3, -2)] = (2\cdot 3)(1, 3, -2) \\ &= (3\cdot 2)(1, 3, -2) = 6(1, 3, -2) \end{aligned}$$

This manipulation leads directly to the result that for all scalars k and m and all vectors $\mathbf{A}$ in R^n,

$$k(m\mathbf{A}) = (km)\mathbf{A} = (mk)\mathbf{A} = m(k\mathbf{A}) \qquad \text{(see Exercise 8)}$$

Every vector line

$$t\ \begin{array}{|cccc|} \hline a_1 & a_2 & \cdots & a_n \\ \hline \end{array}$$
$$= x_1 \quad = x_2 \quad \cdots \quad = x_n$$

contains $\mathbf{0}_n$. To see this, set $t = 0$, giving

$$0\ \begin{array}{|cccc|} \hline a_1 & a_2 & \cdots & a_n \\ \hline \end{array}$$
$$= x_1 \quad = x_2 \quad \cdots \quad = x_n$$

We see that $(x_1, x_2, \ldots, x_n) = (0, 0, \ldots, 0)$ when $t = 0$. Theorem 5.1 extends to n dimensions. Let

$$t\ \begin{array}{|cccc|} \hline a_1 & a_2 & \cdots & a_n \\ \hline \end{array}$$
$$= x_1 \quad = x_2 \quad \cdots \quad = x_n$$

be the vector line α generated by $\mathbf{A} = (a_1, a_2, \ldots, a_n)$,

$$\mathbf{A} = (a_1, a_2, \ldots, a_n) \neq \mathbf{0}_n$$

Then the vector line

$$t\ \begin{array}{|cccc|} \hline ca_1 & ca_2 & \cdots & ca_n \\ \hline \end{array}$$
$$= x_1 \quad = x_2 \quad \cdots \quad = x_n$$

generated by $c\mathbf{A}$ (where $c \neq 0$) is also α.

This result, plus the fact that every vector line in R^n contains the vector

$\mathbf{0}_n$, leads us directly to the result that two distinct vector lines α and β in R^n have only $\mathbf{0}_n$ in common because if they had another vector $\mathbf{C}$ in common, then $\mathbf{C}$ would generate both α and β, contradicting the fact that α and β are different. The geometric picture of this, in two dimensions, shows that two distinct lines through the origin meet only at the origin.

Exercises

1. Perform the multiplication in each of the following:
(a) $3(-1, 2)$.
(b) $-2(0, 2, -1)$.
(c) $0(4, 2, 3, 1)$.
(d) $\frac{1}{2}(6, -8, 0)$.
(e) $\frac{3}{4}(8, 6, -2)$.
(f) $\sqrt{2}(1, 0, \sqrt{2})$.

2. For each vector line represented below, find two vectors that belong to it and two that do not. In each case t is any real number.

(a)

t	8	$-\frac{1}{2}$
	$= x$	$= y$

(b)

t	0	2	-4
	$= x$	$= y$	$= z$

(c)

t	8	2	-1	0
	$= x_1$	$= x_2$	$= x_3$	$= x_4$

(d)

t	3
	$= x$

3. Make a graph of the vector line represented in

(a)

t	-1	3
	$= x$	$= y$

(b)

t	2	1	2
	$= x$	$= y$	$= z$

4. Describe each of the following subsets of vector lines as a ray, a line segment, or a point.

(a)

t	4	8	, $t \geq 1$
	$= x$	$= y$	

(b)

t	3	2	-1	, $t = 3$
	$= x_1$	$= x_2$	$= x_3$	

(c)

t	1
	$= x$

$, -1 \leq t \leq 1$

(d)

t	4	0	-1
	$= x_1$	$= x_2$	$= x_3$

$, 0 \leq t \leq 2$

(e)

t	4	0	2	3
	$= x_1$	$= x_2$	$= x_3$	$= x_4$

$, t \leq 0$

(f)

t	0	0
	$= x$	$= y$

5. Solve for all variables in
(a) $(x, 3, z) = (4, y, 2)$.
(b) $(x + 2, y - 1, z + 3) = (3, 4, 5)$.
(c) $3(x, 2, 4) = 2(8, z, y)$.
(d) $4(x_1, 3, x_3, 5) = 3(6, x_2, 9, x_4)$.
(e) $(a + 3, b - 4) = (b, 2a)$.

6. Designate by a tableau the vector line generated by
(a) $\mathbf{A} = (3, 5)$.
(b) $\mathbf{B} = (0, 1, 3)$.
(c) $\mathbf{C} = (c_1, c_2, 4)$.
(d) $\mathbf{D} = (-2, 0, 5, 2)$.

7. Devise a geometric interpretation in two dimensions for $-\mathbf{A}$, where $\mathbf{A}$ is a nonzero vector.

8. Devise a geometric interpretation in two dimensions for the statement $k(m\mathbf{A}) = (km)\mathbf{A}$.

***9.** Let $\mathbf{A} = (3, -1, 2)$. Describe in tableau form the set of vector points in
(a) Vector line $\overleftrightarrow{\mathbf{0A}}$.
(b) Vector ray $\overrightarrow{\mathbf{0A}}$.
(c) Vector ray $\overrightarrow{\mathbf{A0}}$.
(d) Vector segment $\overline{\mathbf{0A}}$.
(e) Midpoint of vector segment $\overline{\mathbf{0A}}$.

10. (a) Prove there is no k such that $(3, 2, 1) = k(6, 4, 5)$.
(b) Find k such that $(3, 2, 1) = k(30, 20, 10)$.
(c) State a necessary and sufficient condition that there exists a k such that $(a, b, c) = k(d, e, f)$. Generalize to R^n.

11. Let $\mathbf{A} = (2, -3)$, $\mathbf{B} = (x, y)$. Determine whether or not they generate the same line if $(x, y) =$
(a) $(2, 3)$.
(b) $(-2, -3)$.
(c) $(-2, 3)$.
(d) $(-4, 6)$.

5.2 Vector Addition and Vector Planes; Parametric Elimination

In Section 5.1 pictures were used to justify the use of the term "vector line" to describe sets of scalar multiples of a given n-tuple. In this section we justify, pictorially, the term *vector plane* to describe sets of vectors that are sums of scalar multiples of given vectors.

As with matrices, addition of vectors is accomplished by adding corresponding entries. Thus, $(3, 1, 2) + (5, 2, 7) = (8, 3, 9)$, $(5, 6, -1, 4) + (3, -2, 0, 2) = (8, 4, -1, 6)$. If $\mathbf{A} = (a_1, a_2, \ldots, a_n)$ and $\mathbf{B} = (b_1, b_2, \ldots, b_n)$ are any two vectors in R^n, then

$$\mathbf{A} + \mathbf{B} = (a_1 + b_1, a_2 + b_2, \ldots, a_n + b_n)$$

In two dimensions $(3, 1) + (2, 3) = (5, 4)$. In Figure 5.4 $(3, 1)$ and $(2, 3)$ are represented pictorially as solid lines. One dashed line represents the vector in the plane which is equivalent to $(3, 1)$ and has its origin at the terminal point of $(2, 3)$. The result of $(2, 3)$ "followed" by the equivalent of $(3, 1)$ is $(5, 4)$. This can be shown using elementary plane geometry. Thus the "sum" of these two vectors $(3, 1)$ and $(2, 3)$ is a third vector $(5, 4)$, and the three points in the plane $(3, 1)$, $(2, 3)$, and $(5, 4)$, together with $(0, 0)$, determine the parallelogram of Figure 5.4. Any two vectors, not on the same

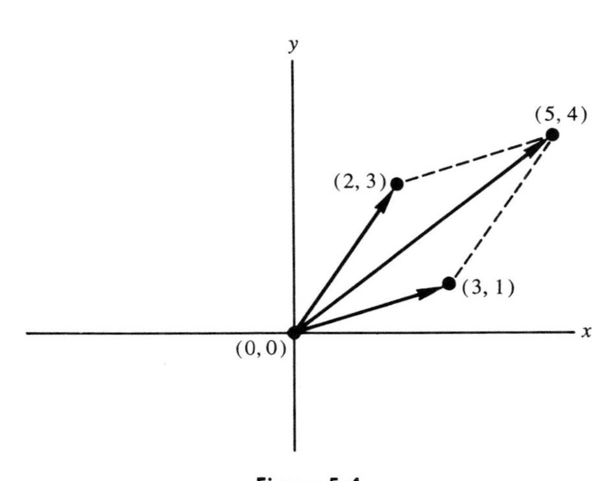

Figure 5.4

vector line when "added together," will produce a third vector which helps to determine (as above) a parallelogram. Figure 5.5 illustrates the two vector lines

$$\alpha:\ r \begin{array}{|cc|}\hline 3 & 1 \\ \hline \end{array} \quad \text{and} \quad \beta:\ s \begin{array}{|cc|}\hline 2 & 3 \\ \hline \end{array}$$
$$\qquad = x \quad = y \qquad\qquad\qquad = x \quad = y$$

Since every point in the plane not on either line can be made the fourth vertex of a parallelogram [(0, 0) is the third] (try it), it is pictorially clear that every point in the plane can be expressed in terms of points on the lines.

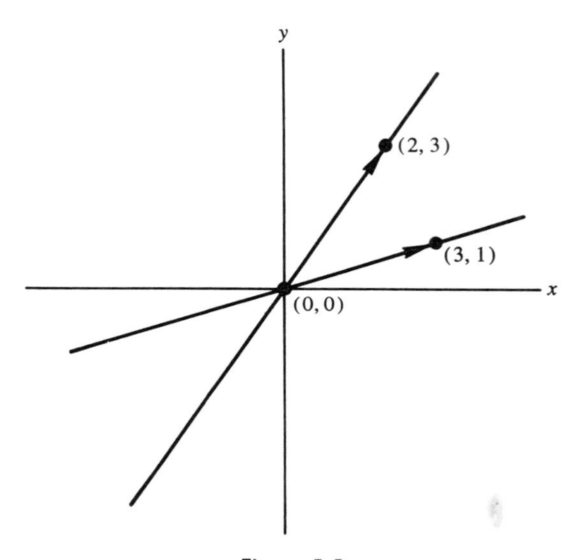

Figure 5.5

[To get all points on the line containing (2, 3), set $s = 0$ and let r vary.] Every point can be expressed as $r(3, 1) + s(2, 3)$, where r and s are real numbers. For instance, $(-4, -13)$ can be expressed as $2(3, 1) + (-5)(2, 3)$. These examples justify the use of *vector plane* to describe the set of vectors that are expressible in the five following ways:

(1) $\{(x, y){:}(x, y) = r(3, 1) + s(2, 3);\ r \text{ and } s \text{ any real numbers}\}$

(2) Letting $\mathbf{X} = (x, y)$:

$\{\mathbf{X}{:}\mathbf{X} = r(3, 1) + s(2, 3);\ r \text{ and } s \text{ any real numbers}\}$

(3) $$\left\{(x, y){:}\ \begin{array}{c|cc|} r & 3 & 1 \\ s & 2 & 3 \\ \hline & = x & = y \end{array} ;\ r \text{ and } s \text{ any real numbers}\right\}$$

(4) $$\left\{(x, y): \left| \begin{array}{l} x = 3r + 2s \\ y = r + 3s \end{array} \right. ;\ r \text{ and } s \text{ any real numbers}\right\}$$

(5)

r	3	1
s	2	3
	$= x$	$= y$

As with vector lines, the tableau (5) is the one we use most often to represent vector planes. The parameters in this case are r and s. Vector lines have one parameter, vector planes two. The vector plane described above is obtained by adding multiples of (3, 1) and (2, 3) together. In the language of linear algebra we are forming *linear combinations* of (3, 1) and (2, 3). In general, if $c_1, c_2, \ldots, c_m$ are any m scalars and $\mathbf{A}_1, \mathbf{A}_2, \ldots, \mathbf{A}_m$ are any m vectors of R^n, then $c_1\mathbf{A}_1 + c_2\mathbf{A}_2 + \cdots + c_m\mathbf{A}_m$ is called a *linear combination* of the $\mathbf{A}$'s. The c's are any scalars. If all c's are zero, we get $\mathbf{0}_n$ regardless of the $\mathbf{A}$'s.

EXAMPLE 1

Let $\mathbf{A} = (3, 2)$, $\mathbf{B} = (1, -4)$. Some linear combinations of $\mathbf{A}$ and $\mathbf{B}$ are

$$1(3, 2) + 0(1, -4) = (3, \quad 2)$$
$$0(3, 2) + 5(1, -4) = (5, -20)$$
$$2(3, 2) - 4(1, -4) = (2, \quad 20)$$

In general a linear combination of $\mathbf{A}$ and $\mathbf{B}$ is $r(3, 2) + s(1, -4)$, where the parameters r and s have specific values. The set of *all* linear combinations of $\mathbf{A}$ and $\mathbf{B}$ is $r(3, 2) + s(1, -4)$, where r and s are any real numbers. Letting (x, y) represent a linear combination of $\mathbf{A}$ and $\mathbf{B}$, we write

$$(x, y) = r(3, 2) + s(1, -4) = (3r + s, 2r - 4s)$$

r and s any real numbers, or else we write

$$x = 3r + s \qquad\qquad y = 2r - 4s$$

The usual tableau with parameters at the left is

r	3	2
s	1	-4
	$= x$	$= y$

The vectors (3, 2) and (1, −4) which form the rows of the tableau *generate* the set of all linear combinations. They are called a *spanning set* of the vector plane.

The geometric interpretation of Example 1 reveals the essential nature of the set of all linear combinations of (3, 2) and (1, −4).

1. If $r = 0$, (x, y) is a point in the line generated by (1, −4).
2. If $s = 0$, (x, y) is a point in the line generated by (3, 2).
3. If neither r nor s is zero, then (x, y) is the fourth vertex of the parallelogram of Figure 5.6.

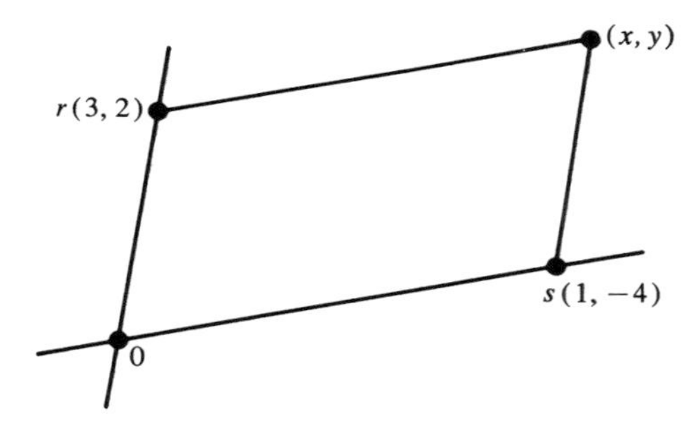

Figure 5.6

For all values of r and s, (x,y), as a point, is in the plane generated by (3, 2) and (1, −4). Also for any point Y in this plane it is possible to determine r and s such that $Y = r(3, 2) + s(1, -4)$.

While the set of all linear combinations of $\mathbf{A} = (3, 2)$ and $\mathbf{B} = (1, -4)$ constitute a vector plane, we must be cautious not to make this assertion for any vectors **A** and **B**. For instance, if $\mathbf{A} = (3, 2)$ and $\mathbf{B} = (4, 6)$, **A** and **B** generate the same vector line and not a vector plane.

In general if **A** and **B** are nonzero vectors in R^n that generate different vector lines, then the set of all linear combinations, $r\mathbf{A} + s\mathbf{B}$, constitutes a *vector plane* in R^n. $\{\mathbf{A}, \mathbf{B}\}$ is a *spanning set* and is said to *generate* the vector plane. A vector plane is designated by π.

EXAMPLE 2

π:				
	p	2	−1	1
	q	1	3	2
		$= x_1$	$= x_2$	$= x_3$

is the set of all linear combinations generated by (2, −1, 1) and (1, 3, 2).

(It is not necessary to use the symbols r and s for parameters; p and q do just as well.)

This tableau represents a vector plane if (2, −1, 1) and (1, 3, 2) generate different vector lines in R^3. To determine that they do, we show that (2, −1, 1) is not a scalar multiple of (1, 3, 2). If (2, −1, 1) is a scalar multiple of (1, 3, 2), then (2, −1, 1) = t(1, 3, 2) for some t. That is,

t	1	3	2
	= 2	= −1	= 1

This tableau can never be true. The first column is true only if $t = 2$. But then columns 2 and 3 are false.

Hence the tableau for π defines a vector plane. To determine vectors in this plane assign values to the parameters p and q, and evaluate (x_1, x_2, x_3). Some vectors for different p and q are as follows:

(p, q)	(0, 0)	(1, 0)	(0, 1)	(8, −2)
(x_1, x_2, x_3)	(0, 0, 0)	(2, −1, 1)	(1, 3. 2)	(14, −14, 4)

Before leaving this example we convince ourselves that not all vectors in R^3 are in π. To do this transform the parametric expression for π to an explicit form using parametric elimination

(1)

p	2	−1	1
q	1*	3	2
	$= x_1$	$= x_2$	$= x_3$

(2)

p	−2	−7	−3*
x_1	1	3	2
	$= q$	$= x_2$	$= x_3$

(3)

x_3	2/3	7/3	−1/3
x_1	−1/3	−5/3	2/3
	$= q$	$= x_2$	$= p$

No longer concerned with parameters p and q, we drop their columns in (3), getting the explicit form

(4)

x_3	$7/3$
x_1	$-5/3$
	$= x_2$

This single equation tells how the components of vectors in π relate to each other. The variables x_1 and x_3 are nonbasic while x_2 is the basic variable. The explicit form is used to determine whether any given vector in R^3 is in π. For instance, $(x_1, x_2, x_3) = (3, 9, 6)$ is in π because it satisfies the equation of the explicit form (4); $(3, 8, 6)$ is not in π because it fails to satisfy that equation.

From now on we obtain the explicit form by deleting the column of the pivot at each exchange.

(1)

p	2	-1	1
q	1*	3	2
	$= x_1$	$= x_2$	$= x_3$

(2)

p	-7	-3*
x_1	3	2
	$= x_2$	$= x_3$

(3)

x_3	$7/3$
x_1	$-5/3$
	$= x_2$

This illustrates the *parametric elimination algorithm*. Later on we use this algorithm in examples with many parameters and vector components.

EXAMPLE 3

The tableau

p	2	-1	1
q	1	3	2
r	-3	5	0
	$= x_1$	$= x_2$	$= x_3$

represents the set S of all linear combinations of (2, -1, 1), (1, 3, 2) , and (-3, 5, 0). Using pivot reduction we obtain the fully reduced tableau (3).

(1)

p	2	-1	1
q	1*	3	2
r	-3	5	0
	$= x_1$	$= x_2$	$= x_3$

(2)

p	-7	-3*
x_1	3	2
r	14	6
	$= x_2$	$= x_3$

(3)

x_3	$7/3$
x_1	$-5/3$
r	0
	$= x_2$

The equation of (3) is not affected by the row of r because of the zero inside the tableau. Thus this row can be removed, showing that S is the vector plane π of Example 2. How can we account for this? It must be that one of the vectors in the initial tableau (1) lies in the vector plane generated by the other two. Indeed $(-3, 5, 0) = -2(2, -1, 1) + 1(1, 3, 2)$. We look into such matters in Section 5.3.

Since vector planes are spanned by two vector lines it is sometimes convenient to express this symbolically. Thus if π is the vector plane spanned by α and β, we write $\pi = \alpha + \beta$.

EXAMPLE 4

Verify that (1, 2, 4, 2) and (2, 3, 6, 5) are in different vector lines α and β. Find the explicit form for $\pi = \alpha + \beta$ a vector plane in R^4.

To verify that (1, 2, 4, 2) and (2, 3, 6, 5) are in different vector lines the reader should show that

t	1	2	4	2
	$= 2$	$= 3$	$= 6$	$= 5$

is inconsistent. Plane π has the parametric representation

(1)

r	1	2	4	2
s	2	3	6	5
	$= x_1$	$= x_2$	$= x_3$	$= x_4$

The parameters r and s are eliminated via parametric elimination.

(1)

r	1*	2	4	2
s	2	3	6	5
	$= x_1$	$= x_2$	$= x_3$	$= x_4$

(2)

x_1	2	4	2
s	-1	-2	1*
	$= x_2$	$= x_3$	$= x_4$

(3)

x_1	4	8
x_4	-1	-2
	$= x_2$	$= x_3$

Replacing (x_1, x_2, x_3, x_4) by (1, 2, 4, 2) or (2, 3, 6, 5) in the explicit tableau is a check that the reduction algorithm is correctly carried out.

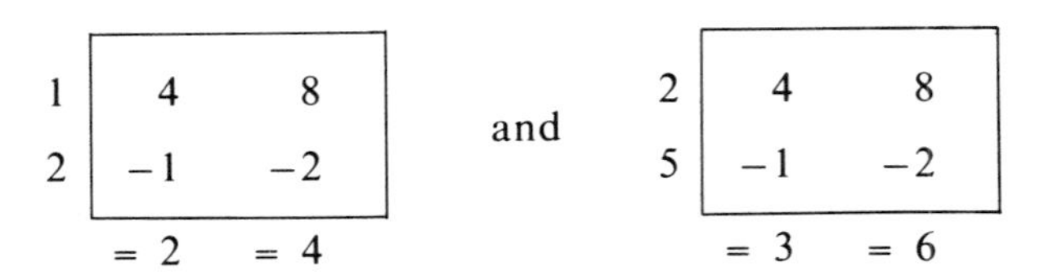

are both true.

Exercises

1. Let $\mathbf{A} = (2, 1, 4)$, $\mathbf{B} = (0, -1, 3)$, and $\mathbf{C} = (1, 2, -1)$. Express each of the following as an ordered triple (3-tuple). By $\mathbf{A} - \mathbf{B}$ we mean $\mathbf{A} + (-\mathbf{B})$. Thus $(2, 1, 4) - (0, -1, 3) = (2, 1, 4) + (0, 1, -3) = (2, 2, 1)$.
   (a) $\mathbf{A} + \mathbf{B}$.
   (b) $(\mathbf{A} + \mathbf{B}) - \mathbf{C}$.
   (c) $\mathbf{A} - \mathbf{C}$.
   (d) $\mathbf{A} - (\mathbf{B} - \mathbf{C})$.
   (e) $\mathbf{A} - \mathbf{B} - \mathbf{C}$.
   (f) $2\mathbf{A} + \mathbf{B} - \mathbf{C}$.
   (g) $2\mathbf{A} + 3\mathbf{B}$.
   (h) $3\mathbf{A} - 2\mathbf{B}$.
   (i) $3\mathbf{A} + 2\mathbf{B} - 2\mathbf{C}$.

2. Express each of the following as an ordered triple:
   (a) $2(3, 2, -1) + 4(0, 5, -2) + (-1, 2, -3)$.
   (b) $5(1, 2, 5) - 2(6, 0, 5) + 3(-1, -2, -3)$.

3. Solve for x, y, and z.
   (a) $2(3, 2, z) + (x, y, 5) = 3(12, 0, 1)$.
   (b) $(x, y, z) + 2(3, 5, 6) = \frac{1}{2}(8, 9, 12)$.
   (c) $x(1, 2, 3) + y(2, -1, 4) + z(-1, 2, -3) = (0, 0, 0)$.

4. Let $\mathbf{I} = (1, 0, 0)$, $\mathbf{J} = (0, 1, 0)$, and $\mathbf{K} = (0, 0, 1)$.
   (a) If $\mathbf{A} = (2, 3, 5)$, show that $\mathbf{A} = 2\mathbf{I} + 3\mathbf{J} + 5\mathbf{K}$.
   (b) If $\mathbf{B} = (x_1, x_2, x_3)$, show that $\mathbf{B} = x_1\mathbf{I} + x_2\mathbf{J} + x_3\mathbf{K}$.
   (c) Express $5\mathbf{I} - 2\mathbf{J} + \frac{1}{2}\mathbf{K}$ as a triple.

5. Find a parametric representation of the vector plane that contains:
   (a) (1, 3, 2) and (−1, 2, 3).
   (b) (1, 1, 1) and (3, 2, 1).
   (c) (−1, 3, 4) and (0, 8, 2).

6. Express each of the following planes in explicit form:

(a)

p	1	3	2
q	2	0	−4
	$= x_1$	$= x_2$	$= x_3$

(b)

p	3	−12	5
q	−1	6	8
	$= x_1$	$= x_2$	$= x_3$

(c)

p	3	2	5
q	2	3	4
	$= x_1$	$= x_2$	$= x_3$

7. Given the vector plane that contains (1, 0, 1) and (0, 1, 4), determine whether or not each of the following vectors is in that plane:
(a) (1, 1, 5). (b) (−1, 3, 8).
(c) (0, 0, 0). (d) (100, 4, 116).
(*Hint:* Use the explicit representation of the plane.)

8. Let **A** = (1, 3, 4, 2) and **B** = (0, 5, 6, 2) be vectors in R^4.
(a) Express, in both parametric and explicit forms, equations of the vector plane containing **A** and **B**.
(b) Find five vectors that are in this plane.
(c) Find two vectors that are not in this plane.
(d) Find a vector line all of whose vectors are in this plane.

9. (a) Show that

p	3	−1	4
q	0	5	6
r	3	4	10
	$= x_1$	$= x_2$	$= x_3$

defines a vector plane in R^3.
(b) Express this plane with a two-row tableau.
(c) Express this plane in explicit form.

10. (a) Show that

p	3	−1	4	2
q	0	5	6	−3
r	3	4	10	−1
	$= x_1$	$= x_2$	$= x_3$	$= x_4$

defines a vector plane in R^4.

(b) Express this plane with a two-row tableau.
(c) Express this plane in explicit form.

11. How many equations in explicit form define a vector plane: (a) In R^3? (b) In R^4? (c) In R^5? (d) In R^n?

12. Discuss the statement: There is only one vector plane in R^2.

5.3 Bases and Dimensions of Vector Subspaces; Pivot Condensation

We return to the vector plane

π:

p	2	−1	1
q	1	3	2
	$= x_1$	$= x_2$	$= x_3$

constructed in Example 2 of Section 5.2. In Example 3 of that section we saw that

p	2	−1	1
q	1	3	2
r	−3	5	0
	$= x_1$	$= x_2$	$= x_3$

defines the same plane despite the fact that its spanning set contains an additional vector. This was explained by the fact that $(-3, 5, 0)$ is in π; that is, $(-3, 5, 0)$ is a linear combination of $(2, -1, 1)$ and $(1, 3, 2)$. We say that $(-3, 5, 0)$ is linearly dependent on $(2, -1, 1)$ and $(1, 3, 2)$. In more general terms whenever one vector can be expressed as a linear combination of others we say that all the vectors in question are *linearly dependent*. Thus the set $\{(-3, 5, 0), (2, -1, 1), (1, 3, 2)\}$ is a linearly dependent set. *Careful:* $\{(2, 0), (1, 0), (0, 1)\}$ is a linearly dependent set since $(2, 0) = 2(1, 0) + 0(0, 1)$. However $(0, 1)$ cannot be expressed as a linear combination of $(1, 0)$ and $(2, 0)$. (Form a tableau, get the fully reduced tableau and see why not.) A set is linearly dependent if *one* of the vectors is expressible as a linear combination of the others. It need not be true that all of them are so expressible (although it may be true).

EXAMPLE 1

$\{(3, 2), (4, 6)\}$ is linearly dependent because $(4, 6) = 2(3, 2)$.

EXAMPLE 2

$\{(3, 2), (1, -4)\}$ is not linearly dependent because there is no r such that $(3, 2) = r(1, -4)$ and no s such that $(1, -4) = s(3, 2)$. To see this suppose that

$$\begin{array}{c|cc|} \cline{2-3} r & 1^* & -4 \\ \cline{2-3} & = 3 & = 2 \end{array}$$

Then by parametric elimination

$$\begin{array}{c|c|} \cline{2-2} 3 & -4 \\ \cline{2-2} & = 2 \end{array}$$

which is false. Also

$$\begin{array}{c|cc|} \cline{2-3} s & 3 & 2^* \\ \cline{2-3} & = 1 & = -4 \end{array}$$

$$\begin{array}{c|c|} \cline{2-2} -4 & 3/2 \\ \cline{2-2} & = 1 \end{array}$$

is false. If a set of vectors is not linearly dependent we call them *linearly independent.*

In these examples it is easy to see whether the sets are linearly dependent or independent. For less obvious cases there is need for more sophisticated methods. Again pivot exchange shows the way.

EXAMPLE 3

Show that $S = \{(2, -1, 1), (1, 3, 2), (-3, 5, 0)\}$ is a linearly dependent set. (We know this but for the moment assume we do not.) To begin, consider the equation

$$p(2, -1, 1) + q(1, 3, 2) + r(-3, 5, 0) = (0, 0, 0)$$

If this equation has a solution (p, q, r) with at least one of p, q, and r not zero, then S is linearly dependent. For the moment assume that r is not zero.

(1) $$p(2, -1, 1) + q(1, 3, 2) + r(-3, 5, 0) = (0, 0, 0)$$

can be written as

(2) $$p(2, -1, 1) + q(1, 3, 2) = -r(-3, 5, 0)$$

and since $r \neq 0$ we can divide by r, getting

(3) $$\frac{-p}{r}(2, -1, 1) + \frac{-q}{r}(1, 3, 2) = (-3, 5, 0)$$

which expresses $(-3, 5, 0)$ as a linear combination of the other two and this makes S a linearly dependent set. [If p or q is not zero the same argument could be set up for $(2, -1, 1)$ or $(1, 3, 2)$.] If $(p, q, r) = (0, 0, 0)$ is the only solution of (1), then none of the vectors can be expressed as linear combinations of the others and the set is linearly independent. Therefore, to demonstrate linear dependence or independence we need only decide the answer to the question: Must p, q, and r all be 0 in (1)? If the answer is yes, S is linearly independent. If the answer is no, we can obtain (3) and S must be linearly dependent. We need not concern ourselves with the actual values of p, q, and r. We proceed by writing the tableau for (1).

(1)

p	2	-1	1
q	1*	3	2
r	-3	5	0
	$= 0$	$= 0$	$= 0$

Now pivot-exchange as indicated.

(2)

p	-2	-7	-3
0	1	3	2
r	3	14	6
	$= q$	$= 0$	$= 0$

The second row of (2) is multiplied by 0; therefore, we delete it since all equations are the same without it.

This leaves

(3)
$$\begin{array}{c|ccc|} \cline{2-4} p & -2 & -7 & -3 \\ r & 3 & 14 & 6 \\ \cline{2-4} \multicolumn{1}{c}{} & \multicolumn{1}{c}{= q} & \multicolumn{1}{c}{= 0} & \multicolumn{1}{c}{= 0} \end{array}$$

From (3), if p and r are both zero, then q must be zero. Therefore, if there is to be a nonzero solution, at least one of p or r must be nonzero, and we need not consider q any further. Since we need not consider it, we drop its column and therefore have

(4)
$$\begin{array}{c|cc|} \cline{2-3} p & -7^{*} & -3 \\ r & 14 & 6 \\ \cline{2-3} \multicolumn{1}{c}{} & \multicolumn{1}{c}{= 0} & \multicolumn{1}{c}{= 0} \end{array}$$

Now pivoting as indicated in (4) yields

(5)
$$\begin{array}{c|cc|} \cline{2-3} 0 & -1/7 & 3/7 \\ r & 2 & 0 \\ \cline{2-3} \multicolumn{1}{c}{} & \multicolumn{1}{c}{= p} & \multicolumn{1}{c}{= 0} \end{array}$$

Using precisely the same argument as above, we can remove the row of the zero, giving

(6)
$$\begin{array}{c|cc|} \cline{2-3} r & 2 & 0 \\ \cline{2-3} \multicolumn{1}{c}{} & \multicolumn{1}{c}{= p} & \multicolumn{1}{c}{= 0} \end{array}$$

Now if $r = 0$, then $p = 0$, so if there is a nonzero solution (p, q, r), r must be nonzero. So we need not consider p and can remove its column, leaving

(7)
$$\begin{array}{c|c|} \cline{2-2} r & 0 \\ \cline{2-2} \multicolumn{1}{c}{} & \multicolumn{1}{c}{= 0} \end{array}$$

We see that r can take on any value whatsoever and still satisfy (7). Let us say $r = 3$. Then from (6) $p = 6$ and from (3) $q = -3$. Thus $(p, q, r) = (6, -3, 3)$ satisfies equation (1), as you can verify. Therefore, S is linearly dependent. We know this since (7) established that r could be nonzero. We did not have to obtain $(p, q, r) = (6, -3, 3)$ as a solution, although it is a check on our work.

The algorithm we have illustrated is called the *pivot condensation algorithm,* and tableau (7) is called the *fully condensed tableau.* The process of eliminating the row of the 0 and the column of the parameter is called *condensation.* Condensation is removal of the row and column of the pivot when performing a pivot exchange on a tableau of the form in this example.

EXAMPLE 4

Determine whether $\{(3, 2, 1, 4), (0, 2, -1, 3), (3, -2, 3, -2)\}$ is linearly dependent or independent.

As in Example 1, form the system

p	3	2	1*	4
q	0	2	−1	3
r	3	−2	3	−2
	= 0	= 0	= 0	= 0

By condensation obtain

q	3*	4	7
r	−6	−8	−14
	= 0	= 0	= 0

and then

r	0	0
	= 0	= 0

The row of zeros in the fully condensed form indicates that r can have any value. Therefore, there is a nontrivial (not all parameters equal to zero) solution to the system and the given set of vectors is linearly dependent.

EXAMPLE 5

Since from

p	1*	2	3	-1
q	2	4	6	-2
r	-1	-2	-3	1
	$= 0$	$= 0$	$= 0$	$= 0$

we get by condensation

q	0	0	0
r	0	0	0
	$= 0$	$= 0$	$= 0$

we conclude that q and r can have any values and therefore that there is a nontrivial solution to the given system. So $\{(1, 2, 3, -1), (2, 4, 6, -2), (-1, -2, -3, 1)\}$ is a linearly dependent set.

In Examples 3, 4, and 5, the fully condensed tableaus contain only zeros on the insides. In each case the conclusion follows that the vectors in the rows of the initial tableau form a linearly dependent set.

EXAMPLE 6

Is $\{3, 2, 1), (1, 2, 3), (4, 4, 2)\}$ linearly dependent? Proceeding as usual yields

(1)

p	3	2	1*
q	1	2	3
r	4	4	2
	$= 0$	$= 0$	$= 0$

q	-8	-4*
r	-2	0
	$= 0$	$= 0$

(2)

(3)

r	-2
	$= 0$

By (3) r can only be 0. From (2) and $r = 0$ flows the conclusion $q = 0$. From (1) and $q = 0$ and $r = 0$ we deduce that $p = 0$. Hence the only solution to (1) is the trivial solution $(p, q, r) = (0, 0, 0)$. Thus the answer to the question is no. The set is linearly independent.

EXAMPLE 7

Determine whether $\{(1, 2), (3, 4), (5, 6)\}$ is linearly dependent or independent.

By condensation we get

(1)

p	1^*	2
q	3	4
r	5	6
	$= 0$	$= 0$

(2)

q	-2^*
r	-4
	$= 0$

(3)

r	-2
	$= q$

From (3) $q = -2r$ is true for all r. [Try some r's and obtain p from (1).] That is, the system has nontrivial solutions, and thus the set of vectors is linearly dependent.

Based on Examples 3–7 we describe the pivot condensation algorithm and the conclusions that can be drawn from the fully condensed tableau. The algorithm is used to determine the linear dependence or independence of a set of vectors S.

1. Form a tableau whose rows are the vectors of S with parameters on the left and 0's at the bottom.
2. Pivot on any nonzero element, removing the row and column of the pivot from the new tableau. This step is called condensation.
3. Repeat step 2, the condensation step, until
   (a) there are all zeros inside the tableau;
   (b) there is exactly one row left;
   (c) there is exactly one column left with more than one row.

This is called the fully condensed tableau.

Conclusions:

1. If 3(a) occurs, then the remaining parameters at the left can be given any values, and thus the set S of vectors is linearly dependent.
2. If 3(b), the set is linearly dependent if there are all zeros inside the tableau [by 3(a)]; otherwise the set is linearly independent. This is true because if the tableau has the element in the ith column, $a_i \neq 0$, then

r	$\cdots$ a_i $\cdots$
	$= 0$

   r must be zero and working backward as in Example 6 will establish the other parameters as being zero also.
3. If 3(c), the set is linearly dependent. If there is a zero inside the column, its parameter on the left can take any value. Otherwise a relationship is established between the remaining parameters, as in Example 7. Since 3(c) *always* results in linear dependence, we save work by

THEOREM 5.2. If a tableau has more rows than columns, then the vectors in its rows form a linearly dependent set. In other words, if a set of vectors S, in R^n, contains more vectors than the number, n, of components in a vector, then S is linearly dependent.

It is interesting to note that, if the inside of a fully condensed tableau has only zeros, then the pivot rank of its initial tableau is equal to the number of condensation steps. Otherwise the pivot rank is one more than the number of condensation steps. This is so because in the nonzero case one more pivot exchange is possible.

The set of all n-tuples together with scalar multiplication and vector addition is called an *n-dimensional vector space*. We state this briefly by just saying that R^n is a vector space.

The set of all linear combinations of a given set of vectors in R^n is a *vector subspace* of R^n. Thus in R^3 vector lines are one-dimensional subspaces and vector planes are two-dimensional subspaces. This raises the question: Precisely what do we mean by dimension? The examples of course indicate what we mean, since a vector line is spanned by one vector, all other vectors in the line being scalar multiples of it. A vector plane is spanned by two linearly independent vectors, all others being sums of scalar multiples of the given two vectors. We shall define dimension precisely a little later.

Recall that a spanning set of vectors for a plane is a set that generates the plane. In Example 2 of Section 5.2, $\{(2, -1, 1), (1, 3, 2)\}$ generates a plane π. $\{(2, -1, 1), (1, 3, 2), (-3, 5, 0)\}$ spans the same plane π. One of these spanning sets has two vectors and the other has three vectors. We use the following definition to distinguish these spanning sets.

DEFINITION 5.1. A set of vectors S in R^n is called a *basis* of a vector subspace of R^n if

(a) S is a spanning set of (generates) the subspace.
(b) S is linearly independent.

There may be many different bases for a subspace. For instance, $\{(2, -1, 1), (1, 3, 2)\}$ is a basis for π and for any nonzero r and s, so is $\{r(2, -1, 1), s(1, 3, 2)\}$, but all bases have in common the property expressed in

THEOREM 5.3. All bases of a given vector subspace V have the same number of vectors.

Proof: We illustrate the proof of this theorem for the case where we supposedly have two different bases, one with three vectors in it, the other with two vectors. The general proof will be clear from the special case. Suppose that V has $\{\mathbf{A}_1, \mathbf{A}_2\}$ as a basis and that it is proposed that $\{\mathbf{B}_1, \mathbf{B}_2, \mathbf{B}_3\}$ is also a basis. Now since $\{\mathbf{A}_1, \mathbf{A}_2\}$ is a basis, every vector in V is expressible as a linear combination of $\mathbf{A}_1$ and $\mathbf{A}_2$. Since $\mathbf{B}_1$, $\mathbf{B}_2$, and $\mathbf{B}_3$ are in V, they are so expressible. That is, there are constants k, t, m, n, p, and q such that

$$\mathbf{B}_1 = k\mathbf{A}_1 + t\mathbf{A}_2 \tag{1}$$

$$\mathbf{B}_2 = m\mathbf{A}_1 + n\mathbf{A}_2 \tag{2}$$

$$\mathbf{B}_3 = p\mathbf{A}_1 + q\mathbf{A}_2 \tag{3}$$

Now consider the vector set $\{k, t), (m, n), (p, q)\}$. By Theorem 5.2 these vectors are linearly dependent. It does not affect the generality of the proof for this special case if we assume that the linear dependence allows us to write

$$(k, t) = s(m, n) + r(p, q)$$

Thus

$$(k, t) = (sm, sn) + (rp, rq)$$

or

$$k = sm + rp \tag{4}$$

$$t = sn + rq \tag{5}$$

Substituting (4) and (5) for k and t in (1) yields

$$\mathbf{B}_1 = (sm + rp)\,\mathbf{A}_1 + (sn + rq)\,\mathbf{A}_2$$

which yields

$$\mathbf{B}_1 = sm\mathbf{A}_1 + sn\mathbf{A}_2 + rp\mathbf{A}_1 + rq\mathbf{A}_2$$

or

$$\mathbf{B}_1 = s(m\mathbf{A}_1 + n\mathbf{A}_2) + \mathbf{r}(p\mathbf{A}_1 + q\mathbf{A}_2) \tag{6}$$

Substituting (2) and (3) in (6) gives

$$\mathbf{B}_1 = s\mathbf{B}_2 + r\mathbf{B}_3$$

which means that $\mathbf{B}_1$ is a linear combination of $\mathbf{B}_2$ and $\mathbf{B}_3$. Thus $\{\mathbf{B}_1, \mathbf{B}_2, \mathbf{B}_3\}$ is not linearly independent and therefore is not a basis. The general procedure to prove the theorem is now clear. This theorem implies the uniqueness of pivot rank, already established in Section 2.5.

The zero vector never belongs to a basis. The zero vector can always be written as a linear combination of any vectors. (Just use 0 as the scalar for each vector.)

EXAMPLE 8

The set of all linear combinations of any set of vectors is a subspace. Thus the set of all linear combinations of

$$S = \{(1, 2, -1, -2), (3, 4, -4, -3), (5, 6, -6, 10)\}$$

is a subspace of R^4. What is a basis for this subspace of R^4?

The answer to this question is obtained by employing another application of the parametric elimination algorithm. All linear combinations of S and thus the subspace spanned by S are given by

(1)

p	1*	2	−1	−2
q	3	4	−4	−3
r	5	6	−6	10
	$= x_1$	$= x_2$	$= x_3$	$= x_4$

Using the algorithm we arrive at the fully condensed tableau (4):

(2)

x_1	2	−1	−2
q	−2	−1*	3
r	−4	−1	20
	$= x_2$	$= x_3$	$= x_4$

(3)

x_1	4	−5
x_3	2	−3
r	−2*	17
	$= x_2$	$= x_4$

(4)

x_1	29
x_3	14
x_2	$-17/2$
	$= x_4$

Now

x_1	1
x_3	0
x_2	0
	$= x_1$

x_1	0
x_3	0
x_2	1
	$= x_2$

x_1	0
x_3	1
x_2	0
	$= x_3$

are all obviously true. If these columns are added to the explicit form (4) they do not affect the solution set. They provide a new tableau:

(5)

x_1	1	0	0	29
x_3	0	0	1	14
x_2	0	1	0	$-17/2$
	$= x_1$	$= x_2$	$= x_3$	$= x_4$

We now declare that the rows of (5) are a basis for the subspace spanned by S. That

$$B = \{(1, 0, 0, 29), (0, 0, 1, 14), (0, 1, 0, -17/2)\}$$

is a basis is clear from the following

1. Since the subspace is spanned by S and S contains three vectors, a basis can have no more than three vectors.
2. Substitution of any vector of S into (5) gives a true statement. For instance, substituting $(5, 6, -6, 10)$ into (5) yields

(6)

5	1	0	0	29
-6	0	0	1	14
6	0	1	0	$-17/2$
	$= 5$	$= 6$	$= -6$	$= 10$

which is true. This shows that $(5, 6, -6, 10)$ is expressible as a linear combination of the vectors of B. In another format

$$(5, 6, -6, 10) = 5(1, 0, 0, 29) + (-6)(0, 0, 1, 14) + 6(0, 1, 0, -17/2)$$

Since all vectors of S are expressible as linear combinations of the vectors of B and any vector **A** in the subspace is expressible as a linear combination of the vectors of S it follows that **A** is expressible as a linear combination of the vectors of B. Thus B spans the subspace.

3. $(1, 0, 0, 29)$ is not a linear combination of the two other vectors of B. This is true because $(0, 0, 1, 14)$ and $(0, 1, 0, -17/2)$ both have 0 in their first position, and no multiples of 0 can add up to 1, which is in the first position of $(1, 0, 0, 29)$. By the same reasoning, no

linear combination of (1, 0, 0, 29) and (0, 0, 1, 14) can yield (0, 1, 0, $-17/2$). (Look at the second component of each of these vectors.) Also no linear combination of (1, 0, 0, 29) and (0, 1, 0, $-17/2$) can yield (0, 0, 1, 14). Therefore, B is a linearly independent set.

Statements 2 and 3 together provide the result; B is a basis.

A basis of this type in which each basis vector has a 1 as one of its components and every other vector in the basis has a 0 in that position is called an *elementary basis*. Elementary bases are always attainable as in Example 8. Apply the parametric elimination algorithm to a spanning set of a subspace and add on to the explicit tableau a unit column for each non-basic variable. The rows of the resulting tableau form an elementary basis.

In Example 8 the basis B contains three vectors. Thus any spanning set with three vectors is a basis. Therefore, S itself is a basis, which is not clear by observation alone. One value of elementary bases is that they are determinable by visual inspection.

EXAMPLE 9

Find an elementary basis for the subspace spanned by

$$S = \{(1, 2, -1, -2), (3, 4, -4, -3), (4, 6, -5, -5)\}$$

The tableau for the subspace spanned by S and the resulting tableaus using parametric elimination are shown below with (4) as the explicit form. The last row of (3) contains all 0's and therefore is removed.

(1)

p	1*	2	−1	−2
q	3	4	−4	−3
r	4	6	−5	−5
	$= x_1$	$= x_2$	$= x_3$	$= x_4$

(2)

x_1	2	−1	−2
q	−2	−1*	3
r	−2	−1	3
	$= x_2$	$= x_3$	$= x_4$

(3)

x_1	4	-5
x_3	2	-3
r	0	0
	$= x_2$	$= x_4$

(4)

x_1	4	-5
x_3	2	-3
	$= x_2$	$= x_4$

Adding the identity columns

x_1	1
x_3	0
	$= x_1$

and

x_1	0
x_3	1
	$= x_3$

to (4) and rearranging columns gives

(5)

x_1	1	4	0	-5
x_3	0	2	1	-3
	$= x_1$	$= x_2$	$= x_3$	$= x_4$

An elementary basis for the subspace spanned by S is $B = \{(1, 4, 0, -5), (0, 2, 1, -3)\}$ as can be confirmed by the methods of Example 8. S is not a basis since it contains three vectors and a basis contains two vectors. Since S spans the subspace it must contain a basis. We assert that the two rows of (1), $(1, 2, -1, -2)$ and $(3, 4, -4, -3)$, corresponding to the nonzero rows of (3) form a basis. It is a valuable exercise for the reader to confirm this fact. That all bases of a subspace V contain the same number of vectors is important to keep in mind. This number is the *dimension* of V, thus making precise the informal discussion of dimension at the beginning of this section.

The dimension of a vector line is 1; that of a vector plane is 2. A vector subspace of R^n generated by a basis with k vectors is a *k-dimensional* subspace. To round things out we say that the dimension of $\{\mathbf{0}_n\}$ is 0.

We close this section with

THEOREM 5.4. Let S be a vector subspace of R^n with basis $B = [\mathbf{A}_1, \mathbf{A}_2, \ldots, \mathbf{A}_k\}$. Then each vector in S is expressible in exactly one way as a linear combination of the vectors of B.

Proof: Let $\mathbf{C}$ be a vector in S and assume two different ways of expressing $\mathbf{C}$ as a linear combination of the vectors in B:

$$\mathbf{C} = c_1\mathbf{A}_1 + c_2\mathbf{A}_2 + \cdots + c_k\mathbf{A}_k = d_1\mathbf{A}_1 + d_2\mathbf{A}_2 + \cdots + d_k\mathbf{A}_k$$

Then

$$(c_1 - d_1)\mathbf{A}_1 + (c_2 - d_2)\mathbf{A}_2 + \cdots + (c_k - d_k)\mathbf{A}_k = \mathbf{0}_n$$

Since $\{\mathbf{A}_1, \mathbf{A}_2, \ldots, \mathbf{A}_k\}$ is linearly independent,

$$c_1 - d_1 = 0, c_2 - d_2 = 0, \ldots, c_n - d_n = 0$$

or

$$c_1 = d_1, c_2 = d_2, \ldots, c_n = d_n$$

Exercises

1. Determine which of the following sets are linearly independent.
   (a) $\{(3, 5, 2), (6, 10, 6), (0, 0, 2)\}$.
   (b) $\{(-1, 2), (3, 4)\}$.
   (c) $\{(6, -1, 0), (3, 0, -1)\}$\
   (d) $\{(1, 1, -1, 1), (1, 0, 1, 1), (1, 2, 1, 1), (2, 1, -1, 0)\}$.
   (e) $\{(2, 1, -1), (1, 3, 2), (0, 0, 1), (3, 4, 2)\}$.
2. Determine which of the following sets of vectors are a basis for a vector plane.
   (a) $\{(3, 2, 1), (0, 2, 4)\}$.
   (b) $\{(3, 2, 1)\}$.
   (c) $\{(0, 0, 0), (8, 6, 2)\}$.
   (d) $\{(1, 2, 5), (-1, 2, -3), (0, 4, 2)\}$.
3. Let $\mathbf{A} = (4, 2, 1, 6)$ be a vector in R^4. Find a vector $\mathbf{B}$ such that
   (a) $\{\mathbf{A}, \mathbf{B}\}$ is a basis of a vector plane in R^4.
   (b) $\{\mathbf{A}, \mathbf{B}\}$ is not a basis of a vector plane in R^4.
4. Find an elementary basis for each of the following vector subspaces. Also determine a basis which is a subset of the spanning set.

(a)

p	3	2	4
q	1	1	-2
	$= x_1$	$= x_2$	$= x_3$

(b)

p	4	1	2	0
q	-1	3	1	2
r	5	-2	1	-2
	$= x_1$	$= x_2$	$= x_3$	$= x_4$

(c)

p	3	4
q	2	1
r	-1	6
	$= x_1$	$= x_2$

(d)

p	2	1	-4
q	-4	-2	8
	$= x_1$	$= x_2$	$= x_3$

(e)

p	1	1	0	0	1
q	1	1	1	0	0
r	0	1	1	1	0
s	0	0	1	1	1
t	1	0	0	1	1
	$= x_1$	$= x_2$	$= x_3$	$= x_4$	$= x_5$

(f)

p	0	0	1
q	0	1	0
r	1	0	0
	$= x_1$	$= x_2$	$= x_3$

5. Let vector **A** be a basis for a vector line in R^n. Describe a method for finding **B** such that {**A**, **B**} is a basis of a vector plane in R^n.

6. Determine whether or not
(a) {(1, 1, 0), (0, 1, 1), (1, 0, 1)} is a basis of R^3.
(b) {(1, 1, 1, 0), (1, 1, 0, 1), (1, 0, 1, 1), (0, 1, 1, 1)} is a basis of R^4.

7. Discuss this statement: The dimension of a vector space is the number of nonbasic variables in an explicit representation of it.

5.4 Vector Spaces and Subspaces in the Abstract

This section rightfully belongs in a text on abstract algebra. For students who have already studied groups and fields this section places vector spaces in a proper abstract algebraic setting. For students who have not studied abstract algebra we recommend skipping this section and perhaps returning to it as independent study after completing the course. The material in this section may be omitted without affecting the continuity of the rest of the book.

The system that consists of the set of vectors in R^n together with vector addition has many properties. Among the properties of this system, denoted $(R^n, +)$, are

1. The sum of any two vectors in R^n is in R^n.
2. Addition of vectors is *associative:*

$$(\mathbf{A} + \mathbf{B}) + \mathbf{C} = \mathbf{A} + (\mathbf{B} + \mathbf{C})$$

3. Addition of vectors is *commutative:*

$$\mathbf{A} + \mathbf{B} = \mathbf{B} + \mathbf{A}$$

4. There is an *additive* identity, namely $\mathbf{0}_n$ such that for each $\mathbf{A}$ in R^n,

$$\mathbf{A} + \mathbf{0}_n = \mathbf{0}_n + \mathbf{A} = \mathbf{A}$$

5. For each $\mathbf{A}$ in R^n there is an *additive inverse,* also in R^n, designated $-\mathbf{A}$, such that

$$\mathbf{A} + (-\mathbf{A}) = (-\mathbf{A}) + \mathbf{A} = \mathbf{0}_n$$

Any system with these properties is called an *Abelian (commutative) group.* The system consisting of all 2×2 matrices together with matrix addition is an Abelian group. The additive identity is

$$\begin{bmatrix} 0 & 0 \\ 0 & 0 \end{bmatrix}$$

and the additive inverse of

$$\begin{bmatrix} a & b \\ c & d \end{bmatrix} \quad \text{is} \quad \begin{bmatrix} -a & -b \\ -c & -d \end{bmatrix}$$

In Section 5.3 we asserted that the set of all n-tuples together with vector addition and scalar multiplication is a vector space. We now know that $(R^n, +)$, the set of all n-tuples together with vector addition, is an Abelian group. Therefore, we have that it is the scalar multiplication which takes the Abelian group and turns it into a vector space. It is easy to see that scalar multiplication has the following properties for all real numbers k and m and for all vectors $\mathbf{A}$ and $\mathbf{B}$ in R^n:

(a) $k(\mathbf{A} + \mathbf{B}) = k\mathbf{A} + k\mathbf{B}$.
(b) $(k + m)\mathbf{A} = k\mathbf{A} + m\mathbf{A}$.
(c) $k(m\mathbf{A}) = (km)\mathbf{A}$.
(d) $1\mathbf{A} = \mathbf{A}$.

Scalar multiplication of 2×2 matrices also has properties (a), (b), (c), and (d). Now since 2×2 matrices with matrix addition form an Abelian group, then 2×2 matrices with matrix addition and scalar multiplication have the same abstract properties (1)–(5) and (a)–(d) as does $(R^n, +)$ together with scalar multiplication. For this reason the set of all 2×2 matrices (with real numbers for entries) is called an abstract vector space, even though there is no physical resemblance between a 2×2 matrix and a vector. There are many other abstract vector spaces. You are asked to look into this matter in the exercises. We now proceed to develop a formal definition for an abstract vector space. Before doing so we list properties of the vector space R^n. These properties hold for any abstract vector space.

Let $\mathbf{A}$, $\mathbf{B}$, $\mathbf{C}$, $\mathbf{D}$, and $\mathbf{X}$ be any vectors in R^n, and let k and m be any scalars.

1. If $\mathbf{A} = (a_1, a_2, \ldots, a_n)$ and $\mathbf{B} = (b_1, b_2, \ldots, b_n)$, then $\mathbf{A} - \mathbf{B} = (a_1 - b_1, a_2 - b_2, \ldots, a_n - b_n)$.
2. $\mathbf{A} - \mathbf{B} = -(\mathbf{B} - \mathbf{A})$.
3. $k(\mathbf{A} + \mathbf{B}) = k\mathbf{A} + k\mathbf{B}$.
4. $(k + m)\mathbf{A} = k\mathbf{A} + m\mathbf{A}$.
5. $\mathbf{A} - \mathbf{0}_n = \mathbf{A}$.
6. $\mathbf{0}_n - \mathbf{A} = -\mathbf{A}$.
7. $\mathbf{A} - \mathbf{A} = \mathbf{0}_n$.
8. $\mathbf{A} - (\mathbf{B} + \mathbf{C}) = \mathbf{A} - \mathbf{B} - \mathbf{C}$.
9. $\mathbf{A} - (\mathbf{B} - \mathbf{C}) = \mathbf{A} - \mathbf{B} + \mathbf{C}$.
10. $\mathbf{X} + \mathbf{A} = \mathbf{B}$ if and only if $\mathbf{X} = \mathbf{B} - \mathbf{A}$.
11. $\mathbf{B} - \mathbf{A} = \mathbf{D} - \mathbf{C}$ if and only if $\mathbf{B} + \mathbf{C} = \mathbf{A} + \mathbf{D}$.

A proof of property 3 is given. It is typical of the others.

Proof of Property 3: Let

$$\begin{aligned}
\mathbf{A} &= (a_1, a_2, \ldots, a_n) \\
\mathbf{B} &= (b_1, b_2, \ldots, b_n) \\
k(\mathbf{A} + \mathbf{B}) &= k(a_1 + b_1, a_2 + b_2, \ldots, a_n + b_n) \\
&= (ka_1 + kb_1, ka_2 + kb_2, \ldots, ka_n + kb_n) \\
&= k\mathbf{A} + k\mathbf{B}
\end{aligned}$$

The construction of the vector space R^n requires two sets of objects: the set of real numbers R (used as scalars) and the set of n-tuples R^n. The set R, together with addition $(+)$ and multiplication $(\cdot)$, is a system called a *field*, which we denote by $(R, +, \cdot)$. The sets R and $R - \{\mathbf{0}\}$ are Abelian groups with respect to addition and multiplication, and multiplication is distributed over addition: $a \cdot (b + c) = a \cdot b + a \cdot c$. The set of n-tuples R^n is an additive Abelian group. The operation scalar multiplication straddles both the field of real numbers and the group of n-tuples. This operation creates the new structure we call vector space. We abstract the properties of the vector space R^n and find out whether other systems have these same abstract properties. If so, then we call that system a vector space. The definition that follows is abstract. In it the field F is not necessarily R, the field of real numbers, and the Abelian group V is not necessarily R^n.

DEFINITION 5.2. Let $(F, +, \cdot)$ be a field and $(V, +)$ an Abelian group. Let $\times$ be such that if k is in F and $\mathbf{A}$ in V, then $k \times \mathbf{A}$ is in V. $(V, +)$ is said to be a *vector space over* $(F, +, \cdot)$ if for every k and m in F and every $\mathbf{A}$ and $\mathbf{B}$ in V:

1. $k \times (\mathbf{A} + \mathbf{B}) = k \times \mathbf{A} + k \times \mathbf{B}$.
2. $(k + m) \times \mathbf{A} = k \times \mathbf{A} + m \times \mathbf{A}$.
3. $k \times (m \times \mathbf{A}) = (k \cdot m) \times \mathbf{A}$.
4. $1 \times \mathbf{A} = \mathbf{A}$.

The following logical consequences of Definition 5.2 are valid for all models (interpretations) of vector spaces.

THEOREM 5.5. For all k in F and all $\mathbf{A}$ in V:

(a) $0 \times \mathbf{A} = \mathbf{0}$.
(b) $k \times \mathbf{0} = \mathbf{0}$.
(c) $-1 \times \mathbf{A} = -\mathbf{A}$.
(d) $(-k) \times \mathbf{A} = -(k \times \mathbf{A})$.
(e) $k \times \mathbf{A} = \mathbf{0}$ implies that $k = 0$ or $\mathbf{A} = \mathbf{0}$.

(In these statements 0 is the additive identity of F, 1 is its multiplicative identity, and $-k$ is the additive inverse of k. Also $\mathbf{0}$ is the additive identity of V.) Hereafter $k \times \mathbf{A}$ will be written $k\mathbf{A}$. The operation is called *scalar multiplication*. Also $k \cdot m$ will be written km.

Proof of (a):

$$c\mathbf{A} + 0\mathbf{A} = (c + 0)\mathbf{A} \qquad \text{(property 2 in Definition 5.2)}$$
$$= c\mathbf{A} \qquad (\text{in } F,\ c + 0 = c)$$

Thus

$$0\mathbf{A} = \mathbf{0}$$

This is typical of the proof of the other parts.

There is no question that $(R^n, +)$ is indeed a vector space over $(R, +, \cdot)$ since the abstract notions came from $(R^n, +)$. However, as an illustration of the procedure to show that a system is a vector space, we proceed to prove the obvious. (From now on we write R^n is a vector space over R, the operations being understood.) We already know that $(R, +, \cdot)$ is a field and that $(R^n, +)$ is an Abelian group. Scalar multiplication has been so defined that for all k in R and $\mathbf{A}$ in R^n, $k\mathbf{A}$ is in R^n. So the only task left is to verify (1)–(4) of Definition 5.2. We have already proved that

1. $k(\mathbf{A} + \mathbf{B}) = k\mathbf{A} + k\mathbf{B}$.
2. $(k + m)\mathbf{A} = k\mathbf{A} + m\mathbf{A}$.
3. $k(m\mathbf{A}) = (km)\mathbf{A}$.
4. $1\mathbf{A} = \mathbf{A}$.

Hence

THEOREM 5.6. R^n is a vector space over R. (When R is the field we will often just say that R^n is a vector space.)

One other model of a vector space is the set of $m \times n$ matrices over R. Another is the set of binomials of the form $ax + by$ (a, b in R) over R. There are still others in the exercises at the end of the section.

The next definition prepares the way for another view of vector lines and vector planes.

DEFINITION 5.3. Let V be a vector space over F and let W be a nonempty subset of V. Then W is a *vector subspace* of V if and only if $(W, +)$ is a vector space over F. [It is understood that "+" in $(W, +)$ is the operation in $(V, +)$.]

If you are given an arbitrary subset of V, it is usually difficult to decide whether this subset determines a subspace. Fortunately the next theorem gives a convenient shortcut.

THEOREM 5.7. A nonempty subset of an (abstract) vector space V over F is a vector subspace if and only if it is closed under addition of V and under scalar multiplication.

Proof: Let W be the nonempty subset. We prove the "if" part first. Assume that W is closed under addition and scalar multiplication. We first prove that $(W, +)$ is an Abelian group, where $+$ is addition in V. Let **A**, **B**, and **C** be any (not necessarily different) vectors of W.

1. By hypothesis $\mathbf{A} + \mathbf{B}$ is in W.
2. $\mathbf{A} + \mathbf{B} = \mathbf{B} + \mathbf{A}$, since this is a property of $(V, +)$. Similarly,
3. $(\mathbf{A} + \mathbf{B}) + \mathbf{C} = \mathbf{A} + (\mathbf{B} + \mathbf{C})$.
4. By hypothesis $k\mathbf{A}$ is in W for all k in F.

By Theorem 5.5, $0\mathbf{A} = \mathbf{0}$, and $\mathbf{0}$ is in W. Since -1 is in F, $(-1)\mathbf{A}$ is in W. But $(-1)\mathbf{A} = -\mathbf{A}$ by Theorem 5.5 (c). The additive inverse of every vector in W is in W. Therefore, $(W, +)$ is an Abelian group. Furthermore, for all k and l in F and **A** and **B** in W:

1. $k(\mathbf{A} + \mathbf{B}) = k\mathbf{A} + k\mathbf{B}$ since **A** and **B** are in V.
2. $(k + l)\mathbf{A} = k\mathbf{A} + l\mathbf{A}$ since **A** is in V.
3. $k(l\mathbf{A}) = (kl)\mathbf{A}$ since **A** is in V.
4. $1\mathbf{A} = \mathbf{A}$.

We conclude that W is a vector space. To prove the "only if" part, assume that W is a vector subspace. It follows immediately from Definition 5.2 that W is closed under addition and scalar multiplication.

COROLLARY 1

(a) $\{\mathbf{0}\}$ is a vector subspace of V.
(b) V is a vector subspace of V.

COROLLARY 2. If W is a vector subspace of V, then the zero vector of V belongs to W.

Theorem 5.7 is useful in proving the basic theorem that follows.

THEOREM 5.8. Let $A = \{\mathbf{A}_1, \mathbf{A}_2, \ldots, \mathbf{A}_n\}$ be a set of vectors in R^n. Then the set S of all linear combinations of vectors in A is a vector subspace of R^n.

Proof: Let

$$\mathbf{C} = c_1\mathbf{A}_1 + c_2\mathbf{A}_2 + \cdots + c_n\mathbf{A}_n$$

and

$$\mathbf{D} = d_1\mathbf{A}_1 + d_2\mathbf{A}_2 + \cdots + d_n\mathbf{A}_n$$

be any two vectors in S. Then

$$\mathbf{C} + \mathbf{D} = (c_1 + d_1)\mathbf{A}_1 + (c_2 + d_2)\mathbf{A}_2 + \cdots + (c_n + d_n)\mathbf{A}_n$$

is also in S. Moreover, $t\mathbf{C} = tc_1\mathbf{A}_1 + tc_2\mathbf{A}_2 + \cdots + tc_n\mathbf{A}_n$ is in S. Thus S is closed under vector addition and scalar multiplication. Hence S is a vector subspace of R^n.

COROLLARY 1. A vector line in R^n is a vector subspace of R^n.

Proof: This follows from Theorem 5.8, for we need only take $n = 1$ and $\mathbf{A}_1$, a nonzero vector.

COROLLARY 2. A vector plane in R^n is a vector subspace of R^n.

Proof: Take $n = 2$, and $\mathbf{A}_1$ and $\mathbf{A}_2$ in different vector lines.

It is clear now that we can construct other kinds of subspaces of R^n by taking $n = 3, \ldots, n$ and choosing $\mathbf{A}_1, \ldots, \mathbf{A}_n$ properly.

THEOREM 5.9. The sum of two vector subspaces of R^n is a vector subspace of R^n.

Proof: This follows from the fact that the sum of two (or more) vector lines is equivalent to the set of all linear combinations of the spanning vectors of those lines.

THEOREM 5.10. The intersection of two vector subspaces of R^n is a vector subspace of R^n. (*Reminder:* The intersection of two sets is the set of all elements, and only those, that are common to both sets.)

Proof: Let the two subspaces be S_1 and S_2, and let $\mathbf{A}_1$ and $\mathbf{A}_2$ be vectors in both sets, that is, in $S_1 \cap S_2$. Since S_1 and S_2 are vector subspaces, $\mathbf{A}_1 + \mathbf{A}_2$, $c\mathbf{A}_1$ and $c\mathbf{A}_2$ are in each, and hence in $S_1 \cap S_2$. Therefore, $S_1 \cap S_2$ is closed under vector addition and scalar multiplication. So $S_1 \cap S_2$ is a vector subspace of R^n.

Exercises

1. Consider the structures: group, field, vector space.
(a) How many operations are there associated with each?
(b) How is a group related to a field?
(c) How do group and field relate to a vector space?
(d) How many distributive relations are there in a field? In a vector space?
(e) How many associative relations are there in a group? Field? Vector space?

In Exercises 2–4 we continue to let $(V, +)$ represent an Abelian group and $(F, +, \cdot)$ a field.

2. For F take Q, the field of rationals. For V take R. Vector addition and scalar multiplication are performed as usual. Prove that R is a vector space over Q.

3. For F take R. For V take B, the set of binomials of the form $ax + by$, where a and b are in R, and x and y are variables over R. Vector addition is ordinary addition of polynomials. Scalar multiplication follows the rule $k(ax + by) = kax + kby$. Prove that B is a vector space over R.

4. For F take R. For V take the set G of mappings from R to R. For vector addition let $(f_1 + f_2)(x) = f_1(x) + f_2(x), f_1, f_2$ in G. For scalar multiplication, let $[af](x) = a \cdot f(x)$, a in R, f in G. Prove that G is a vector space over R.

5. Which of the following, if any, are vector spaces?
(a) Q over R.
(b) Q over Z (the set of integers).
(c) Q over Q.
(d) Z over Q.

6. Prove
(a) Theorem 5.5 (b).
(b) Theorem 5.5 (d).
(c) Theorem 5.5 (e).

In Exercises 7–12 you are asked to prove theorems about a vector space which resemble cancellation laws. In each exercise, k and l are in R, $\mathbf{A}$ and $\mathbf{B}$ are in R^n, and R^n is the vector space over R.

7. If $k\mathbf{A} = l\mathbf{A}$, and $\mathbf{A} \neq \mathbf{0}$, then $k = l$.

8. If $k\mathbf{A} = k\mathbf{B}$, and $k \neq 0$, then $\mathbf{A} = \mathbf{B}$.

9. If $k\mathbf{A} = l\mathbf{B}$, and $k \neq 0$, $\mathbf{A} \neq \mathbf{0}$, then $l \neq 0$.

10. If $k\mathbf{A} = l\mathbf{B}$, and $k \neq 0$, $\mathbf{A} \neq \mathbf{0}$, then $\mathbf{B} \neq \mathbf{0}$.

11. If $k\mathbf{A} = l\mathbf{B}$, $k \neq 0$, $\mathbf{A} \neq \mathbf{B}$, then $k \neq l$.

12. If $k\mathbf{A} = l\mathbf{B}$, $\mathbf{A} \neq \mathbf{0}$, $k \neq l$, then $\mathbf{A} \neq \mathbf{B}$.

13. By showing a counterexample, prove the statement false: "If $k\mathbf{A} = l\mathbf{B}$, $k \neq 0$, $\mathbf{A} \neq \mathbf{0}$, then $\mathbf{A} = \mathbf{B}$."

14. For each set represented below (i) show that it is a vector subspace of R^n and (ii) describe it geometrically.

(a)
$$M = \left\{ (x, y):\ k \begin{array}{|cc|} \hline 3 & 2 \\ \hline \end{array}\ ;\ k \text{ in } R \right\}$$
$$\qquad\qquad\quad = x \quad = y$$

(b)
$$N = \left\{ (x, y, z):\ \begin{array}{c|ccc|} \hline a & 1 & 2 & 3 \\ b & -1 & 2 & -2 \\ \hline \end{array}\ ;\ a, b \text{ in } R \right\}$$
$$\qquad\qquad\quad = x \quad = y \quad = z$$

(c) $Q = \{\mathbf{X}:\mathbf{X} = a(1, -2, 3) + b(-1, 2, -2) + c(0, 0, 1);\ a, b, c,$ in $R\}$.
(d) $T = \{((a + b), 2a, (3a + b));\ a, b,$ in $R\}$.

15. Prove that $\{\mathbf{0}\}$ is a subspace of R^n.

16. Prove that if a subset S of vector space V does not contain $\mathbf{0}$ of V, then S is not a subspace of V.

17. Let α be a vector line in a vector plane π. Prove that
(a) $\mathbf{0} + \alpha = \alpha + \mathbf{0} = \alpha$.
(b) $\alpha + \alpha = \alpha$.
(c) $\alpha + \pi = \pi$.
(d) $\pi + \pi = \pi$.

18. Let V be the set of all 2×2 matrices whose entries are real numbers.
(a) Prove that V is a vector space over R.
(b) Let W_1 be the subset of V of the form

$$\begin{bmatrix} x & 0 \\ 0 & y \end{bmatrix}$$

Prove that W_1 is a vector subspace of V.
(c) Let W_2 be the subset of matrices of V of the form

$$\begin{bmatrix} a & b \\ -a & c \end{bmatrix}$$

Prove that W_2 is a vector subspace of V.
(d) Are $W_1 + W_2$ and $W_1 \cap W_2$ subspaces of V? Why?

5.5 Chapter Review

A vector line in R^n is defined by

t	a_1	a_2	$\cdots$	a_n
	$= x_1$	$= x_2$	$\cdots$	$= x_n$

Examples are given in R^2 and R^3. A vector plane is defined as the set of all linear combinations of two vectors **A** and **B** that generate two different vector lines. In n dimensions a tableau for a vector plane is

p	a_1	a_2	$\cdots$	a_n
q	b_1	b_2	$\cdots$	b_n
	$= x_1$	$= x_2$	$\cdots$	$= x_n$

Parametric elimination is used to get an explicit form for vector lines and planes.

The concepts of linear dependence, linear independence, basis, elementary basis, and dimension are developed. The pivot condensation algorithm is used to determine linear dependence or linear independence. The parametric elimination algorithm is used to obtain elementary bases for vector subspaces.

Among the theorems of the chapter are:

Theorem 5.2. If a tableau has more rows than columns, then the vectors in its rows form a linearly dependent set. In other words, if a set of vectors S in R^n contains more vectors than the number n of components in a vector, then S is linearly dependent.

Theorem 5.3. All bases of a given vector subspace V have the same number of vectors.

Theorem 5.4. Let S be a vector subspace of R^n with basis $B = \{\mathbf{A}_1, \mathbf{A}_2, \ldots, \mathbf{A}_k\}$. Then each vector in S is expressible in exactly one way as a linear combination of the vectors of B.

The optional Section 5.4 deals with abstract vector spaces, some examples of vector spaces other than R^n, and some properties of abstract vector spaces.

Chapter Review Exercises

1. Find the dimensions of:

(a)

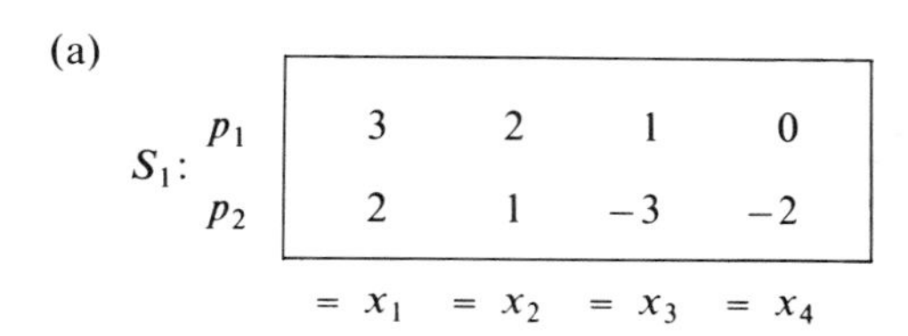

S_1:	3	2	1	0
p_1	3	2	1	0
p_2	2	1	-3	-2
	$= x_1$	$= x_2$	$= x_3$	$= x_4$

(b)

S_2:				
p_3	4	3	5	2
	$= x_1$	$= x_2$	$= x_3$	$= x_4$

(c)

S_3:				
p_4	0	1	0	2
	$= x_1$	$= x_2$	$= x_3$	$= x_4$

(d)

S_4:				
p_5	1	2	3	4
p_6	-1	1	-3	4
p_7	0	1	0	0
	$= x_1$	$= x_2$	$= x_3$	$= x_4$

2. Describe the vector subspace of R^4 whose basis is:
(a) $\{(1, 0, 0, 0)\}$.
(b) $\{(1, 0, 0, 0), (0, 1, 0, 0)\}$.
(c) $\{(1, 0, 0, 0), (0, 1, 0, 0), (0, 0, 1, 0)\}$.
(d) $\{(1, 0, 0, 0), (0, 1, 0, 0), (0, 0, 1, 0), (0, 0, 0, 1)\}$.

3. What is the dimension of $\{(1, 2, 3), (-1, 3, 5), (1, 7, 11), (0, 5, 8), (1, 12, 19)\}$?

4. (a) Why must $\{(1, 0, 0), (0, 1, 0), (0, 0, 1)\}$ be the only elementary basis of R^3?
(b) What conclusions appear to be justifiable about elementary bases for R^n? Why?

***5.** In Example 9 of Section 5.3 we asserted that a basis was extractable from the spanning set S by forming a basis from those vectors corresponding to nonzero rows in the tableaus resulting from parametric elimination. Why is this result true in general? Why is it that if there are no zero rows, then S itself is a basis?

CLA CHAPTER 6

Affine and Euclidean Spaces

In Chapter 5 vector lines, vector planes, and other vector subspaces were analyzed. The terms "vector line" and "vector plane" were justified pictorially. In this chapter we also use pictures to motivate our thinking and justify the use of certain words. It is necessary to keep in mind that the pictures are only motivational. The actual constructive work is with the well-defined manipulations of n-tuples and tableaus.

Figure 6.1 illustrates a line that contains the vectors $\mathbf{A} = (a_1, a_2)$ and $\mathbf{B} = (b_1, b_2)$. This lines does not contain the zero vector and so is not a vector line.

All vector subspaces, including vector lines, contain the zero vector. Therefore, in Chapter 5 there is no discussion of parallelism. The line containing **A** and **B** is parallel to the vector line of Figure 6.2(a). The vector $(\mathbf{B} - \mathbf{A})$ is the fourth vertex of the parallelogram that has **0, A,** and **B** as its three other vertices. This is so because from the geometric notion of addition of vectors (described in Chapter 5) we have $(\mathbf{B} - \mathbf{A}) + \mathbf{A} = \mathbf{B}$.

To describe any vector **X** in the line containing **A** and **B**, construct a

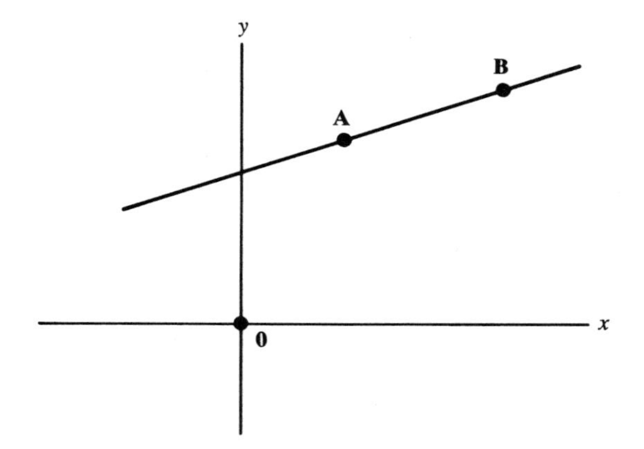

Figure 6.1

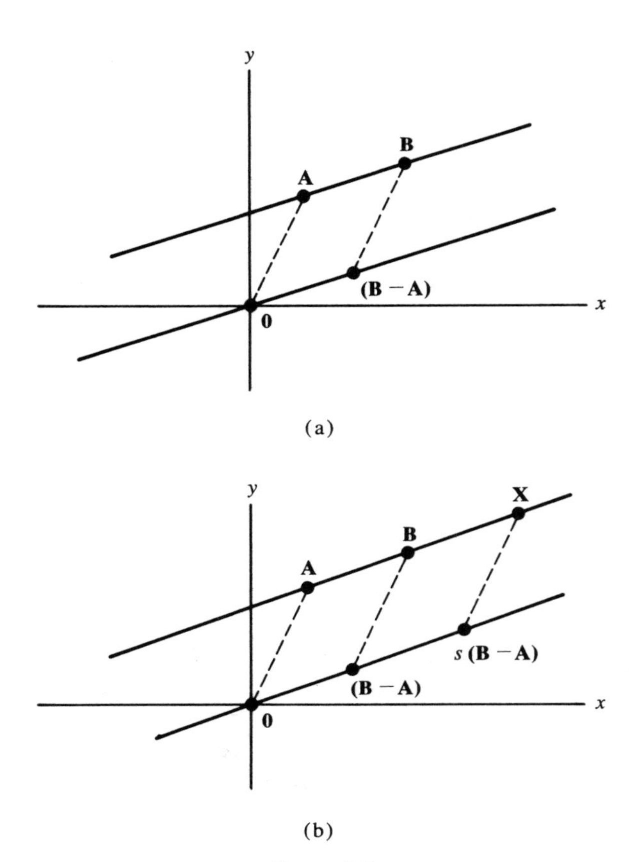

Figure 6.2

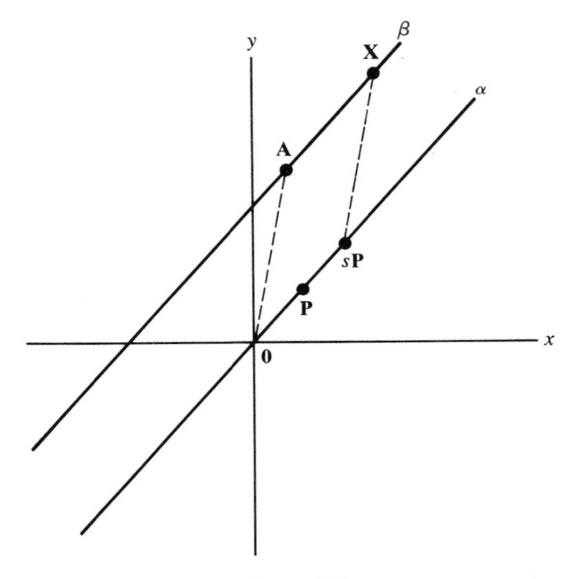

Figure 6.3

side of a parallelogram through **X** that has **0** and **A** as two of the other three vertices. The fourth vertex is in the vector line containing $(\mathbf{B} - \mathbf{A})$ and thus is a multiple of $(\mathbf{B} - \mathbf{A})$, say $s(\mathbf{B} - \mathbf{A})$, where s is a real number.

By vector addition [see Figure 6.2(b)]:

$$s(\mathbf{B} - \mathbf{A}) + \mathbf{A} = \mathbf{X} \tag{1}$$

or, rearranging terms,

$$s\mathbf{B} + (1 - s)\mathbf{A} = \mathbf{X} \tag{2}$$

Any vector **X** in the line which contains both **A** and **B** is described by (2), where s is a real number. Letting $1 - s = t$, (2) is written as

$$s\mathbf{B} + t\mathbf{A} = \mathbf{X}, \qquad s + t = 1 \tag{3}$$

Figure 6.3 illustrates another way to describe lines that are not vector lines. In this figure the given vector line α contains the vector **P**. Vector **A**, not in α, is in a line β parallel to α. All vectors **X** in β are described in terms of **A** and **P** by

$$s\mathbf{P} + \mathbf{A} = \mathbf{X} \tag{4}$$

The vector line α is *translated* by the vector **A** to β. We describe β as the *translate* of α by **A**. In (1), the vector $(\mathbf{B} - \mathbf{A})$ plays the role of **P** in (4).

6.1 Affine Lines

If **A** and **B** are any two distinct vectors not both **0**, then $\beta = \{\mathbf{X}:\mathbf{X} = s\mathbf{A} + t\mathbf{B};\ s + t = 1\}$ is called an *affine line* (generated by **A** and **B**).

EXAMPLE 1

Let $\mathbf{A} = (1, 2)$, $\mathbf{B} = (3, 4)$; then

$$\beta = \{(x_1, x_2) : (x_1, x_2) = s(1, 2) + t(3, 4);\ s + t = 1\} \tag{1}$$

is an affine line in R^2. In tableau form

(2)

β:			¦	
s	1	2	¦	1
t	3	4	¦	1*
	$= x_1$	$= x_2$		$= 1$

The last column, set off by the dashed line, gives the relationship between the parameters.

The affine line β has two parameters, s and t. Vector lines have one parameter. Can affine lines also be described with one parameter? The answer is "yes." To see this, pivot in (2) as indicated. The result is

(3)

β:			¦	
s	-2	-2	¦	-1
1	3	4	¦	1
	$= x_1$	$= x_2$		$= t$

Dropping the last column, which describes t as $t = 1 - s$, gives

(4)

β:		
s	-2	-2
1	3	4
	$= x_1$	$= x_2$

In this form β has one parameter. Substituting $t = 1 - s$ in (1) yields

$$\beta = \{(x_1, x_2) : (x_1, x_2) = s(-2, -2) + (3, 4)\} \tag{5}$$

which is equivalent to (4).

EXAMPLE 2

Let $\mathbf{A} = (-1, 3, -2)$, $\mathbf{B} = (2, 4, -3)$. Write the affine line of R^3 containing **A** and **B** in tableau form with one parameter.

The affine line is

$$\beta = \{(x_1, x_2, x_3) : (x_1, x_2, x_3) = s(-1, 3, -2) + t(2, 4, -3);\ s + t = 1\}$$

Its tableau is

(1)

s	-1	3	-2	1
t	2	4	-3	1^*
	$= x_1$	$= x_2$	$= x_3$	$= 1$

We pivot and eliminate the last column to obtain

(2)

s	-3^*	-1	1
1	2	4	-3
	$= x_1$	$= x_2$	$= x_3$

This is the desired result. Pivoting on (1) to eliminate s instead of t and rearranging rows gives

(3)

t	3^*	1	-1
1	-1	3	-2
	$= x_1$	$= x_2$	$= x_3$

Tableaus (2) and (3) are different. To verify that they represent the same affine line, apply the parametric elimination algorithm to both (2) and (3), pivoting as indicated. From (2) comes the explicit tableau (4):

(4)

x_1	$1/3$	$-1/3$
1	$10/3$	$-7/3$
	$= x_2$	$= x_3$

From (3) comes the explicit tableau (5):

(5)

x_1	$1/3$	$-1/3$
1	$10/3$	$-7/3$
	$= x_2$	$= x_3$

Tableaus (4) and (5) are identical. This tableau is an explicit representation of the affine line β.

EXAMPLE 3

Show that in R^3 the tableaus for the affine line $\beta = \{(x_1, x_2, x_3): (x_1, x_2, x_3) = s(a_1, a_2, a_3) + t(b_1, b_2, b_3); s + t = 1)\}$ are equivalent.

From

(1)

s	a_1	a_2	a_3	1
t	b_1	b_2	b_3	1^*
	$= x_1$	$= x_2$	$= x_3$	$= 1$

we obtain

(2)

s	$a_1 - b_1$	$a_2 - b_2$	$a_3 - b_3$
1	b_1	b_2	b_3
	$= x_1$	$= x_2$	$= x_3$

Also from

(1)

s	a_1	a_2	a_3	1^*
t	b_1	b_2	b_3	1
	$= x$	$= x_2$	$= x_3$	$= 1$

we get

(3)

t	$b_1 - a_1$	$b_2 - a_2$	$b_3 - a_3$
1	a_1	a_2	a_3
	$= x_1$	$= x_2$	$= x_3$

By parametric elimination pivoting on $a_1 - b_1$, we obtain, from (2),

(4)

x_1	$\dfrac{a_2 - b_2}{a_1 - b_1}$	$\dfrac{a_3 - b_3}{a_1 - b_1}$
1	$b_2 - \dfrac{b_1(a_2 - b_2)}{a_1 - b_1}$	$b_3 - \dfrac{b_1(a_3 - b_3)}{a_1 - b_1}$
	$= x_2$	$= x_3$

The reader should, by pivoting on $b_1 - a_1$, show that (3) gives another tableau (5) and then by arithmetic manipulation show that the numbers in the inside of (5) are the same as those of (4). The weakness of this proof lies in the possibility that $b_1 - a_1 = 0$ (which means that $a_l = b_1$). Then division by $b_1 - a_1$ is not allowable and the proof fails. However, this problem is easily overcome, since an identical proof holds if $b_2 - a_2$ is selected as the pivot. If $b_2 - a_2 = 0$ and $b_3 - a_3 = 0$, then $(a_1, a_2, a_3) = (b_1, b_2, b_3)$ and β is just a one-element set that we need not consider.

EXAMPLE 4

Using the result of Example 3, write a one-parameter tableau for the affine line β containing $\mathbf{A} = (3, 2, 4)$ and $\mathbf{B} = (-1, 2, 3)$.

To start, $\mathbf{B} - \mathbf{A} = (-4, 0, -1)$ ($\mathbf{B} - \mathbf{A}$ is not in β). Then from (3) of Example 3 a tableau is

(1)

s	-4	0	-1*
1	3	2	4
	$= x_1$	$= x_2$	$= x_3$

From (2) of Example 3 another tableau is

(2)

s	4	0	1*
1	-1	2	3
	$= x_1$	$= x_2$	$= x_3$

By parametric elimination on (1) we obtain

(3)

x_3	4	0
1	-13	2
	$= x_1$	$= x_2$

and, from (2),

(4)

x_3	4	0
1	-13	2
	$= x_1$	$= x_2$

The explicit representations (3) and (4) are identical, showing that both (1) and (2) are one-parameter tableaus for the affine line β. The reader should show that

(5)

s	-4	0	-1
1	1	2	3
	$= x_1$	$= x_2$	$= x_3$

and

(6)

s	4	0	1
1	3	2	4
	$= x_1$	$= x_2$	$= x_3$

are also one-parameter tableaus for the affine line β.

Let

(1)

$\alpha : s$	a_1	a_2	$\cdots$	a_n
	$= x_1$	$= x_2$	$\cdots$	$= x_n$

be a vector line in R^n. Then the affine line

(2)
$$\beta: \begin{array}{c|cccc|} s & a_1 & a_2 & \cdots & a_n \\ 1 & b_1 & b_2 & \cdots & b_n \\ \hline & = x_1 & = x_2 & \cdots & = x_n \end{array}$$

is a translate of α by $(b_1, b_2, \ldots, b_n)$. (See Figure 6.3.)

EXAMPLE 5

(1)
$$\begin{array}{c|cccc|} s & 1 & 3 & -2 & 6 \\ \hline & = x_1 & = x_2 & = x_3 & = x_4 \end{array}$$

is the tableau of a vector line α in R^4. The affine line β with tableau

(2)
$$\begin{array}{c|cccc|} s & 1 & 3 & -2 & 6 \\ 1 & 5 & 1 & 0 & 3 \\ \hline & = x_1 & = x_2 & = x_3 & = x_4 \end{array}$$

is a translate of α by (5, 1, 0, 3). The vector line α is a translate of α by (0, 0, 0, 0) since α has a tableau

(3)
$$\begin{array}{c|cccc|} s & 1 & 3 & -2 & 6 \\ 1 & 0 & 0 & 0 & 0 \\ \hline & = x_1 & = x_2 & = x_3 & = x_4 \end{array}$$

In the form (3) we see that a vector line is a special kind of affine line.

When an affine line in R^n is written in the one-parameter form

$$\begin{array}{c|cccc|} s & c_1 & c_2 & \cdots & c_n \\ 1 & d_1 & d_2 & \cdots & d_n \\ \hline & = x_1 & = x_2 & \cdots & = x_n \end{array}$$

the vector $\mathbf{C} = (c_1, c_2, \ldots, c_n)$ or a multiple of it is called a *direction vector* of the line. From our examples we know that if an affine line β is generated by **A** and **B**, then **B** − **A** is a direction vector of β [as is **A** − **B** = − 1(**B** − **A**)]. (See Figure 6.2.) In Example 5, (1, 3, −2, 6) is a direction vector of β and α. (−2, −6, 4, −12) is also a direction vector of β and α since (−2, −6, 4, −12) = −2(1, 3, −2, 6).

EXAMPLE 6

Let α be a vector line in R^3 with tableau

(1)

s	2	−3	4
	$= x_1$	$= x_2$	$= x_3$

Then the affine lines with tableaus

(2)

s	2*	−3	4
1	3	5	−4
	$= x_1$	$= x_2$	$= x_3$

and

(3)

s	2*	−3	4
1	5	2	0
	$= x_1$	$= x_2$	$= x_3$

are translates of α by (3, 5, −4) and (5, 2, 0), respectively. But pivoting as indicated on (2) and (3) gives, by parametric elimination (in both cases),

(4)

x_1	$-3/2$	2
1	$19/2$	−10
	$= x_2$	$= x_3$

This shows that the supposedly different affine lines are the same. Therefore, let us give you a word of caution. If two affine lines have the same direction vectors, do not assume that they are distinct before you test them with the parametric elimination algorithm.

The translate of a vector line α by any vector in α is again α.

EXAMPLE 7

(1)

t	-2	-3	1
1	6	9	-3
	$= x_1$	$= x_2$	$= x_3$

is the tableau of an affine line β that is a translate by $(6, 9, -3)$ of the vector line α whose tableau is

(2)

t	-2	-3	1
	$= x_1$	$= x_2$	$= x_3$

Show by parametric elimination that $\alpha = \beta$. The work is left to the reader. Another word of caution: a vector line may have a tableau in the form (1). Parametric elimination, which leads to the explicit form, always tells the story.

All affine lines (including the vector line) in R^n with the same direction vector are called *parallel* to each other. (See Figures 6.2 and 6.3.) If α and β are parallel to each other we write $\alpha \parallel \beta$ (read "α is parallel to β") or $\beta \parallel \alpha$. For any affine lines α, β, δ:

1. $\alpha \parallel \alpha$.
2. If $\alpha \parallel \beta$, then $\beta \parallel \alpha$.
3. If $\alpha \parallel \beta$ and $\beta \parallel \delta$, then $\alpha \parallel \delta$.

EXAMPLE 8

Find the solution set of the system of linear equations in standard form:

(1)

x_1	x_2	x_3	-1	
2	1^*	3	4	$= 0$
3	2	-2	-1	$= 0$

By the methods of Chapter 4 we obtain

(2)

x_1	x_3	-1	
2	3	4	$= -x_2$
-1^*	-8	-9	$= 0$

(3)

	x_3	-1	
	-13	-14	$= -x_2$
	8	9	$= -x_1$

Now (3) is rewritten in the explicit form

(4)

x_3	-8	13
1	9	-14
	$= x_1$	$= x_2$

From (4) we see that $(x_1, x_2, x_3) = (9, -14, 0)$ and $(x_1, x_2, x_3) = (1, -1, 1)$ are two distinct solutions of (1). Also these two solutions (thought of as vectors) generate the affine line whose tableau is

(5)

s	9	-14	0	1
t	1	-1	1	1^*
	$= x_1$	$= x_2$	$= x_3$	1

Pivot as indicated on this affine line to obtain

(6)

s	8	-13	-1^*
1	1	-1	1
	$= x_1$	$= x_2$	$= x_3$

(7)

x_3	-8	13
1	9	-14
	$= x_1$	$= x_2$

Now (7) and (4) are identical and so we have that the solution set to (1) is the affine line with tableau (5). The reader should, by selecting two other solutions of (4), convince himself that (5) indeed represents the affine line which satisfies (1).

A little thought should convince the reader that any system of linear equations whose standard form can be transformed by pivot exchange to an explicit form with one nonbasic variable has as its solution set the set of vectors in an affine line. This line is obtained by selecting any two solutions as generators. Thus the explicit form is the bridge between the parametric form and the standard form.

Exercises

1. Find a parametric representation in two parameters, and then in one parameter, of the affine line generated by:
(a) $(4, 2, 1)$ and $(5, 3, 2)$.
(b) $(2, 5)$ and $(5, 2)$.
(c) $(1, 0, 3, 2)$ and $(6, 1, 2, -1)$.

2. Find an explicit and a standard representation of the affine line in:
(a) Exercise 1(a).
(b) Exercise 1(b).
(c) Exercise 1(c).

3. Given the affine line α generated by $(-1, 2, 4)$ and $(1, 0, 2)$, determine whether or not each of the following vectors is in α.
(a) $(1, 2, 8)$.
(b) $(-1, 4, 10)$.
(c) $(0, 0, 0)$.
(d) What is a direction vector of α?

4. Given $\mathbf{A} = (3, 2, -4)$, find a representation of the affine line containing $\mathbf{A}$ which has direction vector:
(a) $(1, 2, 1)$.
(b) $(-1, 2, 3)$.
(c) $(3, 2, -4)$.

5. Prove that if an affine line in R^n contains vector $\mathbf{A}$ and has direction vector $\mathbf{A}$, then it is a vector line.

6. Show that $\{(3, 2, 1), (1, 2, 3)\}$ and $\{(-1, 2, 5), (2, 2, 2)\}$ generate the same affine line.

7. Prove that the affine lines generated by $\{\mathbf{A}, \mathbf{B}\}$ and $\{n\mathbf{A}, n\mathbf{B}\}$ are parallel if:
(a) $\mathbf{A} = (2, 1), \mathbf{B} = (2, 3)$.
(b) $\mathbf{A} = (1, 2, 3), \mathbf{B} = (-1, 2, -3)$.
(c) $\mathbf{A}$ and $\mathbf{B}$ are distinct vectors in R^n.

8. (a) Recall the geometry proposition: two distinct points determine a line. Formulate the corresponding statement for vectors and affine lines in R^n and prove it.
(b) Recall the geometry proposition (the parallel postulate): through a given point there is one line parallel to a given line. Formulate the corresponding statement for vectors and affine lines in R^n and prove it.

9. Given the vector line α represented by $\mathbf{X} = t\mathbf{A}$, t any real number.

(a) Prove that $\alpha + \mathbf{B}$ is distinct from α if and only if $\mathbf{B}$ is a vector not in α.

(b) Assume that $\mathbf{B}$ is not in α and let α_1 be the translate of α by $\mathbf{B}$. Show that $\alpha_1 + \mathbf{C}$ is distinct from α_1 if and only if $\mathbf{C}$ is not in α_1.

(c) Assume that $\mathbf{C}$ is not in α_1. Let $\alpha_2 = \alpha_1 + \mathbf{C}$. Show that α_1 and α_2 have the same direction vector.

(d) Prove that two affine lines are identical if they have the same direction vector and a vector in common. How does this statement relate to the one in Exercise 8(b)?

10. Given the system S of linear equations

x_1	x_2	x_3	-1	
3	-1	2	4	$= 0$
-2	1	5	2	$= 0$

(a) Find the solution set of S and represent this set as an affine line α in parametric form.

(b) Check that α really represents the solution set of S by use of parametric elimination.

6.2 Affine Planes

Recall from Chapter 5 that vector planes, in fact all vector subspaces, contain the zero vector. We define affine planes in such a way that vector planes are special affine planes and they are the only affine planes containing the zero vector. The word "plane" is used because affine planes in R^3 can be viewed pictorially as in Figure 6.4. If $\mathbf{A}$, $\mathbf{B}$, and $\mathbf{C}$ are vectors and $(\mathbf{B} - \mathbf{A})$ and $(\mathbf{C} - \mathbf{A})$ are linearly independent, then

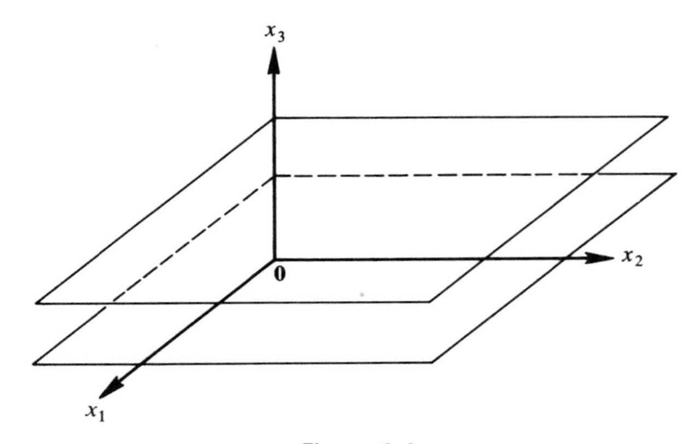

Figure 6.4

$$\pi = \{\mathbf{X}:\mathbf{X} = r\mathbf{A} + s\mathbf{B} + t\mathbf{C};\ r + s + t = 1\}$$

is called an *affine plane* (generated by **A**, **B**, and **C**).

The development of the notion of affine plane follows closely that of affine line and to a great extent is left to the reader. The link among parametric, explicit, and standard form representations is shown in the following examples.

EXAMPLE 1

(1) $$\pi = \{x_1, x_2, x_3) : (x_1, x_2, x_3) = r(1,0,0) + s(0,1,0) + t(0,0,1);\ r + s + t = 1\}$$

is an affine plane in R^3. A tableau for π is

(2)

r	1	0	0	1
s	0	1	0	1
t	0	0	1	1*
	$= x_1$	$= x_2$	$= x_3$	$= 1$

Pivoting as indicated and deleting the last column gives

(3)

r	1*	0	−1
s	0	1	−1
1	0	0	1
	$= x_1$	$= x_2$	$= x_3$

This is the two-parameter form of the affine plane π. The vectors (1, 0, −1) and (0, 1, −1) are *direction vectors* of π.

π_1 is a vector plane with tableau

(4)

r	1	0	−1
s	0	1	−1
	$= x_1$	$= x_2$	$= x_3$

π is a translate of π_1 by (0, 0, 1). π and π_1 are *parallel.*

From (3) by parametric elimination we get

(5)

x_1	0	-1
s	1*	-1
1	0	1
	$= x_2$	$= x_3$

(6)

x_1	-1
x_2	-1
1	1
	$= x_3$

Tableau (6) is the *explicit* form of π. It has two nonbasic variables and one basic variable.

If a tableau for an affine plane π in R^n is

p	c_1	c_2	$\cdots$	c_n
q	d_1	d_2	$\cdots$	d_n
1	e_1	e_2	$\cdots$	e_n
	$= x_1$	$= x_2$	$\cdots$	$= x_n$

then $\mathbf{C} = (c_1, c_2, \ldots, c_n)$, $\mathbf{D} = (d_1, d_2, \ldots, d_n)$, and any linear combinations of $\mathbf{C}$ and $\mathbf{D}$ are *direction vectors* for π. π is a *translate* of π_1 by $(e_1, e_2, \ldots, e_n)$, where π_1 is the vector plane with tableau.

p	c_1	c_2	$\cdots$	c_n
q	d_1	d_2	$\cdots$	d_n
	$= x_1$	$= x_2$	$\cdots$	$= x_n$

EXAMPLE 2

Obtain an explicit form of the affine plane in R^4 generated by $A = (2, 1, 3, -2)$, $B = (8, 7, -9, 8)$, and $C = (5, 4, -3, 3)$. A tableau is

(1)

p	2	1	3	−2	1
q	8	7	−9	8	1
r	5	4	−3	3	1*
	$= x_1$	$= x_2$	$= x_3$	$= x_4$	$= 1$

Pivot to obtain

(2)

p	−3	−3	6	−5
q	3	3*	−6	5
1	5	4	−3	3
	$= x_1$	$= x_2$	$= x_3$	$= x_4$

(3)

p	0	0	0
x_2	1	− 2	$5/3$
1	1	−13	$-11/3$
	$= x_1$	$= x_3$	$= x_4$

From (3) we see that **A**, **B**, and **C** do not generate an affine plane but an affine line. Inspection shows that **B** − **A** = 2(**C** − **A**), which means that (**B** − **A**) and (**C** − **A**) are not linearly independent. Again the explicit form tells the story.

EXAMPLE 3

Show that the solution set of

(1)

x_1	x_2	x_3	x_4	x_5	-1	
1*	0	−1	2	1	2	= 0
0	1	1	−1	2	1	= 0
−1	−1	0	0	2	−2	= 0

is an affine plane of R^5.

We pivot three times to get the explicit form (4) or (5).

(2)

x_2	x_3	x_4	x_5	-1	
0	-1	2	1	2	$= -x_1$
1*	1	-1	2	1	$= 0$
-1	-1	2	3	0	$= 0$

(3)

x_3	x_4	x_5	-1	
-1	2	1	2	$= -x_1$
1	-1	2	1	$= -x_2$
0	1*	5	1	$= 0$

(4)

x_3	x_5	-1	
-1	-9	0	$= -x_1$
1	7	2	$= -x_2$
0	5	1	$= -x_4$

(5)

x_3	1	-1	0
x_5	9	-7	-5
1	0	2	1
	$= x_1$	$= x_2$	$= x_4$

Three distinct solutions of (5) [and therefore of (1)] are (0, 2, 0, 1, 0), $(0, 0, 9, 6, -1)$, and $(8, -4, -1, -4, -1)$. The affine plane is

(6)

r	0	2	0	1	0	1
s	0	0	9	6	-1	1
t	8	-4	-1	-4	1	1
	$= x_1$	$= x_2$	$= x_3$	$= x_4$	$= x_5$	$= 1$

The reader should verify this result by starting with (6) and obtaining (5) as the explicit form.

Exercises

1. Find a parametric representation in three parameters, and another in two parameters, for the affine plane generated by:
(a) $\{(2, 1, 3), (3, 0, 1), (2, -2, 4)\}$.
(b) $\{(1, 2, 5, 1), (3, 3, 2, -1), (5, 4, 0, 4)\}$.
(c) $\{(2, 0, 0), (0, 1, 0), (0, 0, -2)\}$.
(d) $\{(a, 0, 0), (0, b, 0), (0, 0, c)\}$, where a, b, and $c \neq 0$.

2. Find an explicit representation and a standard representation of the plane in:
(a) Exercise 1 (a).
(b) Exercise 1 (b).
(c) Exercise 1 (c).

3. Given the affine plane π generated by $\{(1, -1, 2), (0, 2, 4), (1, 1, 1)\}$, determine whether or not each of the following vectors is in π:
(a) $(2, 2, 7)$.
(b) $(2, 0, 5)$.
(c) $(2, -4, 0)$.

4. Find a parametric, explicit, and standard form representation of the affine plane containing $(2, 5, -1)$ and having direction vectors $(1, 1, 1)$ and $(2, 1, -1)$.

5. Given the affine plane π represented by $2x_1 - 3x_2 + x_3 = 4$, determine which of the following lines are or are not in π.
(a) $\mathbf{X} = t(3, 1, -3) + (2, 1, -3)$.
(b) $\mathbf{X} = t(3, 1, 3) + (2, 1, -3)$.
(c) $\mathbf{X} = t(-3, 1, -3) + (2, 1, 3)$.
(d) $\mathbf{X} = t(2, 3, 5) + (2, 1, 3)$.

6. Given distinct vectors $\mathbf{A}$, $\mathbf{B}$, and $\mathbf{C}$ in R^n, show that all vectors in the affine line generated by $\mathbf{A}$ and $\mathbf{B}$ are in the affine plane generated by $\mathbf{A}$, $\mathbf{B}$, and $\mathbf{C}$.

7. (a) Let α be an affine line and π an affine plane, both in R^3. Devise a definition for "$\alpha \parallel \pi$."
(b) Also devise a definition for "$\alpha \parallel \pi$" if they are in R^4; in R^n.

8. Find a representation of the affine plane π_1 that contains $(1, 2, 3)$ and is parallel to plane π_2 represented by:

(a)

p	-2	1	0
q	3	0	1
1	4	0	0
	$= x_1$	$= x_2$	$= x_3$

(b) Plane π_3 represented by $2x_1 + 3x_2 - x_3 = 5$.

9. Show that the solution set of

x_1	x_2	x_3	x_4	x_5	-1	
1	1	0	0	3	-3	$= 0$
0	-1	-1	1	2	4	$= 0$
-1	0	1	-2	-1	3	$= 0$

is an affine plane of R^5.

6.3 Affine Spaces and Subspaces

If S is any set of vectors of R^n and $\mathbf{A}$ is any vector in R^n, then by $S + \mathbf{A}$ we mean the set of vectors obtained by adding $\mathbf{A}$ to each vector of S. We call this set the $\mathbf{A}$-*translate* of S.

$$S + \mathbf{A} = \{\mathbf{Y} : \mathbf{Y} = \mathbf{X} + \mathbf{A} \text{ for every } \mathbf{X} \text{ in } S\}$$

Thus if

$$S = \{(1, 2, 1), (3, 4, -2), (2, 6, 3)\} \quad \text{and} \quad \mathbf{A} = (1, 1, -1)$$

then

$$S + \mathbf{A} = \{(2, 3, 0), (4, 5, -3), (3, 7, 2)\}$$

If α is a vector line, then $\alpha + \mathbf{A}$ is an affine line. If π is a vector plane, then $\pi + \mathbf{A}$ is an affine plane, the translate of π by $\mathbf{A}$.

EXAMPLE 1

α:	2	-1	3
s	2	-1	3
	$= x_1$	$= x_2$	$= x_3$

is a vector line in R^3. It is a one-dimensional vector subspace of R^3.

If $\mathbf{A} = (1, 3, -4)$, then $\beta = \alpha + \mathbf{A}$ is the affine line with tableau

s	2	-1	3
1	1	3	-4
	$= x_1$	$= x_2$	$= x_3$

and β is a one-dimensional *affine subspace* of R^3.

R^3 is a vector space. It is representable by the tableau

(1)

r	1	0	0
s	0	1	0
t	0	0	1
	$= x_1$	$= x_2$	$= x_3$

If $\mathbf{A} = (a_1, a_2, a_3)$, then $R^3 + \mathbf{A}$ has tableau

(2)

r	1	0	0
s	0	1	0
t	0	0	1
1	a_1	a_2	a_3
	$= x_1$	$= x_2$	$= x_3$

The pivot rank of (2) is 3, as the reader can verify. Thus (2) represents R^3. Therefore, we see that R^3 is also an *affine space*.

EXAMPLE 2

π:

r	2	-1	3	-2	4
s	5	0	0	3	1
	$= x_1$	$= x_2$	$= x_3$	$= x_4$	$= x_5$

is a vector plane in R^5. It is a two-dimensional subspace of the vector space R^5. Its translate by $\mathbf{A} = (3, 1, 4, 1, 1)$ is

π_1:

r	2	-1	3	-2	4
s	5	0	0	3	1
1	3	1	4	1	1
	$= x_1$	$= x_2$	$= x_3$	$= x_4$	$= x_5$

The affine plane π_1 is a two-dimensional affine subspace of the affine space

R^5. We know that the dimensionality of a vector subspace is its pivot rank. The *dimensionality* of an affine subspace which is a translate of the vector subspace is the same as that of the vector subspace. This is clear from the tableaus, since translates merely add a row with a 1 at the left of the tableau. This does not affect the pivot rank.

EXAMPLE 3

S:

r	1	0	0	0	⋯	0
s	0	1	0	0	⋯	0
t	0	0	1	0	⋯	0
u	0	0	0	1	⋯	0
	$= x_1$	$= x_2$	$= x_3$	$= x_4$	⋯	$= x_n$

is a four-dimensional vector subspace of R^n, where n is greater than 4. Thus $S + (1, 1, 1, 1, \ldots, 1)$ is a four-dimensional affine subspace of R^n represented by

r	1	0	0	0	⋯	0
s	0	1	0	0	⋯	0
t	0	0	1	0	⋯	0
u	0	0	0	1	⋯	0
1	1	1	1	1	⋯	1
	$= x_1$	$= x_2$	$= x_3$	$= x_4$	⋯	$= x_n$

In this case if $n = 5$, the vector subspace S is called a *vector hyperplane* of R^5 and the affine subspace an *affine hyperplane* of R^5. In general, for both vector and affine subspaces, if the dimension of the subspace is $n - 1$, it is called a *hyperplane*. Thus lines in R^2 are hyperplanes and planes in R^3 are hyperplanes. Notice that planes in R^4 are not hyperplanes.

If S and T are two sets of vectors in R^n, then $S + T$ is the set of vectors formed by adding each vector in S to every vector in T.

$$S + T = \{\mathbf{X} + \mathbf{Y} : \text{for every } \mathbf{X} \text{ in } S \text{ and every } \mathbf{Y} \text{ in } T\}$$

Thus if

$$S = \{(1, 2), (3, 3)\}, \; T = \{(2, 5), (-1, 2), (1, 4)\}$$

then

$$S + T = \{(3, 7), (0, 4), (2, 6), (5, 8), (2, 5), (4, 7)\}$$

Since in parametric form the tableau of an affine subspace represents addition of vectors, to add two subspaces we need only adjoin these tableaus.

EXAMPLE 4

Let S be a three-dimensional affine subspace of R^4 (a hyperplane) with tableau

(1)

p	2	−4	1	5
q	−5	7	−3	−9
r	8	0	6	−2
1	2	1	2	1
	$= x_1$	$= x_2$	$= x_3$	$= x_4$

Let T be a two-dimensional affine subspace of R^4 (a plane but not a hyperplane) with tableau

(2)

s	5	2	4	−4
t	−9	8	−6	−10
1	2	3	1	1
	$= x_1$	$= x_2$	$= x_3$	$= x_4$

Now $S + T$ has tableau

(3)

p	2	−4	1	5
q	−5	7	−3	−9
r	8	0	6	−2
s	5	2	4	−4
t	−9	8	−6	−10
1	4	4	3	2
	$= x_1$	$= x_2$	$= x_3$	$= x_4$

What kind of subspace is $S + T$? It is not five-dimensional, even though it has five parameters, since it is in R^4 and can have pivot rank at most four. To find the dimension, we need to obtain a basis. We use the methods of

Chapter 5 to obtain an elementary basis. To start, the translate is not involved in determining a basis. (Why?) We consider only the vector subspace with tableau

(4)

p	2	−4	1*	5
q	−5	7	−3	−9
r	8	0	6	−2
s	5	2	4	−4
t	−9	8	−6	−10
	$= x_1$	$= x_2$	$= x_3$	$= x_4$

By parametric elimination, pivoting as indicated, we obtain

(5)

x_3	2	−4	5
q	1*	−5	6
r	−4	24	−32
s	−3	18	−24
t	3	−16	20
	$= x_1$	$= x_2$	$= x_4$

(6)

x_3	6	−7
x_1	−5	6
r	4	−8
s	3	−6
t	−1*	2
	$= x_2$	$= x_4$

(7)

x_3	5
x_1	−4
r	0
s	0
x_2	−2
	$= x_4$

or

(8)

x_3	5
x_1	−4
x_2	−2
	$= x_4$

Now adding unit columns gives

(9)

x_3	0	0	1	5
x_1	1	0	0	−4
x_2	0	1	0	−2
	$= x_1$	$= x_2$	$= x_3$	$= x_4$

From (9) we have that the vector space of (4) is three-dimensional with elementary basis vectors (0, 0, 1, 5), (1, 0, 0, −4), and (0, 1, 0, −2), which are the rows of (9). Thus $S + T$, a translate of (4) by (4, 4, 3, 2) with tableau (3), is also three-dimensional, with the same elementary basis.

We know that an affine subspace is the solution set of a system of linear equations. Therefore, the intersection of two affine subspaces is the intersection of the solution sets of their respective systems of linear equations.

We demonstrate this in Example 5 by obtaining and combining the standard forms for two affine subspaces [tableau (10)]. We then get the explicit form of this intersection [tableau (16)].

EXAMPLE 5

Let U be the affine plane of R^5 with tableau

(1)

p	2*	−4	1	5	−3
q	−5	7	−3	−9	0
1	1	1	−1	2	1
	$= x_1$	$= x_2$	$= x_3$	$= x_4$	$= x_5$

and V the three-dimensional affine subspace of R^5 with tableau

(2)

r	8	0	6	-2	16
s	5	2*	4	-4	0
t	-8	8	-6	0	-32
1	2	1	1	-1	-1
	$= x_1$	$= x_2$	$= x_3$	$= x_4$	$= x_5$

An explicit form (4) for (1) is obtained by pivoting as indicated:

(3)

x_1	-2	$1/2$	$5/2$	$-3/2$
q	-3*	$-1/2$	$7/2$	$-15/2$
1	3	$-3/2$	$-1/2$	$5/2$
	$= x_2$	$= x_3$	$= x_4$	$= x_5$

(4)

x_1	$5/6$	$1/6$	$7/2$
x_2	$1/6$	$-7/6$	$5/2$
1	-2	3	-5
	$= x_3$	$= x_4$	$= x_5$

A standard form for (4) is

(5)

x_1	x_2	x_3	x_4	x_5	-1	
$5/6$	$1/6$	-1	0	0	2	$= 0$
$1/6$	$-7/6$	0	-1	0	-3	$= 0$
$7/2$	$5/2$	0	0	-1	5	$= 0$

An explicit form (8) for (2) is obtained by parametric elimination.

(6)

r	8*	6	-2	16
x_2	$5/2$	2	-2	0
t	-28	-22	16	-32
1	$-1/2$	-1	1	-1
	$= x_1$	$= x_3$	$= x_4$	$= x_5$

(7)

x_1	$3/4$	$-1/4$	2
x_2	$1/8$	$-11/8$	-5
t	-1^*	9	24
1	$-5/8$	$7/8$	0
	$= x_3$	$= x_4$	$= x_5$

(8)

x_1	$13/2$	20
x_2	$-1/4$	-2
x_3	-9	-24
1	$-19/4$	-15
	$= x_4$	$= x_5$

A standard form for (8) is

(9)

x_1	x_2	x_3	x_4	x_5	-1	
$13/2$	$-1/4$	-9	-1	0	$19/4$	$= 0$
20	-2	-24	0	-1	15	$= 0$

The intersection of the affine subspaces U and V with tableaus (1) and (2) is the intersection of the solution sets of (5) and (9): in other words, the set of solutions satisfying both (5) and (9). Combining (5) and (9) gives (10), and so the solution set of (10) is the intersection of the affine subspaces U and V.

(10)

x_1	x_2	x_3	x_4	x_5	-1	
$5/6$	$1/6$	-1^*	0	0	2	$= 0$
$1/6$	$-7/6$	0	-1	0	-3	$= 0$
$7/2$	$5/2$	0	0	-1	5	$= 0$
$13/2$	$-1/4$	-9	-1	0	$19/4$	$= 0$
20	-2	-24	0	-1	15	$= 0$

An explicit form for (10) is (16), obtained from the following sequence of tableaus:

(11)

x_1	x_2	x_4	x_5	-1	
$-5/6$	$-1/6$	0	0	-2	$= -x_3$
$1/6$	$-7/6$	-1^*	0	-3	$= 0$
$7/2$	$5/2$	0	-1	5	$= 0$
-1	$-7/4$	-1	0	$-53/4$	$= 0$
0	-6	0	-1	-33	$= 0$

(12)

x_1	x_2	x_5	-1	
$-5/6$	$-1/6$	0	-2	$= -x_3$
$-1/6$	$7/6$	0	3	$= -x_4$
$7/2$	$5/2$	-1^*	5	$= 0$
$-7/6$	$-7/12$	0	$-41/4$	$= 0$
0	-6	-1	-33	$= 0$

(13)

x_1	x_2	-1	
$-5/6$	$-1/6$	-2	$= -x_3$
$-1/6$	$7/6$	3	$= -x_4$
$-7/2$	$-5/2$	-5	$= -x_5$
$-7/6$	$-7/12^*$	$-41/4$	$= 0$
$-7/2$	$-17/2$	-38	$= 0$

(14)

x_1	-1	
$-1/2$	$13/14$	$= -x_3$
$-5/2$	$-35/2$	$= -x_4$
$3/2$	$545/14$	$= -x_5$
2	$123/7$	$= -x_2$
$27/2^*$	$1554/14$	$= 0$

(15)

-1	
$^{955}/_{189}$	$= -x_3$
$^{590}/_{189}$	$= -x_4$
$^{239}/_{9}$	$= -x_5$
$^{203}/_{189}$	$= -x_2$
$^{1559}/_{189}$	$= -x_1$

(16)

1	$^{1559}/_{189}$	$^{203}/_{189}$	$^{955}/_{189}$	$^{590}/_{189}$	$^{239}/_{9}$
	$= x_1$	$= x_2$	$= x_3$	$= x_4$	$= x_5$

In the form (15) or (16) we see that the solution set to (10) consists of a single vector. To round out the theory we call such a one-element solution set an *affine subspace of dimension zero.* (If the one vector is the zero vector we call the subspace a *vector subspace of dimension zero.*) Thus U intersects V in just one vector.

By direct computation the reader is invited to show that the dimension of $U + V$ is 5 and that in Example 4 the dimension of the intersection of S and T (written $S \cap T$) is 2. These results are summarized in the following table:

	Dimension
S	3
T	2
$S + T$	3
$S \cap T$	2
U	2
V	3
$U + V$	5
$U \cap V$	0

These results illustrate the following theorem.

THEOREM 6.1. Let S and T be any intersecting affine subspaces of R^n with s the dimension of S and t the dimension of T. Let a be the dimension of $S + T$ and b the dimension of $S \cap T$. Then $s + t = a + b$.

Since vector subspaces of R^n are affine subspaces containing the zero vector, this theorem applies to all vector subspaces. In general, if S and T are any sets of vectors, $S \cap T$ is the set of vectors common to both S and T.

We have illustrated many ideas in the preceding sections and now take a few moments to express these ideas in ways that might be overlooked.

1. Vector subspaces are affine subspaces that contain the zero vector.
2. A set of vectors in R^n is a vector subspace if it is closed under linear combinations (Chapter 5). It is an affine subspace if it is closed under linear combinations, where the sum of the scalars is 1.
3. Parametric and standard representations of subspaces (the word "subspace" by itself means that it applies to both vector and affine) are convertible one to the other via their explicit forms. A parametric representation converts to an explicit representation by parametric elimination. A standard representation converts to an explicit representation by pivot reduction.
4. The dimension of a subspace is the number of nonbasic variables in the explicit representation.
5. Every row (except the row of the 1) in a parametric representation of a subspace represents a line. Thus a subspace is a sum of lines. The dimension of the subspace is the number of lines in the sum with linearly independent generating vectors.
6. Each equation in standard form has an explicit form with $n - 1$ nonbasic variables. Thus the solution set for an equation [by (4)] is a hyperplane. The solution set for a system of linear equations is the intersection of the solution sets of the individual equations. Therefore, the solution set for a system of linear equations is the intersection of hyperplanes. The number of distinct hyperplanes that form the intersection is equal to the number of basic variables in the explicit representation.
7. The sum of two subspaces is the subspace of their adjoined parametric representations.
8. The intersection of two subspaces is the subspace of the solution set of their adjoined standard forms.

Exercises

1. Given the affine line

α:

t	3	-1	2
1	0	2	3
	$= x_1$	$= x_2$	$= x_3$

and the affine line

β:

r	4	-2	8
1	1	2	1
	$= x_1$	$= x_2$	$= x_3$

find a representation for and describe the nature of: (a) $\alpha + \beta$; (b) $\alpha \cap \beta$.

2. Let λ be the affine line

s	2	-1	0
1	1	2	1
	$= x_1$	$= x_2$	$= x_3$

and α and β the affine lines in Exercise 1. Find a representation and describe the nature of:

(a) $\alpha + \lambda$.
(b) $\beta + \lambda$.
(c) $\alpha + \beta + \lambda$.
(d) $\alpha \cap \lambda$.
(e) $\beta \cap \lambda$.
(f) $\alpha \cap \beta \cap \lambda$.

3. Given affine plane

π:

p	1	2	3
q	0	-1	-2
1	2	1	-1
	$= x_1$	$= x_2$	$= x_3$

and affine line

α:

r	1	0	-1
1	2	1	-1
	$= x_1$	$= x_2$	$= x_3$

represent and describe:
(a) $\alpha + \pi$.
(b) $\alpha \cap \pi$.
(c) Find dimensions of α, π, $\alpha + \pi$, and $\alpha \cap \pi$ and verify Theorem 6.1.

4. Given affine line

β:

t	1	1	1
1	3	-1	2
	$= x_1$	$= x_2$	$= x_3$

and plane π in Exercise 3, represent and describe:
(a) $\beta + \pi$.
(b) $\beta \cap \pi$.
(c) Verify that Theorem 6.1 does not apply to β and π. Why not?

5. Show that

α:

p	1	2	3
1	1	-1	2
	$= x_1$	$= x_2$	$= x_3$

and β:

q	1	4	1
1	-1	1	2
	$= x_1$	$= x_2$	$= x_3$

have no vectors in common and are not parallel. Such affine lines are called *skew* lines.

6. Given affine plane

π:

p	1	-1	0	2
q	2	0	1	-1
1	1	2	3	4
	$= x_1$	$= x_2$	$= x_3$	$= x_4$

and affine line

α:				
r	0	2	1	-5
1	4	3	2	1
	$= x_1$	$= x_2$	$= x_3$	$= x_4$

(a) Show that $\alpha + \pi$ is a two-dimensional affine subspace.
(b) Show that $\alpha \cap \pi = \phi$, a zero-dimensional affine subspace.
(c) Verify that Theorem 6.1 is not valid for α and π. Explain.

7. Given affine hyperplane

H:				
p	1	0	0	1
q	0	1	1	1
r	0	1	1	0
1	1	2	3	4
	$= x_1$	$= x_2$	$= x_3$	$= x_4$

and affine plane

π:				
s	1	1	1	2
t	3	-1	0	-1
1	2	1	3	4
	$= x_1$	$= x_2$	$= x_3$	$= x_4$

Find the dimensions of H, π, $H + \pi$, and $H \cap \pi$ and verify Theorem 6.1.

6.4 Inner Products and Euclidean Spaces

The purpose of this section is to investigate what happens to the vector space R^n (and to its vector subspaces) when we superimpose on it an operation called *inner product*. Inner product is needed to discuss distance and perpendicularity for R^2 and R^3, and as usual we extend our definitions to R^n. We motivate our discussion with pictures, but the true algebraic work of this section begins with Definition 6.1.

An economist, studying relations between demand, d, and supply, s, records the data as time passes and obtains a set of ordered pairs of numbers (d, s). He then plots them as points on a graph in an attempt to discover relations between them. For this purpose he may use axes that are perpendicular to each other and calibrated with the same unit length. These perpendicular axes form a *rectangular coordinate system* in two dimensions. [See Figure 6.5(a).] The unit length on the d axis is OI and on the s axis OJ. The notion of rectangular coordinate system extends to three dimensions, where we have three mutually perpendicular axes whose base points O, I, J, and K are such that $OI = OJ = OK = 1$. [See Figure 6.5(b).]

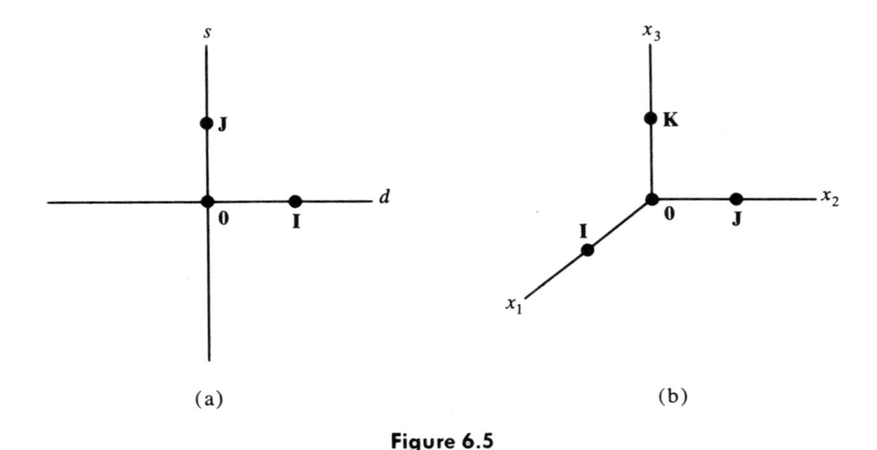

Figure 6.5

Similarly, even though we cannot draw pictures, we conceive of an n-dimensional rectangular coordinate system as one with n mutually perpendicular axes, all intersecting at a common point O and all marked off with a unit distance.

The definition of inner product, which follows, is motivated by the following consideration. In Figure 6.6(a) we wish to find the distance from **A** to **B**. Figure 6.6(b) shows the point **C** which forms a right triangle with **A** and **B**. Then by the Pythagorean theorem the distance d from **A** to **B** is given by

$$\begin{aligned} d &= \sqrt{(b_1 - a_1)^2 + (b_2 - a_2)^2} \\ &= \sqrt{(b_1 - a_1)(b_1 - a_1) + (b_2 - a_2)(b_2 - a_2)} \end{aligned}$$

Why this distance formula motivates Definition 6.1 will be self-explanatory.

DEFINITION 6.1. Let

$$\mathbf{A} = (a_1, a_2, \ldots, a_n) \qquad \text{and} \qquad \mathbf{B} = (b_1, b_2, \ldots, b_n)$$

be vectors in R^n. The *inner product* of **A** and **B**, denoted $\mathbf{A}\cdot\mathbf{B}$ (read **A** dot **B**), is $a_1b_1 + a_2b_2 + \cdots + a_nb_n$.

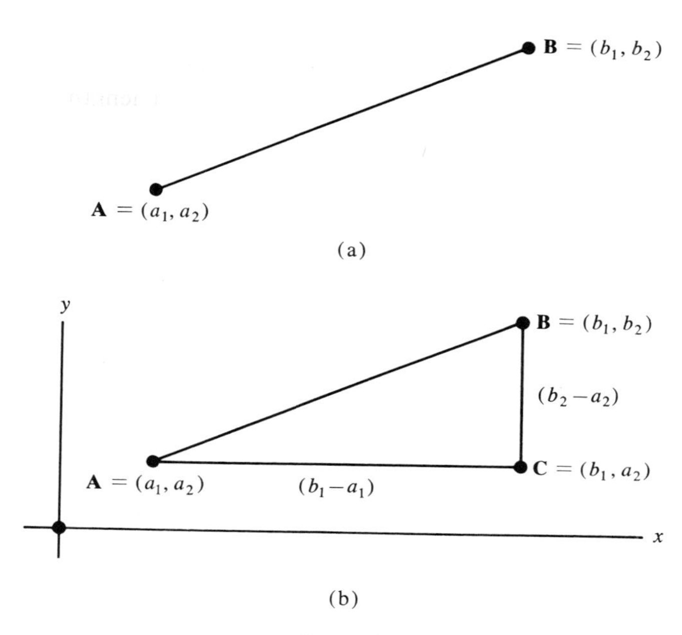

Figure 6.6

EXAMPLE 1

(a) Let $\mathbf{X} = (3, 4)$ and $\mathbf{Y} = (2, 7)$; then $\mathbf{X}\cdot\mathbf{Y} = 3\cdot 2 + 4\cdot 7 = 34$.
(b) Let $\mathbf{A} = (1, 3, 3)$, $\mathbf{B} = (2, 0, -4)$, and $\mathbf{C} = (2, 1, -1)$.

$$\mathbf{A}\cdot\mathbf{B} = 1\cdot 2 + 3\cdot 0 + 3(-4) = -10$$
$$\mathbf{A}\cdot\mathbf{C} = 1\cdot 2 + 3\cdot 1 + 3(-1) = 2$$
$$\mathbf{B}\cdot\mathbf{C} = 2\cdot 2 + 0\cdot 1 - 4(-1) = 8$$

(c) Let $\mathbf{C} = (1, 2, 3, 4)$ and $\mathbf{D} = (1, -1, 2, -2)$.

$$\mathbf{C}\cdot\mathbf{D} = 1\cdot 1 + 2(-1) + 3\cdot 2 + 4(-2) = -3$$
$$\mathbf{C}\cdot\mathbf{0}_4 = 1\cdot 0 + 2\cdot 0 + 3\cdot 0 + 4\cdot 0 = 0$$
$$\mathbf{D}\cdot\mathbf{C} = 1\cdot 2 - 1(2) + 2(3) - 2(4) = -3 = \mathbf{C}\cdot\mathbf{D}$$

(d) Let $\mathbf{X} = (x_1, x_2, x_3)$ and $\mathbf{A} = (3, 4, -2)$; then $\mathbf{A}\cdot\mathbf{X} = 3x_1 + 4x_2 - 2x_3$.

(e) Let $\mathbf{A} = (2, 4, 1)$ and $\mathbf{B} = (1, 2)$; then $\mathbf{A}\cdot\mathbf{B}$ is not defined since $\mathbf{A}$ is in R^3 and $\mathbf{B}$ is in R^2.

In Definition 6.1, $\mathbf{A}\cdot\mathbf{B}$ is a scalar, *not* a vector. The inner product definition assigns to each pair of vectors in R^n a scalar. (An inner product is also called a *dot product* and a *scalar product*. However, we have reserved "scalar product" to describe the product of a scalar and a vector, and therefore will not use it in this connection.) When the n-dimensional vector space R^n has this inner product defined on it, then it is called *Euclidean n-space* and instead of writing R^n we write E^n. Since all the theorems that we have proved up to this point are independent of the inner product definition, they hold for E^n as well as for R^n.

Recall that $a_1x_1 + a_2x_2 + a_3x_3 = a_0$ is a standard representation for a plane in R^3 with $\mathbf{A} = (a_1, a_2, a_3)$ and $\mathbf{X} = (x_1, x_2, x_3)$. Thus the plane equation is $\mathbf{A}\cdot\mathbf{X} = a_0$.

Similarly, an equation, in standard form, for any hyperplane $a_1x_1 + a_2x_2 + \cdots + a_nx_n = a_0$ of R^n is $\mathbf{A}\cdot\mathbf{X} = a_0$, where $\mathbf{A} = (a_1, a_2, \ldots, a_n)$ and $\mathbf{X} = (x_1, x_2, \ldots, x_n)$.

The following theorem, verifiable by the reader, lists basic properties of inner products.

THEOREM 6.2. For all vectors, $\mathbf{A}$, $\mathbf{B}$, $\mathbf{C}$, and $\mathbf{D}$ in E^n and for all scalars r,

(a) $\mathbf{A}\cdot\mathbf{B} = \mathbf{B}\cdot\mathbf{A}$ (commutativity).
(b) $\mathbf{A}\cdot(\mathbf{B} + \mathbf{C}) = \mathbf{A}\cdot\mathbf{B} + \mathbf{A}\cdot\mathbf{C}$ (distributivity).
(b^1) $(\mathbf{A} + \mathbf{B})\cdot(\mathbf{C} + \mathbf{D}) = \mathbf{A}\cdot\mathbf{C} + \mathbf{A}\cdot\mathbf{D} + \mathbf{B}\cdot\mathbf{C} + \mathbf{B}\cdot\mathbf{D}$.
(c) $(r\mathbf{A})\cdot\mathbf{B} = r(\mathbf{A}\cdot\mathbf{B}) = \mathbf{A}\cdot(r\mathbf{B})$.
(d) $(-\mathbf{A})\cdot\mathbf{B} = -(\mathbf{A}\cdot\mathbf{B}) = \mathbf{A}\cdot(-\mathbf{B})$.
(e) $(-\mathbf{A})\cdot(-\mathbf{B}) = \mathbf{A}\cdot\mathbf{B}$.

$\mathbf{A}\cdot\mathbf{B}\cdot\mathbf{C}$ has no meaning since $(\mathbf{A}\cdot\mathbf{B})$ is a scalar and Definition 6.1 has nothing to say about an inner product of a scalar and a vector. However, $(\mathbf{A}\cdot\mathbf{B})\,\mathbf{C}$ is a scalar multiple of $\mathbf{C}$ and is quite different from $(\mathbf{B}\cdot\mathbf{C})\,\mathbf{A}$, a scalar multiple of $\mathbf{A}$.

EXAMPLE 2

If $\mathbf{A} = (3, 1, -1)$, $\mathbf{B} = (2, 2, 2)$, and $\mathbf{C} = (-1, 1, 4)$, then $\mathbf{A}\cdot\mathbf{B} = 6$ and $\mathbf{B}\cdot\mathbf{C} = 8$. Therefore,

$$(\mathbf{A}\cdot\mathbf{B})\,\mathbf{C} = 6(-1, 1, 4) = (-6, 6, 24)$$

and

$$(\mathbf{B}\cdot\mathbf{C})\,\mathbf{A} = 8(3,\ 1,\ -1) = (24,\ 8,\ -8)$$

If $\mathbf{A} = (a_1, a_2, a_3)$, then

$$\mathbf{A}\cdot\mathbf{A} = a_1^2 + a_2^2 + a_3^2$$

or

$$\mathbf{A}\cdot\mathbf{A} = (a_1 - 0)^2 + (a_2 - 0)^2 + (a_3 - 0)^2$$

Thus

$$\sqrt{\mathbf{A}\cdot\mathbf{A}} = \sqrt{(a_1 - 0)^2 + (a_2 - 0)^2 + (a_3 - 0)^2}$$

which is the distance from $\mathbf{0}_3$ to $\mathbf{A}$. (See Figure 6.6.)

For any $\mathbf{A}$ in E^n when $n = 2$ or $n = 3$, $\sqrt{\mathbf{A}\cdot\mathbf{A}}$ is the distance of $\mathbf{A}$ from the origin. If $n > 3$, then the notion of distance has no meaning, but certainly $\sqrt{\mathbf{A}\cdot\mathbf{A}}$ still can be computed. We call $\sqrt{\mathbf{A}\cdot\mathbf{A}}$ the *norm* of $\mathbf{A}$ and write

$$\|\mathbf{A}\| = \sqrt{\mathbf{A}\cdot\mathbf{A}}$$

or

$$\|\mathbf{A}\|^2 = \mathbf{A}\cdot\mathbf{A}$$

EXAMPLE 3

If $\mathbf{A} = (1,\ -1,\ 2)$, then

$$\|\mathbf{A}\| = \sqrt{(1)^2 + (-1)^2 + (2)^2} = \sqrt{6}$$

If $\mathbf{I} = (1,\ 0,\ 0,\ 0)$, then

$$\|\mathbf{I}\| = \sqrt{1^2 + 0^2 + 0^2 + 0^2} = 1$$

If $\mathbf{B} = (1/2,\ -1/2,\ 1/\sqrt{2},\ 0,\ 0)$, then

$$\|\mathbf{B}\| = \sqrt{1/4 + 1/4 + 1/2} = 1$$

A vector, such as $\mathbf{I}$ and $\mathbf{B}$ above, whose norm is 1 is called a *unit vector*.

EXAMPLE 4

If $\mathbf{A} = (2,\ 3,\ 4)$, then

$$\|\mathbf{A}\| = \sqrt{4 + 9 + 16} = \sqrt{29}$$

Now

$$\frac{1}{\sqrt{29}}(2, 3, 4) = \left(\frac{2}{\sqrt{29}}, \frac{3}{\sqrt{29}}, \frac{4}{\sqrt{29}}\right)$$

$$\left\|\left(\frac{2}{\sqrt{29}}, \frac{3}{\sqrt{29}}, \frac{4}{\sqrt{29}}\right)\right\| = \sqrt{4/29 + 9/29 + 16/29} = \sqrt{1} = 1$$

Verify for all $\mathbf{A} \neq \mathbf{0}_n$ that $\mathbf{A}/\|\mathbf{A}\|$ is a unit vector. Also verify that for any vector $\mathbf{A}$ in E^n and any scalar t:

$$\|t\mathbf{A}\| = |t|\ \|\mathbf{A}\| \qquad (|t| \text{ the absolute value of } t)$$

In two or three dimensions the distance from $\mathbf{0}$ to $\mathbf{B} - \mathbf{A}$ is the same as the distance from $\mathbf{A}$ to $\mathbf{B}$. (See Figure 6.2.) Thus in two or three dimensions, $\|\mathbf{B} - \mathbf{A}\| = \|\mathbf{A} - \mathbf{B}\|$ is the distance from $\mathbf{A}$ to $\mathbf{B}$ (or $\mathbf{B}$ to $\mathbf{A}$). If

$$\mathbf{A} = (a_1, a_2, \ldots, a_n) \qquad \text{and} \qquad \mathbf{B} = (b_1, b_2, \ldots, b_n)$$

Then

$$\mathbf{A} - \mathbf{B} = (a_1 - b_1, a_2 - b_2, \ldots, a_n - b_n)$$

and

$$\|\mathbf{A} - \mathbf{B}\| = \sqrt{(a_1 - b_1)^2 + (a_2 - b_2)^2 + \cdots + (a_n - b_n)^2}$$

This is the plane distance formula,

$$\sqrt{(a_1 - b_1)^2 + (a_2 - b_2)^2}$$

when $n = 2$, and the three-dimensional space distance formula,

$$\sqrt{(a_1 - b_1)^2 + (a_2 - b_2)^2 + (a_3 - b_3)^2}$$

when $n = 3$.

EXAMPLE 5

How far apart are $\mathbf{A} = (1, 2, 4, -1)$ and $\mathbf{B} = (0, -1, 2, -3)$? The distance between $\mathbf{A}$ and $\mathbf{B}$ is

$$\|\mathbf{B} - \mathbf{A}\| = \sqrt{(0 - 1)^2 + (-1 - 2)^2 + (2 - 4)^2 + (-3 - (-1))^2}$$

$$= \sqrt{18}$$

Exercises

1. Let $\mathbf{A} = (2, -4, 2)$, $\mathbf{B} = (2, 1, 1)$, and $\mathbf{C} = (1, -1, -3)$.
(a) Calculate $\mathbf{A}\cdot\mathbf{B}$, $\mathbf{A}\cdot\mathbf{C}$, and $\mathbf{B}\cdot\mathbf{C}$.
(b) Express as a triple $(\mathbf{A}\cdot\mathbf{B})\,\mathbf{C}$.
(c) Express as a triple $(\mathbf{A}\cdot\mathbf{B})\,\mathbf{C} + (\mathbf{A}\cdot\mathbf{C})\,\mathbf{B}$.

2. (a) Determine k so that $(k, 2, 8)\cdot(5, -k, 6) = 0$.
(b) Determine the values of x for which $(x, x - 1, x, 0)\cdot(3x, x, 3, 1) = 0$.

3. Let $\mathbf{A} = (1, 1, -2)$ and $\mathbf{B} = (2, -1, -1)$.
(a) Calculate $\|\mathbf{A}\|$ and $\|\mathbf{B}\|$.
(b) Calculate $\mathbf{A}\cdot\mathbf{B}$.

4. Find the length of the vector:
(a) $(1, 2, 2)$.
(b) $(3, 4, 12)$.
(c) $(a, a, a) - (0, 1, -1)$.

5. Using the definition of inner product, prove that: for $\mathbf{A}$ and $\mathbf{B}$ in E^n,
(a) $\mathbf{A}\cdot\mathbf{B} = \mathbf{B}\cdot\mathbf{A}$.
(b) $(-\mathbf{A})\cdot\mathbf{B} = -(\mathbf{A}\cdot\mathbf{B}) = \mathbf{A}\cdot(-\mathbf{B})$.
(c) $(-\mathbf{A})\cdot(-\mathbf{B}) = \mathbf{A}\cdot\mathbf{B}$.

6. Prove that for $\mathbf{A}$, $\mathbf{B}$, and $\mathbf{C}$ in E^n, $\mathbf{A}\cdot(\mathbf{B} + \mathbf{C}) = \mathbf{A}\cdot\mathbf{B} + \mathbf{A}\cdot\mathbf{C}$.

7. Prove that for $\mathbf{A}$ and $\mathbf{B}$ in E^n and for r any real number, $(r\mathbf{A})\cdot\mathbf{B} = r(\mathbf{A}\cdot\mathbf{B}) = \mathbf{A}\cdot(r\mathbf{B})$.

8. A cube has vertices at $\mathbf{0} = (0,0,0)$, $\mathbf{I} = (1,0,0)$, $\mathbf{J} = (0,1,0)$, and $\mathbf{K} = (0, 0, 1)$ (see the figure).

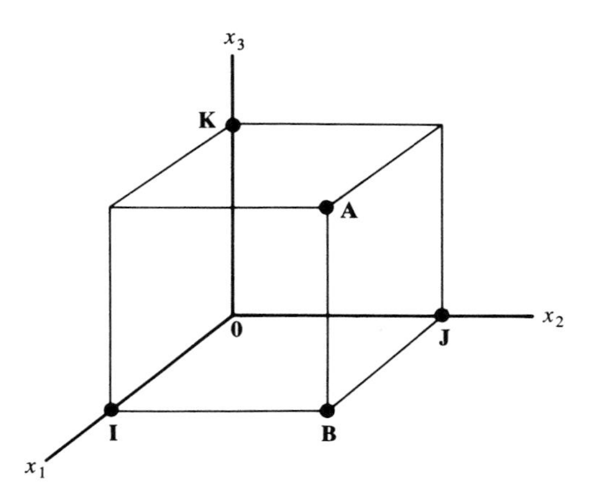

(a) Represent $\mathbf{A}$ and $\mathbf{B}$ as triples.
(b) Calculate the lengths of $\mathbf{0A}$ and $\mathbf{0B}$.

9. If **A** is any vector in E^3, and **I**, **J**, and **K** are as given in Exercise 8, prove that $\mathbf{A} = (\mathbf{A}\cdot\mathbf{I})\mathbf{I} + (\mathbf{A}\cdot\mathbf{J})\mathbf{J} + (\mathbf{A}\cdot\mathbf{K})\mathbf{K}$.

10. Let **A** be a vector E^n. Prove that:
(a) $\mathbf{A}\cdot\mathbf{A} \geq 0$.
(b) If $\mathbf{A}\cdot\mathbf{A} = 0$, $\mathbf{A} = \mathbf{0}$.

11. By the law of cosines the cosine of the angle θ between two vector lines is given by $\cos\theta = (\mathbf{A}\cdot\mathbf{B}/\|\mathbf{A}\|\ \|\mathbf{B}\|)$, where **A** and **B** are generators of the lines. Find the cosine of the angle θ between

α: r	1	2	-1	2
	$= x_1$	$= x_2$	$= x_3$	$= x_4$

and

β: s	1	-1	-1	0
	$= x_1$	$= x_2$	$= x_3$	$= x_4$

6.5 Orthogonal Vectors

Again we appeal to a pictorial representation to motivate our use of words.

In Figure 6.7 suppose that **B** is at a vertex of an isoceles triangle with **A** and $-\mathbf{A}$ at the other two vertices. The distance from **A** to **B** is $\|\mathbf{B} - \mathbf{A}\|$

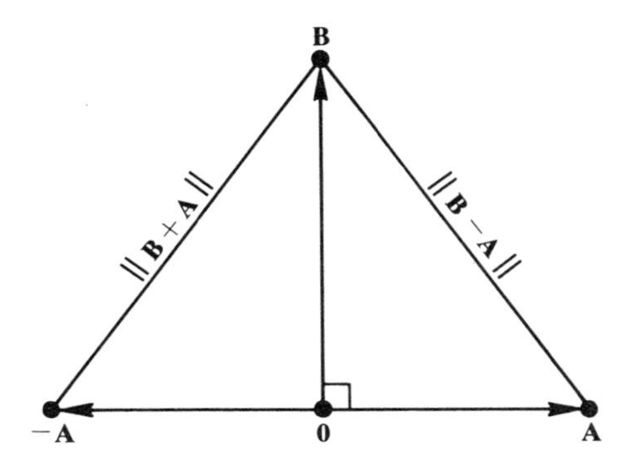

Figure 6.7

and the distance from $-\mathbf{A}$ to **B** is $\|\mathbf{B} - (-\mathbf{A})\| = \|\mathbf{B} + \mathbf{A}\|$. Since the triangle is isoceles, these two distances are equal. Thus

$$\|\mathbf{B} + \mathbf{A}\| = \|\mathbf{B} - \mathbf{A}\|$$

or, upon squaring both sides,

$$(\mathbf{B} + \mathbf{A})\cdot(\mathbf{B} + \mathbf{A}) = (\mathbf{B} - \mathbf{A})\cdot(\mathbf{B} - \mathbf{A})$$

Thus

$$\mathbf{B}\cdot\mathbf{B} + 2\mathbf{A}\cdot\mathbf{B} + \mathbf{A}\cdot\mathbf{A} = \mathbf{B}\cdot\mathbf{B} - 2\mathbf{A}\cdot\mathbf{B} + \mathbf{A}\cdot\mathbf{A}$$

yielding

$$4\mathbf{A}\cdot\mathbf{B} = 0 \qquad \text{or} \qquad \mathbf{A}\cdot\mathbf{B} = 0$$

Contrary to our usual notion of multiplication, an inner product equal to zero does not imply that either of the vectors is zero (although that is possible) but rather that in two or three dimensions their vector lines are perpendicular. In dimensions greater than three, perpendicularity has no geometric meaning but certainly $\mathbf{A}\cdot\mathbf{B} = 0$ is possible. We coin another word for this condition.

Vectors $\mathbf{A}$ and $\mathbf{B}$ in E^n are called *orthogonal* if and only if $\mathbf{A}\cdot\mathbf{B} = 0$. The vectors may be zero vectors. It follows that $\mathbf{0}_n$ and any vector in E^n are orthogonal.

THEOREM 6.3. Let $\mathbf{A}$ and $\mathbf{B}$ be nonzero vectors in E^n such that $\mathbf{A}\cdot\mathbf{B} = 0$. If $\mathbf{C}$ is in the vector line generated by $\mathbf{A}$ and $\mathbf{D}$ is in the vector line generated by $\mathbf{B}$, then $\mathbf{C}\cdot\mathbf{D} = 0$.

Proof: Since $\mathbf{C}$ is in the vector line generated by $\mathbf{A}$, there exists a scalar t such that $\mathbf{C} = t\mathbf{A}$. Similarly there exists a scalar s such that $\mathbf{D} = s\mathbf{B}$. Then $\mathbf{C}\cdot\mathbf{D} = (t\mathbf{A})\cdot(s\mathbf{B}) = (ts)(\mathbf{A}\cdot\mathbf{B}) = (ts)(0) = 0$.

Theorem 6.3 defines orthogonality between vector lines. Two vector lines are *orthogonal* if every vector in one is orthogonal to every vector in the other. Two affine lines are *orthogonal* if their corresponding vector lines are orthogonal. Two affine lines are orthogonal (whether they intersect or not) if their direction vectors are orthogonal.

EXAMPLE 1

α:					
	s	1	2	0	-3
	1	3	1	0	-2
		$= x_1$	$= x_2$	$= x_3$	$= x_4$

β:					
	t	0	3	4	2
	1	4	0	0	3
		$= x_1$	$= x_2$	$= x_3$	$= x_4$

A direction vector of α is $(1, 2, 0, -3)$. A direction vector of β is $(0, 3, 4, 2)$. Since $(1, 2, 0, -3)\cdot(0, 3, 4, 2) = 0$, α is orthogonal to β. We write $\alpha \perp \beta$.

Notice that orthogonality of affine lines is determined by translating them to the origin and checking the orthogonality of their corresponding vector lines. Thus vectors in the orthogonal affine lines need not, themselves, be orthogonal. For instance, in Example 1, setting $s = 1$ shows that $(4, 3, 0, -5)$ is in α, and setting $t = 2$ shows that $(4, 6, 8, 7)$ is in β. But $(4, 3, 0, -5)\cdot(4, 6, 8, 7) = -1$.

Two vector sets are ***orthogonal*** to each other if each vector in one set is orthogonal to every vector in the other set. If the sets have many elements, it may be difficult to check orthogonality. But if the sets are vector subspaces, then

THEOREM 6.4. Let vector subspaces S and T in E^n have respective bases $A = \{\mathbf{A}_1, \mathbf{A}_2, \ldots, \mathbf{A}_k\}$ and $B = \{\mathbf{B}_1, \mathbf{B}_2, \ldots, \mathbf{B}_r\}$ such that A and B are orthogonal to each other. Then S and T are orthogonal to each other.

Proof: S is the set of all linear combinations of the A's, namely $a_1\mathbf{A}_1 + a_2\mathbf{A}_2 + \cdots + a_k\mathbf{A}_k$, and T is the set of all linear combinations of the B's, namely $b_1\mathbf{B}_1 + b_2\mathbf{B}_2 + \cdots + b_r\mathbf{B}_r$. The inner product of a vector in S and a vector in T is

$$(1)\quad (a_1\mathbf{A}_1 + a_2\mathbf{A}_2 + \cdots + a_k\mathbf{A}_k)\cdot(b_1\mathbf{B}_1 + b_2\mathbf{B}_2 + \cdots + b_r\mathbf{B}_r)$$

Each term in the expansion of this product has a factor $\mathbf{A}_i\cdot\mathbf{B}_j$, which is 0. Therefore, (1), which is the sum of the zero terms, is zero. Thus S and T are orthogonal to each other.

THEOREM 6.5. Let $A = \{\mathbf{A}_1, \mathbf{A}_2, \ldots, \mathbf{A}_n\}$ be a basis of a vector subspace S in E^m. Suppose $\mathbf{B}$ is a vector in S orthogonal to each vector in A; then $\mathbf{B} = 0$.

Proof: Since $\mathbf{B}$ is in S,

$$\mathbf{B} = a_1\mathbf{A}_1 + a_2\mathbf{A}_2 + \cdots + a_n\mathbf{A}_n$$

$$\mathbf{B}\cdot\mathbf{B} = a_1\mathbf{A}_1\cdot\mathbf{B} + a_2\mathbf{A}_2\cdot\mathbf{B} + \cdots + a_n\mathbf{A}_n\cdot\mathbf{B}$$

Since $\mathbf{B}\cdot\mathbf{A}_i = 0$ for all i, $\mathbf{B}\cdot\mathbf{B} = 0$. But $\mathbf{B}\cdot\mathbf{B} = 0$ means that $\mathbf{B} = 0$.

COROLLARY. If a nonzero vector $\mathbf{B}$ is orthogonal to every vector in a basis of a vector subspace S, then $\mathbf{B}$ is not in S.

Figure 6.8 illustrates the notions that follow. In this figure $\mathbf{A}$ generates the vector line α. An affine plane that contains $\mathbf{A}$ is π. Two vectors, $\mathbf{B}$ and $\mathbf{C}$, of π are such that $\mathbf{A}$, $\mathbf{B}$, and $\mathbf{C}$ generate π. Therefore, $\mathbf{B} - \mathbf{A}$ and $\mathbf{C} - \mathbf{A}$ are direction vectors of π. To say that α is orthogonal to π means that

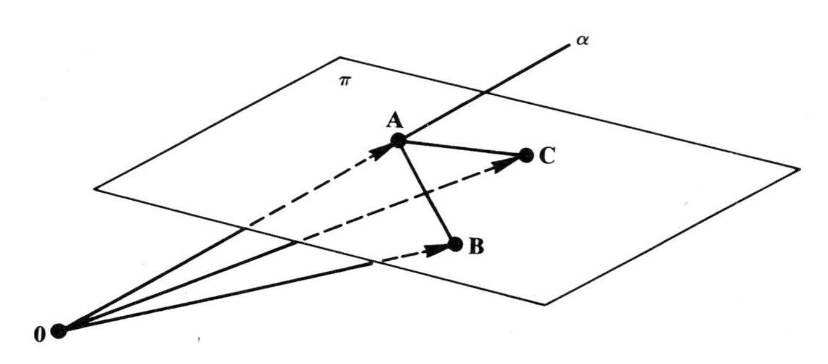

Figure 6.8

$\mathbf{A}\cdot(\mathbf{B} - \mathbf{A}) = 0$ and $\mathbf{A}\cdot(\mathbf{C} - \mathbf{A}) = 0$. Thus any multiple of $\mathbf{A}$ (an element of α) is orthogonal to any linear combination of $\mathbf{B} - \mathbf{A}$ and $\mathbf{C} - \mathbf{A}$. That is,

$$(r\mathbf{A})\cdot[s(\mathbf{B} - \mathbf{A}) + p(\mathbf{C} - \mathbf{A})] = 0$$

The reader should check this statement.

EXAMPLE 2

By Theorem 6.4 the vector line

(1)
$$\alpha\colon\ \begin{array}{r|ccc|}\hline s & 3 & 2 & -1 \\ \hline \end{array}$$
$$= x_1 \quad = x_2 \quad = x_3$$

and the vector plane

(2)
$$\pi\colon\ \begin{array}{r|ccc|}\hline p & -1 & 2 & 1 \\ q & 5 & -7 & 1 \\ \hline \end{array}$$
$$= x_1 \quad = x_2 \quad = x_3$$

are orthogonal since $(3, 2, -1)\cdot(-1, 2, 1) = 0$ and $(3, 2, -1)\cdot(5, -7, 1) = 0$. The vector line α is also orthogonal to

(3)
$$\pi_1\colon\ \begin{array}{r|ccc|}\hline p & -1 & 2 & 1 \\ q & 5 & -7 & 1 \\ 1 & a_1 & a_2 & a_3 \\ \hline \end{array}$$
$$= x_1 \quad = x_2 \quad = x_3$$

which is a translate of π by (a_1, a_2, a_3). The affine line, α_1, a translate of α, is also orthogonal to π and π_1.

The vector line α or any of its translates are called *normal lines* to π or any of its translates. The vector $(3, 2, -1)$ is called a *normal vector* to π or any of its translates.

We derive a formula to describe the affine plane π which contains the vector $\mathbf{A} = (a_1, a_2, a_3)$ and has $\mathbf{A}$ as a normal vector. Let $\mathbf{X}$ be any vector of π other than $\mathbf{A}$. Then $\mathbf{X} - \mathbf{A}$ is a direction vector of π. Therefore,

$$\mathbf{A}\cdot(\mathbf{X} - \mathbf{A}) = 0 \tag{1}$$

or

$$\mathbf{A}\cdot\mathbf{X} - \mathbf{A}\cdot\mathbf{A} = 0 \tag{2}$$

In E^3 (2) is

(3)

x_1	x_2	x_3	-1	
a_1	a_2	a_3	$\mathbf{A}\cdot\mathbf{A}$	$= 0$

which is the standard form for an affine plane.

EXAMPLE 3

What is the standard form for the affine plane normal to $\mathbf{A} = (3, 2, -1)$ which contains $(3, 2, -1)$?

$$\mathbf{A}\cdot\mathbf{A} = (3, 2, -1)(3, 2, -1) = 14$$

Therefore, from (3) the standard form is

(4)

x_1	x_2	x_3	-1	
3	2	-1	14	$= 0$

The components of the normal vector appear in the standard form tableau as coefficients. This plane is the same as that of Example 2. To see this, obtain a standard form from (3) of Example 2 [letting $(a_1, a_2, a_3) = (3, 2, -1)$]. The result is

(5)

x_1	x_2	x_3	-1	
$-3/2$	-1	$1/2$	-7	$= 0$

Why are (4) and (5) the same?

Two affine planes π and π_1 are parallel if two of the direction vectors of π_1 are linear combinations of two of the direction vectors of π. Then in standard form the tableaus for π and π_1 have normal vectors differing by a constant multiple. (The reader should show this for examples in E^3.) Thus, as our geometric intuition indicates, parallel planes have the same normal line.

The set of vectors $\{(1, 0, 0), (0, 1, 0), (0, 0, 1)\}$ is such that each member is orthogonal to every other member.

In E^n a set of vectors $\{\mathbf{A}_1, \mathbf{A}_2, \ldots, \mathbf{A}_m\}$ is called an *orthogonal set* if for each pair of vectors,

$$\mathbf{A}_i, \mathbf{A}_j, \qquad i \neq j, \qquad \mathbf{A}_i \cdot \mathbf{A}_j = 0$$

Of special interest are the bases of vector subspaces that are also orthogonal sets. (We shall see that every vector subspace of E^n has an orthogonal basis.)

Let π be the vector plane generated by a basis $\{\mathbf{A}_1, \mathbf{A}_2\}$. An orthogonal basis for π, $\{\mathbf{B}_1, \mathbf{B}_2\}$, must satisfy three conditions.

1. $\mathbf{B}_1$ and $\mathbf{B}_2$ are linearly independent.
2. $\mathbf{B}_1 \cdot \mathbf{B}_2 = 0$ [In an exercise you will be asked to show that for nonzero vectors, $\mathbf{B}_1 \cdot \mathbf{B}_2 = 0$ implies linear independence. Thus to show (1), just show (2).]
3. $\mathbf{B}_1$ and $\mathbf{B}_2$ are in π. That is, they are linear combinations of $\mathbf{A}_1$ and $\mathbf{A}_2$.

To convert the basis $\{\mathbf{A}_1, \mathbf{A}_2\}$ to an orthogonal basis $\{\mathbf{B}_1, \mathbf{B}_2\}$ do the following. Let

$$\mathbf{B}_1 = \mathbf{A}_1 \qquad \text{and} \qquad \mathbf{B}_2 = r\mathbf{A}_1 + \mathbf{A}_2$$

Then

$$\mathbf{B}_1 \cdot \mathbf{B}_2 = \mathbf{A}_1 \cdot (r\mathbf{A}_1 + \mathbf{A}_2) = r\mathbf{A}_1 \cdot \mathbf{A}_1 + \mathbf{A}_1 \cdot \mathbf{A}_2$$

Now suppose that

$$\mathbf{B}_1 \cdot \mathbf{B}_2 = 0$$

Then

$$r\mathbf{A}_1 \cdot \mathbf{A}_1 + \mathbf{A}_1 \cdot \mathbf{A}_2 = 0$$

which means that

$$r\mathbf{A}_1 \cdot \mathbf{A}_1 = -\mathbf{A}_1 \cdot \mathbf{A}_2$$

or since

$$\mathbf{A}_1 \cdot \mathbf{A}_1 \neq 0$$

then

$$r = -\frac{\mathbf{A}_1 \cdot \mathbf{A}_2}{\mathbf{A}_1 \cdot \mathbf{A}_1}$$

Thus

$$\mathbf{B}_2 = \mathbf{A}_2 - \frac{\mathbf{A}_1 \cdot \mathbf{A}_2}{\mathbf{A}_1 \cdot \mathbf{A}_1} \mathbf{A}_1$$

The desired orthogonal basis is

$$\left\{ \mathbf{A}_1, \mathbf{A}_2 - \frac{\mathbf{A}_1 \cdot \mathbf{A}_2}{\mathbf{A}_1 \cdot \mathbf{A}_1} \mathbf{A}_1 \right\}$$

EXAMPLE 4

Let π be the vector plane generated by the basis $\{(1, 2, 3), (-1, 0, 4)\}$. Let $\mathbf{B}_1 = (1, 2, 3)$. Then

$$\begin{aligned} \mathbf{B}_2 &= (-1, 0, 4) - \frac{(1, 2, 3) \cdot (-1, 0, 4)}{(1, 2, 3) \cdot (1, 2, 3)} (1, 2, 3) \\ &= (-1, 0, 4) - \frac{11}{14}(1, 2, 3) \\ &= \left(\frac{-25}{14}, \frac{-22}{14}, \frac{23}{14} \right) = \frac{1}{14}(-25, -22, 23) \end{aligned}$$

If two vectors are orthogonal, so are their multiples. Thus, an orthogonal basis is

$$\{(1, 2, 3), (-25, -22, 23)\} \qquad \text{(verify this)}$$

The method above is used to derive orthogonal bases for k-dimensional Euclidean subspaces. It is known as the *Gram–Schmidt process.* We show this for a basis having three vectors and generalize the pattern that emerges.

Let Euclidean subspace S have a basis $\{\mathbf{A}_1, \mathbf{A}_2, \mathbf{A}_3\}$ where the $\mathbf{A}$'s are in E^n, $n \geq 3$. To find an orthogonal basis $\{\mathbf{B}_1, \mathbf{B}_2, \mathbf{B}_3\}$, let

$$\mathbf{B}_1 = \mathbf{A}_1 \qquad \text{and} \qquad \mathbf{B}_2 = \mathbf{A}_2 - \frac{\mathbf{B}_1 \cdot \mathbf{A}_2}{\mathbf{B}_1 \cdot \mathbf{B}_1} \mathbf{A}_1$$

as before. Then $\mathbf{B}_1 \cdot \mathbf{B}_2 = 0$. $\mathbf{B}_3$ must satisfy three conditions.

1. $\mathbf{B}_3 = r\mathbf{B}_1 + s\mathbf{B}_2 + \mathbf{A}_3$, thus making $\mathbf{B}_3$ a linear combination of $\mathbf{A}_1, \mathbf{A}_2$, and $\mathbf{A}_3$.
2. $\mathbf{B}_1 \cdot \mathbf{B}_3 = 0$.
3. $\mathbf{B}_2 \cdot \mathbf{B}_3 = 0$.

From conditions 1 and 2

$$\begin{aligned}\mathbf{B}_1 \cdot \mathbf{B}_3 &= \mathbf{B}_1 \cdot (r\mathbf{B}_1 + s\mathbf{B}_2 + \mathbf{A}_3)\\ &= r\mathbf{B}_1 \cdot \mathbf{B}_1 + s\mathbf{B}_1 \cdot \mathbf{B}_2 + \mathbf{B}_1 \cdot \mathbf{A}_3 = 0\end{aligned}$$

Since $\mathbf{B}_1 \cdot \mathbf{B}_2 = 0$ and $\mathbf{B}_1 \cdot \mathbf{B}_1 \neq 0$, $r = \mathbf{B}_1 \cdot \mathbf{A}_3 / \mathbf{B}_1 \cdot \mathbf{B}_1$.

From conditions 1 and 3,

$$\mathbf{B}_2 \cdot \mathbf{B}_3 = \mathbf{B}_2 \cdot (r\mathbf{B}_1 + s\mathbf{B}_2 + \mathbf{A}_3) = r\mathbf{B}_2 \cdot \mathbf{B}_1 + s\mathbf{B}_2 \cdot \mathbf{B}_2 + \mathbf{B}_2 \cdot \mathbf{A}_3$$

Since $\mathbf{B}_2 \cdot \mathbf{B}_1 = 0$ and $\mathbf{B}_2 \cdot \mathbf{B}_2 \neq 0$, $s = -\mathbf{B}_2 \cdot \mathbf{A}_3 / \mathbf{B}_2 \cdot \mathbf{B}_2$. Thus

$$\mathbf{B}_3 = \mathbf{A}_3 - \frac{\mathbf{B}_2 \cdot \mathbf{A}_3}{\mathbf{B}_2 \cdot \mathbf{B}_2}\mathbf{B}_2 - \frac{\mathbf{B}_1 \cdot \mathbf{A}_3}{\mathbf{B}_1 \cdot \mathbf{B}_1}\mathbf{B}_1$$

EXAMPLE 5

Let S be a vector subspace in E^4 with basis $\{(1,0,1,0), (1,1,3,0), (1,2,0,0)\}$. An orthogonal basis $\{\mathbf{B}_1, \mathbf{B}_2, \mathbf{B}_3\}$ can be found as follows, by the Gram–Schmidt process.

$$\mathbf{B}_1 = (1, 0, 1, 0)$$

$$\begin{aligned}\mathbf{B}_2 &= (1, 1, 3, 0) - \frac{(1, 0, 1, 0) \cdot (1, 1, 3, 0)}{(1, 0, 1, 0) \cdot (1, 0, 1, 0)}(1, 0, 1, 0)\\ &= (1, 1, 3, 0) - 2(1,0,1,0)\\ &= (-1, 1, 1, 0)\end{aligned}$$

$$\begin{aligned}\mathbf{B}_3 &= (1, 2, 0, 0) - \frac{(-1, 1, 1, 0) \cdot (1, 2, 0, 0)}{(-1, 1, 1, 0) \cdot (-1, 1, 1, 0)}(-1, 1, 1, 0)\\ &\qquad - \frac{(1, 0, 1, 0) \cdot (1, 2, 0, 0)}{(1, 0, 1, 0) \cdot (1, 0, 1, 0)}(1, 0, 1, 0)\\ &= (1, 2, 0, 0) - \tfrac{1}{3}(-1, 1, 1, 0) - \tfrac{1}{2}(1, 0, 1, 0)\\ &= (\tfrac{5}{6}, \tfrac{5}{3}, -\tfrac{5}{6}, 0) = \tfrac{5}{6}(1, 2, -1, 0)\end{aligned}$$

Therefore,

$$\{\mathbf{B}_1, \mathbf{B}_2, \mathbf{B}_3\} = \{(1, 0, 1, 0), (-1, 1, 1, 0), (1, 2, -1, 0)\}$$

If each vector of an orthogonal basis is replaced by a corresponding unit vector, the basis is called *orthonormal.*

EXAMPLE 6

The orthogonal basis in Example 5 is converted into an orthonormal basis by dividing each vector by its norm. Thus (1, 0, 1, 0) is divided by $\sqrt{2}$, and so on. The resulting orthonormal basis is

$$\left\{\left(\frac{1}{\sqrt{2}}, 0, \frac{1}{\sqrt{2}}, 0\right), \left(-\frac{1}{\sqrt{3}}, \frac{1}{\sqrt{3}}, \frac{1}{\sqrt{3}}, 0\right), \left(\frac{1}{\sqrt{6}}, \frac{2}{\sqrt{6}}, \frac{-1}{\sqrt{6}}, 0\right)\right\}$$

Exercises

1. (a) Find a representation in standard form of the affine plane containing (1, 2, 1) that is orthogonal to (1, −1, 2).
(b) Find the distance from **0** to this plane.

2. Find an equation for the affine plane perpendicular to the segment between (2, −1, 3) and (4, 3, 1) through its midpoint. The midpoint formula is

$$\mathbf{M} = \left(\frac{a_1 + b_1}{2}, \frac{a_2 + b_2}{2}, \frac{a_3 + b_3}{2}\right)$$

3. Find an equation for the affine plane through (1, 2, 5) and parallel to the plane represented by

x_1	x_2	x_3	-1	
3	−1	4	7	= 0

4. Let $\mathbf{A} = (1, 1, 1)$ and let α be the vector line with tableau

t	−1	2	3
	$= x_1$	$= x_2$	$= x_3$

Find vector **B** in α such that $\mathbf{A} - \mathbf{B}$ is orthogonal to α. This locates the point in α nearest to **A**, and $\|\mathbf{A} - \mathbf{B}\|$ is the distance from **A** to α.

5. Let $\mathbf{A} = (1, 1, 1)$ and let α be the affine line

s	1	2	2
1	0	1	0
	$= x_1$	$= x_2$	$= x_3$

Find vector **B** in α such that $\mathbf{A} - \mathbf{B}$ is orthogonal to α. This locates the point in α nearest to **A** and $\|\mathbf{A} - \mathbf{B}\|$ is the distance from **A** to α.

6. Given affine lines α_1 and α_2,

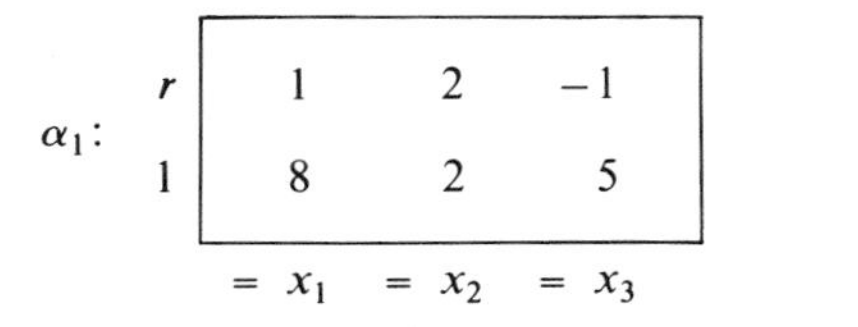

α_1:

r	1	2	-1
1	8	2	5
	$= x_1$	$= x_2$	$= x_3$

α_2:

s	2	1	0
1	1	2	5
	$= x_1$	$= x_2$	$= x_3$

find a vector **A** in α_1 and a vector **B** in α_2 such that the direction vector $\mathbf{A} - \mathbf{B}$ is orthogonal to both α_1 and α_2. The norm of $\mathbf{A} - \mathbf{B}$ is the (shortest) distance between α_1 and α_2. Calculate that distance.

7. Let the affine hyperplane π be given by

x_1	x_2	x_3	x_4	-1	
1	1	-3	-1	5	$= 0$

Find the "distance" between $(1, 2, -1, 3)$ and the hyperplane. [Hint: Express the hyperplane parallel to π that contains $(1, 2, -1, 3)$.]

8. Show that $\{(1, 0, 1, 0), (-1, 1, 1, 0), (a, 2a, -a, b)\}$ is an orthogonal set for all a, b in R.

9. Let $\{\mathbf{A}_1, \mathbf{A}_2, \ldots, \mathbf{A}_k\}$ be an orthogonal set of vectors in E^n and $c_1, c_2, \ldots, c_n$ any nonzero scalars. Prove that $\{c_1\mathbf{A}_1, c_2\mathbf{A}_2, \ldots, c_n\mathbf{A}_n\}$ is an orthogonal set.

10. Given that a basis of a vector plane is $\{(1, -3, 2), (4, 2, -3)\}$, find an orthogonal basis for the plane that has $(1, -3, 2)$ as one of its vectors.

11. Let $\{\mathbf{A}_1, \mathbf{A}_2, \mathbf{A}_3\}$ be a set of nonzero vectors in E^n.
(a) Prove it is not an orthogonal set if it is linearly dependent.
(b) Show that $\{\mathbf{A}_1, \mathbf{A}_2, \mathbf{A}_3\}$ is linearly independent if it is orthogonal.

12. Convert each of the following bases of vector subspaces into orthogonal bases.
(a) $\{(1, 2), (3, 4)\}$.
(b) $\{(1, 0, 1), (1, 1, 1)\}$.

(c) $\{(1, -3, 2), (4, 2, -3)\}$.
(d) $\{(1, 0, 1, 0), (1, 0, 5, 0), (0, 2, 3, 1)\}$.

13. Convert each of the following bases (see Exercise 12) into orthonormal bases.
(a) $\{(1, 2), (3, 4)\}$.
(b) $\{(1, 0, 1), (1, 1, 1)\}$.
(c) $\{(1, 0, 1, 0), (1, 0, 5, 0), (0, 2, 3, 1)\}$.

6.6 Orthogonal Complements

In this section we examine vector subspaces P and R in E^n that are orthogonal to each other, and, in addition, have the property that $P + R = E^n$. The following three examples illustrate subspaces that have both of these properties or have one of them and not the other.

EXAMPLE 1

Let α and β be vector lines with respective tableaus

(1)

t	1	0	0
	$= x_1$	$= x_2$	$= x_3$

(2)

r	0	1	0
	$= x_1$	$= x_2$	$= x_3$

Since $(1, 0, 0)\cdot(0, 1, 0) = 0$, it is true that $\mathbf{A}\cdot\mathbf{B} = 0$, for every vector $\mathbf{A}$ in α and every vector $\mathbf{B}$ in β. $\alpha + \beta$ is given by

(3)

s	1	0	0
t	0	1	0
	$= x_1$	$= x_2$	$= x_3$

and is a vector plane (dimension two) in E^3. Note, $\alpha + \beta \neq E^3$.

EXAMPLE 2

Let α be a vector line with a tableau

(1)

t	1	2	1
	$= x_1$	$= x_2$	$= x_3$

and π a vector plane with a tableau

(2)

r	0	1	0
s	0	0	1
	$= x_1$	$= x_2$	$= x_3$

Then $\alpha + \pi$ has a tableau

(3)

t	1	2	1
r	0	1	0
s	0	0	1
	$= x_1$	$= x_2$	$= x_3$

$\alpha + \pi = E^3$ since $\{(1, 2, 1), (0, 1, 0), (0, 0, 1)\}$ is a basis of E^3. To see that this is a basis, use the pivot condensation test.

t	1*	2	1
r	0	1	0
s	0	0	1
	$= 0$	$= 0$	$= 0$

r	1*	0
s	0	0
	$= 0$	$= 0$

s	1
	$= 0$

The three vectors are linearly independent and the result follows. These basis vectors are not orthogonal since $(1, 2, 1) \cdot (0, 1, 0) \neq 0$.

EXAMPLE 3

Let α have a tableau

(1)

t	1	0	0
	$= x_1$	$= x_2$	$= x_3$

and π have a tableau

(2)

r	0	1	0
s	0	0	1
	$= x_1$	$= x_2$	$= x_3$

Then $\alpha + \pi$ has a tableau

(3)

t	1	0	0
r	0	1	0
s	0	0	1
	$= x_1$	$= x_2$	$= x_3$

Since the rows of (3) are linearly independent, $\alpha + \pi = E^3$. Every vector in α is orthogonal to every vector in π.

α and π of Example 3 are called *orthogonal complements* in E^3. The subspaces of Examples 1 and 2 are not orthogonal complements. In Example 1 the subspaces are orthogonal but are not complements; that is, their sum is not the entire space. In Example 2 the subspaces are complements but they are not orthogonal.

In general, let S and T be vector subspaces of E^n. Then S and T are called *orthogonal complements* in E^n if

1. $S + T = E^n$ (S and T are complements).
2. For all $\mathbf{A}$ in S and $\mathbf{B}$ in T, $\mathbf{A} \cdot \mathbf{B} = 0$ (S and T are orthogonal to each other).

The orthogonal complement of S is designated $S^\perp$ (read "S orthogonal"). In the above definition $T = S^\perp$ and $S = T^\perp$.

EXAMPLE 4

$\{\mathbf{0}_n\}$ and E^n are orthogonal complements.

Now suppose that a subspace of E^3 is given, say α with tableau

(1)

t	1	2	3
	$= x_1$	$= x_2$	$= x_3$

and we seek $\alpha^\perp$, assuming that it exists and is unique. If $\alpha^\perp = \{(x_1, x_2, x_3)\}$, the orthogonality condition requires that for all (x_1, x_2, x_3) in $\alpha^\perp$,

(2) $$(x_1, x_2, x_3)\cdot t(1, 2, 3) = 0$$

We know that (2) is true for $t = 0$. If $t \neq 0$, then

(3)

x_1	x_2	x_3	
1	2	3	$= 0$

must hold for all (x_1, x_2, x_3). This is a standard representation of a vector plane π. By pivoting we see that (3) has a parametric representation

(4)

r	-2	1	0
s	-3	0	1
	$= x_1$	$= x_2$	$= x_3$

A basis for $\alpha^\perp$ is $\{(-2, 1, 0), (-3, 0, 1)\}$. A basis for α is $\{(1, 2, 3)\}$. $\alpha + \pi$ has a tableau

(5)

t	1	2	3
r	-2	1	0
s	-3	0	1
	$= x_1$	$= x_2$	$= x_3$

The linear independence of the basis in (5) is shown by pivot condensation:

(6)

r	-2	1^*	0
s	-3	0	1
t	1	2	3
	$= 0$	$= 0$	$= 0$

(7)

s	-3	1^*
t	5	3
	$= 0$	$= 0$

(8)

t	14
	$= 0$

Therefore, $\alpha + \alpha^\perp$ has dimension three and thus $\alpha + \alpha^\perp = E^3$.

We now demonstrate a truly remarkable fact. Pivoting on the 1 of (1) yields an explicit form for α:

(9)

x_1	2	3
	$= x_2$	$= x_3$

From (9) we obtain a standard form for α:

(10)

x_1	x_2	x_3	
2	-1	0	$= 0$
3	0	-1	$= 0$

which is the same as

(11)

x_1	x_2	x_3	
-2	1	0	$= 0$
-3	0	1	$= 0$

The inside of (11) is the same as the inside of (4)! Thus the two tableaus can be combined into one with α in standard representation and $\alpha^\perp$ in parametric representation.

(12)

	y_1	y_2	y_3	
r	-2	1	0	$= 0$
s	-3	0	1	$= 0$
	$= x_1$	$= x_2$	$= x_3$	

The tableau (1) with α in parametric form is also a dual tableau when $\alpha^\perp$ is presented in a standard form:

(13)

	y_1	y_2	y_3	
t	1	2	3	$= 0$
	$= x_1$	$= x_2$	$= x_3$	

This result is generally true. We express it as

THEOREM 6.6. Let S and T be vector subspaces in E^n that are represented in the same tableau, one (S) in parametric form, the other (T) in standard form. Then S and T are orthogonal complements.

Proof: We offer a proof for the case $n = 5$ and S having dimension three. Assume, with no loss of generality, that the tableau has been fully reduced, with the unit columns appearing first as follows:

T:

(1)

		y_1	y_2	y_3	y_4	y_5	
	p	1	0	0	a_1	a_2	$= 0$
S:	q	0	1	0	b_1	b_2	$= 0$
	r	0	0	1	c_1	c_2	$= 0$
		$= x_1$	$= x_2$	$= x_3$	$= x_4$	$= x_5$	

Recall the key equation from Section 1.3. From the key equation

$$x_1y_1 + x_2y_2 + x_3y_3 + x_4y_4 + x_5y_5 = (-p)(0) + (-q)(0) + (-r)(0) = 0$$

or letting

$$\mathbf{X} = (x_1, x_2, x_3, x_4, x_5), \qquad \mathbf{Y} = (y_1, y_2, y_3, y_4, y_5)$$

$$\mathbf{X} \cdot \mathbf{Y} = 0$$

which means that every vector $\mathbf{Y}$ in T is orthogonal to every vector $\mathbf{X}$ in S. Thus S and T are orthogonal to each other.

S has a basis $\{(1, 0, 0, a_1, a_2), (0, 1, 0, b_1, b_2), (0, 0, 1, c_1, c_2)\}$; the vectors appear in the rows of (1). To find a basis for T, convert the standard form of (1) to parametric form as

(2)

s	$-a_1$	$-b_1$	$-c_1$	1	0
t	$-a_2$	$-b_2$	$-c_2$	0	1
	$= y_1$	$= y_2$	$= y_3$	$= y_4$	$= y_5$

Then a basis of T is $\{(-a_1, -b_1, -c_1, 1, 0), (-a_2, -b_2, -c_2, 0, 1)\}$. To see that $S + T = E^5$, use condensation to show that the five vectors $(1, 0, 0, a_1, a_2)$, $(0, 1, 0, b_1, b_2)$, $(0, 0, 1, c_1, c_2)$, $(-a_1, -b_1, -c_1, 1, 0)$, and $(-a_2, -b_2, -c_2, 0, 1)$ are linearly independent.

EXAMPLE 5

Find $S^\perp$ if S has a tableau

(1)

r	1	2	0	1
s	3	2	1	0
	$= x_1$	$= x_2$	$= x_3$	$= x_4$

Adding a standard representation to the tableau (1) gives

(2)

	y_1	y_2	y_3	y_4	
r	1	2	0	1	$= 0$
s	3	2	1	0	$= 0$
	$= x_1$	$= x_2$	$= x_3$	$= x_4$	

$S^\perp$ is the standard representation. (Check this.)

EXAMPLE 6

Find $\pi^\perp$ if π in E^4 is represented by

(1)

x_1	x_2	x_3	x_4	
2	1	-1	1	$= 0$

Adding the parametric representation on the left and bottom of (1) yields

(2)

	x_1	x_2	x_3	x_4	
p	2	1	-1	1	$= 0$
	$= y_1$	$= y_2$	$= y_3$	$= y_4$	

$\pi^\perp$ is in parametric form. Consideration of the dual tableaus indicates that every k-dimensional vector space in E^n has an orthogonal complement whose dimension is $n - k$. Thus the orthogonal complement of a line is a hyperplane (and vice versa).

We have concerned ourselves with orthogonal complements of vector subspaces. It is also possible to find a corresponding relationship for affine subspaces. This is done by moving the origin of E^n to any vector in a given affine subspace and proceeding as for vector subspaces.

The last theorem of this chapter leads to an unexpected and pretty relation between the sum and intersection of two vector subspaces and orthogonal complements.

THEOREM 6.7. Let S and T be two vector subspaces in E^n. Then $(S + T)^\perp = S^\perp \cap T^\perp$.

Proof: Using the parametric representation of S and T, $S + T$ is represented in its columns by

		y_1	y_2	$\cdots$	y_n	
S	p_1	a_{11}	a_{12}	$\cdots$	a_{1n}	$= 0$
	p_2	a_{21}	a_{22}	$\cdots$	a_{2n}	$= 0$
	$\vdots$	$\vdots$	$\vdots$		$\vdots$	$\vdots$
	p_r	a_{r1}	a_{r2}	$\cdots$	a_{rn}	$= 0$
T	q_1	b_{11}	b_{12}	$\cdots$	b_{1n}	$= 0$
	$\vdots$	$\vdots$	$\vdots$		$\vdots$	$\vdots$
	q_s	b_{s1}	b_{s2}	$\cdots$	b_{sn}	$= 0$
		$= x_1$	$= x_2$	$\cdots$	$= x_n$	

Therefore, the $r + s$ rows represent $(S + T)^{\perp}$. Now from another viewpoint the first r rows represent $S^{\perp}$ and the last s rows represent $T^{\perp}$. The $r + s$ rows therefore also represent $S^{\perp} \cap T^{\perp}$. Thus $(S + T)^{\perp} = S^{\perp} \cap T^{\perp}$.

Exercises

1. Find the orthogonal complement of the vector subspaces with tableaus:

(a)

t	3	2
	$= x_1$	$= x_2$

(b)

t	3	2	-1
	$= x_1$	$= x_2$	$= x_3$

(c)

r	-1	-2	-3
s	1	0	2
	$= x_1$	$= x_2$	$= x_3$

(d)

r	1	0	1	0
s	2	1	0	3
	$= x_1$	$= x_2$	$= x_3$	$= x_4$

2. Find the orthogonal complement of each of the vector hyperplanes represented by tableaus:

(a)

x_1	x_2	
1	2	$= 0$

(b)

x_1	x_2	x_3	
4	-2	-3	$= 0$

(c)

x_1	x_2	x_3	x_4	
1	1	-1	-1	$= 0$

3. Prove that the bases of S and $S^{\perp}$, where S is a vector subspace in E^n, constitute a linearly independent set.

4. Find the set of all vectors in E^n that are orthogonal to both $(1, 1, 0, -1)$ and $(0, 2, 3, -1)$.

5. Prove Theorem 6.5 for the special case $n = 4$ with S having dimension two.

6. Interpret the statement "Every k-dimensional vector subspace in E^n has an orthogonal complement whose dimension is $n - k$" if:
(a) $n = 2, k = 1$.
(b) $n = 3, k = 1$.
(c) $n = 3, k = 2$.

7. Can a vector subspace and its orthogonal complement have the same dimension? Discuss your answer.

8. Interpret Theorem 6.6 if S and T are in E^3 and:
   (a) are two distinct lines.
   (b) are a vector plane π and a vector line α.

6.7 Chapter Review

1. By adding vector $\mathbf{A}$ of R^n to each vector in a vector subspace S of R^n, the result $S + \mathbf{A}$ is an affine subspace. If $S = R^n$, then $R^n + \mathbf{A} = R^n$. Thus R^n is called a vector space or an affine space. Two affine subspaces are said to be parallel if they are translates of the same vector subspace.
   (a) Vector subspaces are affine subspaces that contain the zero vector.
   (b) A set of vectors in R^n is a vector subspace if it is closed under linear combinations (Chapter 5). It is an affine subspace if it is closed under linear combinations, where the sum of the scalars is 1.
   (c) Parametric and standard representations of subspaces (the word "subspace" by itself means that it applies to both vector and affine) are convertible one to the other via their explicit forms. A parametric representation converts to an explicit representation by parametric elimination. A standard representation converts to an explicit representation by pivot reduction.
   (d) The dimension of a subspace is the number of nonbasic variables in the explicit representation.
   (e) Every row (except the row of the 1) in a parametric representation of a subspace represents a line. Thus a subspace is a sum of lines. The dimension of the subspace is the number of lines in the sum with linearly independent generating vectors.
   (f) Each equation in standard form has an explicit form with $n - 1$ nonbasic variables. Thus the solution set for an equation [by (d)] is a hyperplane. The solution set for a system of linear equations is the intersection of the solution sets of the individual equations. Therefore, the solution set for a system of linear equations is the intersection of hyperplanes. The number of distinct hyperplanes that form the intersection is equal to the number of basic variables in the explicit representation.
   (g) The sum of two subspaces is the subspace of their adjoined parametric representations.
   (h) The intersection of two subspaces is the subspace of the solution set of their adjoined standard forms.
2. By introducing the definition of an inner product, R^n is converted into E^n and the notion of orthogonality and length of segment are introduced.

3. Orthogonal sets, orthogonal subspaces, orthogonal bases, and orthonormal bases of subspaces are studied.
4. Orthogonal complements are two subspaces whose sum is E^n and which are orthogonal to each other. It applies the notion of orthogonality to two subspaces whose dimensions are k and $n - k$. Orthogonal complements are representable in the same tableau, one in parametric form, the other in standard form.

   Finally, for all vector subspaces S and T

$$(S + T)^{\perp} = S^{\perp} \cap T^{\perp}$$

Chapter Review Exercises

1. Find a parametric representation for:
   (a) The affine line containing $(3, 4, -2)$ and $(1, 0, 2)$.
   (b) The affine plane containing $(1, 2, 3)$, $(1, 0, -1)$, and $(2, 1, -1)$.
   (c) The intersection, if any, of the line in (a) and the plane in (b).

2. Show that the two affine planes represented by

p	2	1	-3
q	0	-2	1
1	3	5	6
	$= x_1$	$= x_2$	$= x_3$

and

r	4	0	-5
s	2	-3	-1
1	-1	3	-8
	$= x_1$	$= x_2$	$= x_3$

are parallel.

3. Given that $\mathbf{A} = (1, 2, 3)$, $\mathbf{B} = (1, 0, -1)$, and $\mathbf{C} = (2, 1, -1)$:
   (a) Find a representation for the affine line through $\mathbf{A}$ that is parallel to the affine line that contains $\mathbf{B}$ and $\mathbf{C}$.
   (b) Find a representation of the vector line that is normal to the affine line containing $\mathbf{B}$ and $\mathbf{C}$.

(c) Find a representation of the vector line that is normal to the affine plane containing **A**, **B**, and **C**.

4. Show that (3, 2, 1), (−2, 4, 0), and (13, −2, 3) are on the same affine line, without finding a representation of that line.

5. Prove that for all affine lines, α, β, and λ in R^n:
(a) $\alpha \parallel \alpha$.
(b) If $\alpha \parallel \beta$, then $\beta \parallel \alpha$.
(c) If $\alpha \parallel \beta$ and $\beta \parallel \lambda$, then $\alpha \parallel \lambda$.

6. Given affine hyperplane π in E^4 having the tableau

x_1	x_2	x_3	x_4	-1	
1	1	−3	2	5	= 0

(a) Find an equation for the hyperplane that contains (3, −1, 2, 7) and is parallel to π.
(b) Find the length of the normal vector to π that is in π.
(c) Find the distance from (3, 2, −1, 7) to π.

7. Show that (2, 1, −3) and (0, 3, 1) are orthogonal and find a third nonzero vector that is orthogonal to each.

8. Convert the basis {(1, 1, 0), (0, 4, 1), (2, 2, 3)} to:
(a) an orthogonal basis.
(b) an orthonormal basis.

9. Represent the orthogonal complement of the subspace represented by:

(a)

r	3	2	5
s	0	1	3
	$= x_1$	$= x_2$	$= x_3$

(b)

x_1	x_2	x_3	x_4	
3	2	−1	1	= 0

(c)

r	1	2	3	4
p	2	1	4	−2
q	0	5	6	8
	$= x_1$	$= x_2$	$= x_3$	$= x_4$

CHAPTER 7

Mappings

This chapter is a continuation of the study of mappings that began in Chapter 2. In this chapter three types of mappings, linear, affine, and Euclidean, are distinguished. As might be expected, they are closely related to vector spaces (linear spaces), affine spaces, and Euclidean spaces. A word of caution. In Chapter 2 tableaus represent mappings. In Chapters 5 and 6 tableaus represent spaces and subspaces. In this chapter tableaus represent both mappings and spaces and so care must be taken when reading this chapter to distinguish mappings from spaces.

7.1 Linear Mappings

Recall that a mapping assigns to each member of its domain exactly one member of its codomain.

EXAMPLE 1

Consider $M: R^2 \to R^3$ with $(x_1, x_2) \mapsto (x_1 + x_2, x_1 - x_2, 2x_1 + 3x_2)$. M is expressible as

(1) $$(x_1, x_2) \mapsto (x_1 + x_2, x_1 - x_2, 2x_1 + 3x_2)$$

Clearly implicit in this formulation is the domain R^2 and codomain R^3.

(2) $$M((x_1, x_2)) = (x_1 + x_2, x_1 - x_2, 2x_1 + 3x_2)$$

(3)

x_1	x_2	
1	1	$= y_1$
1	-1	$= y_2$
2	3	$= y_3$

Tableau (3) has equations expressed in rows. In this chapter we use tableaus for mappings with equations in columns. This parametric form gives the nonbasic variables x_1 and x_2 the role of parameters. It simplifies reading the image of the mappings.

(4)

x_1	1	1	2
x_2	1	-1	3
	$= y_1$	$= y_2$	$= y_3$

Mappings are also expressible as matrix equations. Letting

$$\mathbf{A} = \begin{bmatrix} 1 & 1 & 2 \\ 1 & -1 & 3 \end{bmatrix}$$

$$\mathbf{X} = \begin{bmatrix} x_1 \\ x_2 \end{bmatrix}$$

$$\mathbf{Y} = \begin{bmatrix} y_1 \\ y_2 \\ y_3 \end{bmatrix}$$

We can express the mapping as

(5) $$\mathbf{A}^T\mathbf{X} = \mathbf{Y}$$

or

(6)
$$\begin{bmatrix} 1 & 1 \\ 1 & -1 \\ 2 & 3 \end{bmatrix} \begin{bmatrix} x_1 \\ x_2 \end{bmatrix} = \begin{bmatrix} y_1 \\ y_2 \\ y_3 \end{bmatrix}$$

or

(7)
$$\mathbf{X}^T\mathbf{A} = \mathbf{Y}^T$$

or

(8)
$$\begin{bmatrix} x_1 & x_2 \end{bmatrix} \begin{bmatrix} 1 & 1 & 2 \\ 1 & -1 & 3 \end{bmatrix} = \begin{bmatrix} y_1 & y_2 & y_3 \end{bmatrix}$$

The domain of M is R^2, a vector space. The image of M is found by eliminating the x's in tableau (4), as though they were parameters.

(1)

x_1	1*	1	2
x_2	1	−1	3
	$= y_1$	$= y_2$	$= y_3$

Pivoting gives

(2)

y_1	1	2
x_2	−2	1*
	$= y_2$	$= y_3$

(3)

y_1	5
y_3	−2
	$= y_2$

(4)

y_1	1	5	0
y_3	0	−2	1
	$= y_1$	$= y_2$	$= y_3$

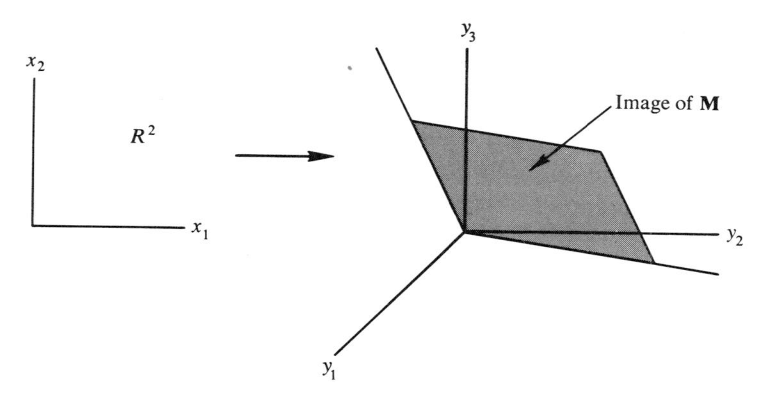

Figure 7.1

The image of M has basis $\{(1, 5, 0)(0, -2, 1)\}$. The image is a vector plane in R^3—a vector subspace of R^3. The linear mapping M is illustrated in Figure 7.1.

EXAMPLE 2

In Example 1 we found, by pivoting, that the image of the mapping M is a vector subspace of R^3. Now let us add to the tableau of M a row with a 1 at the left, pivot, and see what happens. Let N be the mapping $N:R^2 \rightarrow R^3$ with tableau

(1)

x_1	1*	1	2
x_2	1	-1	3
1	2	-3	-2
	$= y_1$	$= y_2$	$= y_3$

Using parametric elimination, we pivot as indicated.

(2)

y_1	1	2
x_1	-2	1*
1	1	-6
	$= y_2$	$= y_3$

(3)

y_1	5
y_3	-2
1	13
	$= y_2$

From (3) the image of N is an affine plane in R^3. Its graph is a plane that does not contain the origin. It is a translate by 13 of the vector plane of Example 1.

Mappings of the type shown in Example 1, that is, mappings without a constant row (or constant column if written in the form used in Chapter 2), were shown many years ago to have two distinctive properties. These are

1. $M(c\mathbf{X}) = cM(\mathbf{X})$.
2. $M(\mathbf{X} + \mathbf{Y}) = M(\mathbf{X}) + M(\mathbf{Y})$.

These two properties of M are illustrated in Examples 3 and 4.

EXAMPLE 3

Let $M: R^2 \to R^3$, $(x_1, x_2) \mapsto (x_1 + x_2,\ x_1 - x_2,\ 2x_1 + 3x_2)$ be the mapping of Example 1. Let $\mathbf{X} = (x_1, x_2)$ be any vector in R^2 and c any scalar. Then

$$c\mathbf{X} = c(x_1, x_2) = (cx_1, cx_2)$$

The image of $c\mathbf{X}$ under M (using the second notation in Example 1), $M(c\mathbf{X})$, is

$$\begin{aligned} M(c\mathbf{X}) &= (cx_1 + cx_2, cx_1 - cx_2, 2cx_1 + 3cx_2) \\ &= c(x_1 + x_2, x_1 - x_2, 2x_1 + 3x_2) \\ &= cM(x_1, x_2) \end{aligned}$$

In words, under M the image of a scalar times a vector is equal to the scalar times the image of the vector. Scalar multiplication is preserved under M. Symbolically, $M(c\mathbf{X}) = cM(\mathbf{X})$. This is called the *homogeneity property*. To illustrate, let $(x_1, x_2) = (2, 3)$ and $c = 4$. Then $c(x_1, x_2) = 4(2, 3) = (8, 12)$. From Example 1

$$\begin{aligned} M(2, 3) &= (5, -1, 13) \\ M(4(2, 3)) &= M(8, 12) = (20, -4, 52) = 4(5, -1, 3) = 4M(2, 3) \end{aligned}$$

EXAMPLE 4

Let $\mathbf{A} = (a_1, a_2)$ and $\mathbf{B} = (b_1, b_2)$ be any vectors in R^2 and let M be as above.

$$\begin{aligned}
\mathbf{A} + \mathbf{B} &= (a_1 + b_1, a_2 + b_2) \\
M(\mathbf{A}) &= M(a_1, a_2) = (a_1 + a_2, a_1 - a_2, 2a_1 + 3a_2) \\
M(\mathbf{B}) &= M(b_1, b_2) = (b_1 + b_2, b_1 - b_2, 2b_1 + 3b_2) \\
M(\mathbf{A} + \mathbf{B}) &= M(a_1 + b_1, a_2 + b_2) = (a_1 + b_1 + a_2 + b_2, \\
&\quad a_1 + b_1 - a_2 - b_2, 2a_1 + 2b_1 + 3a_2 + 3b_2) \\
&= (a_1 + a_2, a_1 - a_2, 2a_1 + 3a_2) \\
&\quad + (b_1 + b_2, b_1 - b_2, 2b_1 + 3b_2) \\
&= M(\mathbf{A}) + M(\mathbf{B})
\end{aligned}$$

Thus the image of the sum of **A** and **B** is the sum of the images of **A** and **B**. This is called the *additive property*.

To illustrate, let $\mathbf{A} = (-2, 3)$ and $\mathbf{B} = (4, 7)$. Then

$$\begin{aligned}
M(\mathbf{A}) &= M(-2, 3) = (1, -5, 5) \\
M(\mathbf{B}) &= M(4, 7) = (11, -3, 29) \\
\mathbf{A} + \mathbf{B} &= (2, 10) \\
M(\mathbf{A} + \mathbf{B}) &= M(2, 10) = (12, -8, 34) = M(\mathbf{A}) + M(\mathbf{B})
\end{aligned}$$

DEFINITION 7.1. Let $M : R^m \to R^n$ be a mapping. M is called a *linear mapping* if and only if for all **X** and **Y** in R^m and all scalars c,

1. $M(\mathbf{X}) + M(\mathbf{Y}) = M(\mathbf{X} + \mathbf{Y})$ (the additive property).
2. $M(c\mathbf{X}) = cM(\mathbf{X})$ (the homogeneity property).

Example 1 illustrates a linear mapping M. However, not all mappings are linear mappings. For instance, mapping N of Example 2 is not a linear mapping. To see this, let $\mathbf{X} = (x_1, x_2)$. Then

$$N(c\mathbf{X}) = N(cx_1, cx_2) = (cx_1 + cx_2 + 2, cx_1 - cx_2 - 3, 2cx_1 + 3cx_2 - 2)$$

But

$$\begin{aligned}
cN(x_1, x_2) &= c(x_1 + x_2 + 2, x_1 - x_2 - 3, 2x_1 + 3x_2 - 2) \\
&= (cx_1 + cx_2 + 2c, cx_1 - cx_2 - 3c, 2cx_1 + 3cx_2 - 2c)
\end{aligned}$$

Thus

$$N(c\mathbf{X}) \neq cN(\mathbf{X}) \qquad \text{except when } c = 1$$

Therefore, N does not have the homogeneity property and is not a linear mapping. Does N have the additive property?

EXAMPLE 5

Let $P: R^2 \to R^1, (x_1, x_2) \mapsto \sqrt[3]{x_1^3 + y_1^3}$. Let $\mathbf{A} = (a_1, a_2)$. Then $c\mathbf{A} = (ca_1, ca_2)$.

$$\begin{aligned} P(\mathbf{A}) &= \sqrt[3]{a_1^3 + a_2^3} \\ P(c\mathbf{A}) &= \sqrt[3]{c^3a_1^3 + c^3a_2^3} = c\sqrt[3]{a_1^3 + a_2^3} \\ &= cP(\mathbf{A}) \end{aligned}$$

P has the homogeneity property. Let $\mathbf{B} = (b_1, b_2)$. Then

$$\begin{aligned} P(\mathbf{A}) + P(\mathbf{B}) &= \sqrt[3]{a_1^3 + a_2^3} + \sqrt[3]{b_1^3 + b_2^3} \\ \text{But } P(\mathbf{A} + \mathbf{B}) &= P(a_1 + b_1, a_2 + b_2) \\ &= \sqrt[3]{(a_1 + b_1)^3 + (a_2 + b_2)^3} \\ P(\mathbf{A}) + P(\mathbf{B}) &\neq P(\mathbf{A} + \mathbf{B}) \text{ [try } \mathbf{A} = (1, 2), \mathbf{B} = (2, 2)] \end{aligned}$$

Thus P is not a linear mapping.

The following properties of linear mappings are often used. We know that $c\mathbf{X}_1 + d\mathbf{X}_2$ is a linear combination of $\mathbf{X}_1$ and $\mathbf{X}_2$ and $cM(\mathbf{X}_1) + dM(\mathbf{X}_2)$ is (using the same scalars) a linear combination of $M(\mathbf{X}_1)$ and $M(\mathbf{X}_2)$. It follows from the definition of linear mapping that $M: R^m \to R^n$ is a linear mapping if and only if for all $\mathbf{X}_1$ and $\mathbf{X}_2$ in R^m, and all scalars c and d,

$$M(c\mathbf{X}_1 + d\mathbf{X}_2) = cM(\mathbf{X}_1) + dM(\mathbf{X}_2)$$

This is extendable to any linear combination. Thus if $a_1, a_2, \ldots, a_s$ are constants, then for a linear mapping M,

$$M(a_1\mathbf{X}_1 + a_2\mathbf{X}_2 + \cdots + a_s\mathbf{X}_s) = a_1M(\mathbf{X}_1) + a_2M(\mathbf{X}_2) + \cdots + a_sM(\mathbf{X}_s)$$

It follows that if $T: R^m \to R^n$ in a linear mapping, then for every $\mathbf{X}$ in R^m,

$$T(-\mathbf{X}) = -T(\mathbf{X}) \qquad \text{and} \qquad T(\mathbf{0}_m) = \mathbf{0}_n$$

[since $T(\mathbf{A} + (-\mathbf{A})) = T(\mathbf{A}) + T(-\mathbf{A})$].

EXAMPLE 6

Let $T: R^2 \to R^2$ be a linear mapping such that $T(1, 2) = (3, 4)$ and $T(-2, 1) = (1, 2.)$ These data are sufficient to find the rule of T. Since $(1, 2)$ and $(-2, 1)$ are linearly independent, all vectors $\mathbf{X} = (x_1, x_2)$ in R^2 can be expressed as linear combinations of them.

That is

(1) $$(x_1, x_2) = u(1, 2) + v(-2, 1)$$

In tableau form, (1) is

(2)

u	1*	2
v	-2	1
	$= x_1$	$= x_2$

(3)

x_1	1	2
v	2	5*
	$= u$	$= x_2$

(4)

x_1	$\frac{1}{5}$	$-\frac{2}{5}$
x_2	$\frac{2}{5}$	$\frac{1}{5}$
	$= u$	$= v$

(Notice we are not using parametric elimination but rather are solving for u and v.) Since T is a linear mapping we get, from (1),

(5) $$T(x_1, x_2) = uT(1, 2) + vT(-2, 1)$$

Substituting for u and v in (5) using the values of (4) and using the given values for $T(1, 2)$ and $T(-2, 1)$ yields

(6) $$\begin{aligned} T(x_1, x_2) &= (\tfrac{1}{5}x_1 + \tfrac{2}{5}x_2)(3, 4) + (-\tfrac{2}{5}x_1 + \tfrac{1}{5}x_2)(1, 2) \\ &= (\tfrac{1}{5}x_1 + \tfrac{7}{5}x_2, 2x_2) = (y_1, y_2) \end{aligned}$$

In tableau form the linear mapping T is

(7)

x_1	$1/5$	0
x_2	$7/5$	2
	$= y_1$	$= y_2$

EXAMPLE 7

Let $T: R^3 \to R^3$ be a linear mapping such that $T(1, 0, 0) = (1, 2, 3)$, $T(0, 1, 0) = (2, 3, 1)$, and $T(0, 0, 1) = (0, -2, 0)$. Since $(x_1, x_2, x_3) = x_1(1, 0, 0) + x_2(0, 1, 0) + x_3(0, 0, 1)$,

$$\begin{aligned} T(x_1, x_2, x_3) &= x_1 T(1, 0, 0) + x_2 T(0, 1, 0) + x_3 T(0, 0, 1) \\ &= x_1(1, 2, 3) + x_2(2, 3, 1) + x_3(0, -2, 0) \\ &= (x_1 + 2x_2, 2x_1 + 3x_2 - 2x_3, 3x_1 + x_2) \end{aligned}$$

In tableau form

x_1	1	2	3
x_2	2	3	1
x_3	0	-2	0
	$= y_1$	$= y_2$	$= y_3$

Observe that the respective images of (1, 0, 0), (0, 1, 0), and (0, 0, 1) appear as consecutive rows of the tableau. A matrix equation for T is

$$\begin{bmatrix} x_1 & x_2 & x_3 \end{bmatrix} \begin{bmatrix} 1 & 2 & 3 \\ 2 & 3 & 1 \\ 0 & -2 & 0 \end{bmatrix} = \begin{bmatrix} y_1 & y_2 & y_3 \end{bmatrix}$$

Example 7 suggests

THEOREM 7.1. Let $T: R^m \to R^n$ be a linear mapping. If

$$\begin{aligned} T(1, 0, 0, \ldots, 0) &= (a_1, a_2, \ldots, a_n) \\ T(0, 1, 0, \ldots, 0) &= (b_1, b_2, \ldots, b_n) \\ &\vdots \\ T(0, 0, \ldots, 1) &= (k_1, k_2, \ldots, k_n) \end{aligned}$$

then

(1)

$$\begin{array}{c|cccc|} & & & & \\ x_1 & a_1 & a_2 & \cdots & a_n \\ x_2 & b_1 & b_2 & \cdots & b_n \\ \vdots & \vdots & \vdots & & \vdots \\ x_m & k_1 & k_2 & \cdots & k_n \\ \hline \end{array}$$
$$= y_1 \quad = y_2 \quad \cdots = y_n$$

is the tableau for T.

The proof is an extension of that of Example 7. In words, Theorem 7.1 states that any linear mapping has a tableau where rows are the images of the *natural basis* of the domain. In matrix language (1) is

(2) $$\begin{bmatrix} x_1 & x_2 & \cdots & x_m \end{bmatrix} \begin{bmatrix} a_1 & a_2 & \cdots & a_n \\ b_1 & b_2 & \cdots & b_n \\ \vdots & \vdots & & \vdots \\ k_1 & k_2 & \cdots & k_n \end{bmatrix} = \begin{bmatrix} y_1 & y_2 & \cdots & y_n \end{bmatrix}$$

It is also true that if any mapping is expressible as (1) or (2) of Theorem 7.1, then it is linear. In mapping language let $T: R^m \to R^n$ and let $\mathbf{A}$ be the $m \times n$ matrix such that for all $\mathbf{X}$ in R^m,

$$\mathbf{XA} = \mathbf{Y} = T(\mathbf{X})$$

($\mathbf{X}$ and $\mathbf{Y}$ are one-rowed matrices). Then T is a linear mapping.

To see this observe that for all $\mathbf{X}_1$ and $\mathbf{X}_2$ in R^m,

$$\begin{aligned} T(\mathbf{X}_1) &= \mathbf{X}_1\mathbf{A} \\ T(\mathbf{X}_2) &= \mathbf{X}_2\mathbf{A} \\ T(\mathbf{X}_1) + T(\mathbf{X}_2) &= \mathbf{X}_1\mathbf{A} + \mathbf{X}_2\mathbf{A} = (\mathbf{X}_1 + \mathbf{X}_2)\mathbf{A} \\ &= T(\mathbf{X}_1 + \mathbf{X}_2) \end{aligned}$$

Also

$$\begin{aligned} T(c\mathbf{X}) &= (c\mathbf{X})\mathbf{A} = c(\mathbf{XA}) \\ &= cT(\mathbf{X}) \end{aligned}$$

Therefore, T is a linear mapping.

This result, combined with Theorem 7.1, shows that there is a one-to-one correspondence between the set of linear mappings $R^m \to R^n$ and the set of $m \times n$ matrices, in which the rows of the matrices are the images of the natural basis vectors.

Exercises

1. For each of the following mappings, state the domain, range, and express its rule both in tableau form (equations in columns) and as a matrix equation.

(a) $(x, y) \mapsto (2x, 3y)$.

(b) $(x, y) \mapsto (2x, 3y, x - y)$.

(c) $(x_1, x_2, x_3) \mapsto (x_1 + x_2 + x_3, 2x_2 - x_3)$

(d) $x \mapsto (2x, 3x, 4x)$.

2. Determine whether or not each of the following is a linear mapping.

(a) $(x_1, x_2) \mapsto (x_1 + x_2, x_1 - x_2)$

(b) $(x_1, x_2) \mapsto (x_1 + x_2 + x_3)$.

(c) $(x_1, x_2, x_3) \mapsto (0, x_2)$.

(d) $(x_1, x_2) \mapsto (x_1^2, x_2)$.

(e) $(x_1, x_2) \mapsto (0, 0)$.

(f) $(x_1, x_2) \mapsto (1, 1)$.

3. Given the mapping M with rule

$$\begin{bmatrix} x_1 & x_2 \end{bmatrix} \begin{bmatrix} 1 & 3 & 5 \\ 2 & 4 & 6 \end{bmatrix} = \begin{bmatrix} y_1 & y_2 & y_3 \end{bmatrix}$$

(a) What are the domain and codomain of M?

(b) Show that M is a linear mapping.

(c) Evaluate M at $(x_1, x_2) = (0, 0), (1, 1), (3, -1), (a, 2a)$.

4. Show that $L : R^2 \to R^3$ with

x_1	1	0	1
x_2	1	1	-1
	$= y_1$	$= y_2$	$= y_3$

is a linear mapping by showing that it satisfies the definition of a linear mapping.

5. Show that $S: R^2 \to R^2$ with $(x, y) \mapsto \sqrt[5]{x^5 + y^5}$ has the homogeneity property but not the additive property.

6. Given that $L: R^2 \to R^3$ is a linear mapping, find its rule if:
(a) $L(1, 0) = (1, 2, 3)$ and $L(0, 1) = (-1, 1, -2)$.
(b) $L(1, 2) = (1, 0, 1)$ and $L(2, 1) = (0, 1, 1)$.

7. Given that $M : R^3 \to R^2$ is a linear mapping, find its rule if
(a) $M(1, 0, 0) = (2, 3)$, $M(0, 1, 0) = (1, 5)$, $M(0, 0, 1) = (4, 2)$.
(b) $M(1, 1, 0) = (1, 1)$, $M(1, 0, 1) = (2, 2)$, $M(2, 1, 0) = (3, 3)$.

8. Show that $(x_1, x_2, x_3, x_4) \mapsto (x_1 + x_2, x_3 + x_4)$ is a linear mapping $L : R^4 \to R^2$.

9. Let $L : R^3 \to R^2$ be a linear mapping with rule $(x_1, x_2, x_3) \mapsto (x_1, x_2, a)$, where a is a constant. Find a.

10. Let $T: R^m \to R^n$ be a linear mapping and let $\mathbf{A}_1$, $\mathbf{A}_2$, and $\mathbf{A}_3$ be in R^m.
(a) Show that $T(\mathbf{A}_1 + \mathbf{A}_2 + \mathbf{A}_3) = T(\mathbf{A}_1) + T(\mathbf{A}_2) + T(\mathbf{A}_3)$.
(b) Show that $T(a_1\mathbf{A}_1 + a_2\mathbf{A}_2 + a_3\mathbf{A}_3) = a_1T(\mathbf{A}_1) + a_2T(\mathbf{A}_2) + a_3T(\mathbf{A}_3)$.

7.2 Surjections, Injections, and Bijections

In Chapter 2 we discussed image and kernel sets of tableaus. Chapter 2 uses tableaus with equations read in rows, and this chapter uses tableaus with equations read in columns. This requires a modification of the schema of Chapter 2. Since the tableaus in this chapter are transposes of tableaus in Chapter 2, this modification is effected by transposing the schema. The result is

(1)

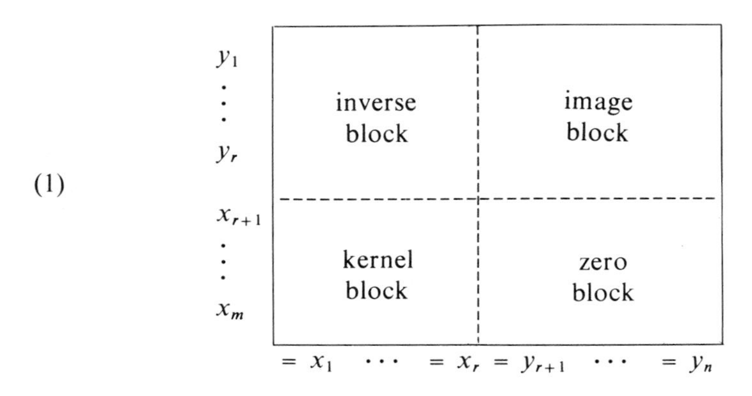

We of course recognize this as a schema for a linear mapping.

EXAMPLE 1

The linear mapping with tableau

$$
(1)\qquad
\begin{array}{c|cccc|}
\hline
x_1 & 4 & -4 & 5 & 2 \\
x_2 & 1 & 5 & 2 & -4 \\
x_3 & -6 & -10 & -9 & 8 \\
x_4 & 6 & -2 & 8 & 0 \\
x_5 & -3 & -9 & -5 & 7 \\
\hline
 & = y_1 & = y_2 & = y_3 & = y_4
\end{array}
$$

has a fully exchanged tableau

$$
(2)\qquad
\begin{array}{c|cccc|}
\hline
x_1 & -3/4 & 0 & 1/2 & 0 \\
y_3 & 5/4 & -4 & -21/2 & 6 \\
x_3 & 1/4 & 0 & -3/2 & -2 \\
y_1 & -3/2 & 5 & -14 & -8 \\
y_4 & 1/4 & -2 & 3/2 & 1 \\
\hline
 & = x_4 & = y_2 & = x_2 & = x_5
\end{array}
$$

On rearranging rows and columns so that the zeros that prevent further pivot exchanges appear in the lower right block, (2) becomes

$$
(3)\qquad
\begin{array}{c|ccc:c|}
\hline
y_1 & -14 & -3/2 & -8 & 5 \\
y_3 & -21/2 & 5/4 & 6 & -4 \\
y_4 & 3/2 & 1/4 & 1 & -2 \\
\hdashline
x_1 & 1/2 & -3/4 & 0 & 0 \\
x_3 & -3/2 & 1/4 & -2 & 0 \\
\hline
 & = x_2 & = x_4 & = x_5 & = y_2
\end{array}
$$

The image set is determined by

(4)

y_1	5
y_3	-4
y_4	-2
	$= y_2$

The kernel set is determined by

(5)

x_1	$\frac{1}{2}$	$-\frac{3}{4}$	0
x_3	$-\frac{3}{2}$	$\frac{1}{4}$	-2
	$= x_2$	$= x_4$	$= x_5$

Adding unit columns to (4) and (5), the image set and the kernel set become, respectively,

(6)

y_1	1	5	0	0
y_3	0	-4	1	0
y_4	0	-2	0	1
	$= y_1$	$= y_2$	$= y_3$	$= y_4$

and

(7)

x_1	1	$\frac{1}{2}$	0	$-\frac{3}{4}$	0
x_3	0	$-\frac{3}{2}$	1	$\frac{1}{4}$	-2
	$= x_1$	$= x_2$	$= x_3$	$= x_4$	$= x_5$

We conclude from (6) that the image set of the linear mapping is a vector subspace of R^4 of dimension three and that $\{(1, 5, 0, 0), (0, -4, 1, 0), (0, -2, 0, 1)\}$ is an elementary basis of this subspace. From (7) the kernel is a vector subspace of R^5 of dimension two with an elementary basis $\{(1, \frac{1}{2}, 0, -\frac{3}{4}, 0), (0, -\frac{3}{2}, 1, \frac{1}{4}, -2)\}$.

The sum of the dimensions of the image set and kernel is equal to the dimension of the domain of the linear mapping. The general result is

THEOREM 7.2. Let $L: R^m \to R^n$ be a linear mapping whose initial tableau has pivot rank r. Then

1. The image set of L is a vector subspace of R^n.
2. The kernel set of L is a vector subspace of R^m.
3. The dimension of the image set is r.
4. The dimension of the kernel set is $m - r$.

Proof: The proof follows the same line of reasoning as used in Example 1.

In Chapter 2 we observed that every fully exchanged tableau fit into one of four categories:

1. All basic variables are exchanged but some nonbasic variables are not.
2. All nonbasic variables are exchanged but some basic variables are not.
3. All basic and all nonbasic variables are exchanged.
4. Some basic variables and some nonbasic variables are not exchanged

There is a much more general language of mapping into which these four categories fit. We place our linear mappings into this general language so that they may be viewed in a broader framework.

DEFINITION 7.2. Let $T: R^m \to R^n$ be a linear mapping. Then T is a *surjection* if its image set is its codomain (category 1). A surjection is also called an *onto mapping,* suggesting that T "covers" its entire codomain.

DEFINITION 7.3. Let $T:R^m \to R^n$ be a linear mapping. Then T is an *injection* if for each $\mathbf{Y}$ in the image set of T there exists exactly one $\mathbf{X}$ in R^m such that $T(\mathbf{X}) = \mathbf{Y}$ (category 2). An injection is also called a *one-to-one mapping.* Every $\mathbf{X}$ in R^m has exactly one image. If T is injective, there exists exactly one preimage for every image; thus the description, one-to-one, fits.

If all the x's in a tableau are exchanged to the bottom, then the linear mapping $L: R^m \to R^n$ is an injection; since setting the y's equal to zero forces the x's to be zero, the kernel is the zero vector. If in the fully exchanged tableau not all x's are exchanged to the bottom the kernel contains more than the zero vector, as is clear from the schema, and also the mapping is not an injection. Thus the statements "L is injective" and "the kernel of L is $\mathbf{0}_m$" imply each other.

DEFINITION 7.4. Let $T: R^m \to R^n$ be a linear mapping. Then T is a *bijection* if it is both surjective and injective (category 3). A bijection is also called an *invertible mapping*, suggesting that a bijection has an inverse mapping, $R^n \to R^m$, which is also bijective.

Using the new language of Definitions 7.2, 7.3, and 7.4, categories 1–4 are restated.

A linear mapping T belongs to one of the following:

1. T is a surjection but not an injection.
2. T is an injection but not a surjection.
3. T is a bijection.
4. T is neither a surjection nor an injection.

The following examples illustrate Definitions 7.2, 7.3, and 7.4.

EXAMPLE 2

$T_1 : R^3 \to R^2$ with tableau

(1)

x_1	2	1*
x_2	3	2
x_3	−2	1
	$= y_1$	$= y_2$

(2)

y_2	2	1
x_2	−1*	−2
x_3	−4	−1
	$= y_1$	$= x_1$

(3)

y_2	2	−3
y_1	−1	2
x_3	−4	7
	$= x_2$	$= x_1$

Since in (3) all the y's have been exchanged (have become nonbasic variables), T_1 is a surjection. The image set of T_1 is its codomain. On the other hand, all x's cannot become basic variables. Therefore, T_1 is not an injection, which means that T_1 is in category 1. There just are not enough columns for all the x's. For our choice of pivots, one of them (x_3) continues to be nonbasic. This implies that, for a given (y_1, y_2) and an arbitrarily chosen x_3, there corresponds x_1 and an x_2. That is, a given image (y_1, y_2) has many preimages (x_1, x_2, x_3) in R^3.

EXAMPLE 3

$T_2 : R^2 \to R^3$ with tableau

(1)

x_1	2	3	−2
x_2	1*	4	0
	$= y_1$	$= y_2$	$= y_3$

(2)

x_1	−2	−5	−2*
y_1	1	4	0
	$= x_2$	$= y_2$	$= y_3$

(3)

y_3	1	$5/2$	$-1/2$
y_1	1	4	0
	$= x_2$	$= y_2$	$= x_1$

Tableau (3) shows that not all the y's become nonbasic variables. Hence T_2 is not surjective. Only y_1 and y_3 can be assigned values arbitrarily and then

(4) $$y_2 = 4y_1 + \tfrac{5}{2}y_3$$

is determined.

Only those values of (y_1, y_2, y_3) which satisfy (4) are in the image set of T_2. Since the solution set of (4) is a vector plane, not every vector of R^3 is an image. On the other hand, all the x's have become basic variables. Thus T_2 is an injection and is in category 2.

For each (y_1, y_2, y_3) in the image set of T_2 there is exactly one preimage, (x_1, x_2).

EXAMPLE 4

$T_3 : R^3 \rightarrow R^3$ with tableau

(1)

x_1	1*	2	3
x_2	3	4	9
x_3	1	0	1
	$= y_1$	$= y_2$	$= y_3$

(2)

y_1	1	2	3
x_2	-3	-2*	0
x_3	-1	-2	-2
	$= x_1$	$= y_2$	$= y_3$

(3)

y_1	-2	1	3
y_2	$3/2$	$-1/2$	0
x_3	2	-1	-2*
	$= x_1$	$= x_2$	$= y_3$

(4)

y_1	1	$-1/2$	$3/2$
y_2	$3/2$	$-1/2$	0
y_3	-1	$1/2$	$-1/2$
	$= x_1$	$= x_2$	$= x_3$

All y's have become nonbasic and all x's have become basic variables. Thus T_3 is both surjective and injective, hence bijective, and belongs to category 3. Maps in category 3 are the only ones that have inverses. The inverse of T_3 (denoted T_3^{-1}) has the fully exchanged tableau (4) as its initial tableau. Note again that the method for finding the inverse of a square matrix corresponds to that of finding the inverse of an invertible mapping.

EXAMPLE 5

$T_4 : R^3 \rightarrow R^3$ with tableau

(1)

x_1	1*	2	3
x_2	3	2	1
x_3	1	1	1
	$= y_1$	$= y_2$	$= y_3$

(2)

y_1	-1	2	-1
x_2	1	-4	0
x_3	1	-1	2
	$= x_1$	$= y_2$	$= y_3$

(3)

y_1	1	2	3
x_2	-3	-4	-8
y_2	-1	-1*	-2
	$= x_1$	$= x_3$	$= y_3$

Since x_2 and y_3 cannot be exchanged, it follows that all the x's cannot become basic and all the y's cannot become nonbasic variables. Hence T is neither surjective nor injective, and belongs to category 4.

In the four preceding examples, we observe again, that the pivot rank r of the initial tableau determines the nature of a linear mapping. Summarizing and generalizing gives

THEOREM 7.3. Let $T:R^m \to R^n$ be a linear mapping whose initial tableau has pivot rank r. Then

(a) T is a surjection if and only if $n = r$.
(b) T is an injection if and only if $m = r$.
(c) T is a bijection if and only if $m = n = r$.
(d) T is neither a surjection nor an injection if and only if $m > r$ and $n > r$.

Exercises

For each mapping as defined in Exercises 1–11, determine whether it is surjective, injective, bijective, or none of these.

1.

x_1	1	2
x_2	3	4
	$= y_1$	$= y_2$

2.

x_1	1	2	−1
x_2	3	6	0
	$= y_1$	$= y_2$	$= y_3$

3.

x_1	2	−1
x_2	1	2
x_3	0	3
	$= y_1$	$= y_2$

4.

x_1	1	−2	2
x_2	−4	8	−8
x_3	3	−6	6
	$= y_1$	$= y_2$	$= y_3$

5.

x_1	1	−2	2
x_2	4	8	−8
x_3	3	−6	6
	$= y_1$	$= y_2$	$= y_3$

6.

x_1	1	−2	2
x_2	4	8	−8
x_3	3	6	6
	$= y_1$	$= y_2$	$= y_3$

7.

x_1	2	1	0	3
x_2	1	4	2	1
	$= y_1$	$= y_2$	$= y_3$	$= y_4$

8.

x	1	2	3
	$= y_1$	$= y_2$	$= y_3$

9.

x_1	1
x_2	2
x_3	3
	$= y$

10.

x_1	1	4	2	5
x_2	0	4	12	12
x_3	1	3	−1	2
	$= y_1$	$= y_2$	$= y_3$	$= y_4$

11.

x_1	2	0	4
x_2	-1	3	5
x_3	2	1	5
x_4	6	0	8
	$= y_1$	$= y_2$	$= y_3$

12. Given mapping $M: R^m \to R^n$ for which $M(\mathbf{0}_m) \neq \mathbf{0}_n$, show that M is not linear.

13. Let $\{\mathbf{A}, \mathbf{B}\}$ be a basis for R^2 and let $M: R^2 \to R^n$ be a linear mapping. Show that either
    (a) $M(\mathbf{A})$ and $M(\mathbf{B})$ are linearly independent; or
    (b) The image set of M has dimension one; or
    (c) The image set of M is $\{\mathbf{0}_n\}$.

14. Let $M: R^3 \to R^3$ be a linear mapping. Prove that if the kernel of M is a vector line, then the image of M is a vector plane.

15. Let $M: R^n \to R^n$ be a linear mapping. Prove that:
    (a) If M is injective, then the image of M is R^n.
    (b) If the image of M is R^n, then the kernel of M is $\{\mathbf{0}_n\}$.

16. Let $M: R^m \to R^n$ be a linear mapping with $\mathbf{A}$ in the image set of M such that $M(\mathbf{B}) = \mathbf{A}$. Show that $\mathbf{B}$ plus the kernel of M is a solution of the equation $M(\mathbf{X}) = \mathbf{A}$.

7.3 Affine Mappings

In this section we study mappings called affine mappings. They correspond to the tableaus with constant columns of Chapter 2. In the previous two sections we saw that linear mappings mapped vector spaces to vector subspaces. In this section we see that affine mappings map affine spaces to affine subspaces. Also we see that affine mappings have the same type of relationship to tableaus of affine subspaces that linear mappings have to tableaus of vector subspaces. In addition we see that affine mappings relate to linear mappings as affine subspace tableaus do to vector subspace tableaus.

DEFINITION 7.5. A mapping $F: R^m \to R^n$ is called an *affine mapping* if for all $\mathbf{X}$ and $\mathbf{Z}$ in R^m and any scalars s and t such that $s + t = 1$,

$$F(s\mathbf{X} + t\mathbf{Z}) = sF(\mathbf{X}) + \mathrm{t}\mathrm{F}(\mathbf{Z})$$

Since $s + t = 1, s = 1 - t$, so the above equation is

$$F((1 - t)\mathbf{X} + t\mathbf{Z}) = (1 - t)F(\mathbf{X}) + \mathrm{t}\mathrm{F}(\mathbf{Z})$$

EXAMPLE 1

Let F be the mapping $F: R^2 \to R^3$ such that

$$(x_1, x_2) \mapsto (x_1 - x_2 + 2, 2x_1 + 4x_2 - 1, 3x_1 - 5x_2 + 3)$$

We verify that F is affine with $\mathbf{X} = (x_1, x_2) = (2, -1)$ and $\mathbf{Z} = (1, 3)$.

1. $F(2, -1) = (5, -1, 14)$.
2. $(1 - t)F(2, 1) = (1 - t)(5, -1, 14) = (5 - 5t, -1 + t, 14 - 14t)$.
3. $F(1, 3) = (0, 13, -9)$.
4. $tF(1, 3) = t(0, 13, -9) = (0, 13t, -9t)$.
5. $(1 - t)\mathbf{X} + t\mathbf{Z} = (2 - t, -1 + 4t)$.
6. $F((1 - t)\mathbf{X} + t\mathbf{Z}) = F(2 - t, \ -1 + 4t) = (5 - 5t, \ -1 + 14t, \ 14 - 23t)$. (Verify this.)
7. $(1 - t)F(\mathbf{X}) + tF(\mathbf{Z}) = (1 - t)F(2, -1) + tF(1, 3) = (5 - 5t, \ -1 + 14t, \ 14 - 23t)$.

Since the equations in 6 and 7 are equal, this verifies that F is affine.

Affine subspaces are translates of vector subspaces. Our intuition tells us to expect similar relations between affine mappings and linear mappings. In this case the difference between intuition and fact is a little labor, just enough to prove the following theorem.

THEOREM 7.4. Let F and L be mappings $R^m \to R^n$ such that for all $\mathbf{X}$ in R^m,

$$F(\mathbf{X}) = L(\mathbf{X}) + F(\mathbf{0}_m) \tag{1}$$

[Letting $\mathbf{A} = F(\mathbf{0}_m)$, we write $F(\mathbf{X}) = L(\mathbf{X}) + \mathbf{A}$.] Then if L is a linear mapping, F is an affine mapping, and conversely if F is an affine mapping then L is a linear mapping.

Proof: To prove the first contention suppose that L is a linear mapping. This means that for $\mathbf{X}$ and $\mathbf{Y}$ in R^m,

$$L((1 - t)\mathbf{X} + t\mathbf{Y}) = (1 - t)L(\mathbf{X}) + tL(\mathbf{Y}) \tag{2}$$

From (1),

$$F((1 - t)\mathbf{X} + t\mathbf{Y}) = L((1 - t)\mathbf{X} + t\mathbf{Y}) + F(\mathbf{0}_m) \tag{3}$$

Substituting from (2) into (3) gives

$$F((1 - t)\mathbf{X} + t\mathbf{Y}) = (1 - t)L(\mathbf{X}) + tL(\mathbf{Y}) + F(\mathbf{0}_m) \tag{4}$$

From (1),

$$L(\mathbf{X}) = F(\mathbf{X}) - F(\mathbf{0}_m),\ L(\mathbf{Y}) = F(\mathbf{Y}) - F(\mathbf{0}_m) \tag{5}$$

Now substitute (5) into (4) to obtain

$$F((1 - t)\mathbf{X} + t\mathbf{Y}) = (1 - t)(F(\mathbf{X}) - F(\mathbf{0}_m)) + t(F(\mathbf{Y}) - F(\mathbf{0}_m)) + F(\mathbf{0}_m) \tag{6}$$

or, after appropriate cancellations,

$$F((1 - t)\mathbf{X} + t\mathbf{Y}) = (1 - t)F(\mathbf{X}) + tF(\mathbf{Y}) \tag{7}$$

which proves that F is an affine mapping.

To prove the second contention we assume that F is an affine mapping and show that L is a linear mapping. To show this we have two tasks:

(a) Show that $L(p\mathbf{X}) = pL(\mathbf{X})$.
(b) Show that $L(\mathbf{X} + \mathbf{Y}) = L(\mathbf{X}) + L(\mathbf{Y})$.

To show (a), since F is an affine mapping,

$$\begin{aligned} F(p\mathbf{X}) &= F(p\mathbf{X} + (1 - p)\mathbf{0}_m) = pF(\mathbf{X}) + (1 - p)F(\mathbf{0}_m) \\ &= p(F(\mathbf{X}) - F(\mathbf{0}_m)) + F(\mathbf{0}_m) \end{aligned} \tag{8}$$

or

$$F(p\mathbf{X}) - F(\mathbf{0}_m) = pF(\mathbf{X}) - F(\mathbf{0}_m) \tag{9}$$

From (5) we obtain

$$L(p\mathbf{X}) = pL(\mathbf{X}) \tag{10}$$

the desired result. To show (b) we start with

$$L(\mathbf{X} + \mathbf{Y}) = L[2(\tfrac{1}{2}\mathbf{X} + \tfrac{1}{2}\mathbf{Y})] \tag{11}$$

Therefore, from (10),

$$L(\mathbf{X} + \mathbf{Y}) = 2L(\tfrac{1}{2}\mathbf{X} + \tfrac{1}{2}\mathbf{Y}) \tag{12}$$

From (5),

$$L(\tfrac{1}{2}\mathbf{X} + \tfrac{1}{2}\mathbf{Y}) = F(\tfrac{1}{2}\mathbf{X} + \tfrac{1}{2}\mathbf{Y}) - F(\mathbf{0}_m) \tag{13}$$

which when substituted into (12) gives

$$L(\mathbf{X} + \mathbf{Y}) = 2F(\tfrac{1}{2}\mathbf{X} + \tfrac{1}{2}\mathbf{Y}) - 2F(\mathbf{0}_m) \tag{14}$$

Since F is an affine mapping and $\frac{1}{2} + \frac{1}{2} = 1$, (14) becomes

$$L(\mathbf{X} + \mathbf{Y}) = 2[\tfrac{1}{2}F(\mathbf{X}) + \tfrac{1}{2}F(\mathbf{Y})] - 2F(\mathbf{0}_m) \tag{15}$$

or

$$L(\mathbf{X} + \mathbf{Y}) = [F(\mathbf{X}) - F(\mathbf{0}_m)] + [F(\mathbf{Y}) - F(\mathbf{0}_m)] \tag{16}$$

which by (5) yields

$$L(\mathbf{X} + \mathbf{Y}) = L(\mathbf{X}) + L(\mathbf{Y}) \tag{17}$$

the desired result.

We already know that any linear mapping is expressible as a tableau. The significance of Theorem 7.4 is that every affine mapping is also so expressible. The tableau of the affine mapping has an additional row $F(\mathbf{0}_m)$. Formally stated we have

THEOREM 7.5. Let $F:R^m \to R^n$ be an affine mapping and $L:R^m \to R^n$ a linear mapping such that $F(\mathbf{X}) = L(\mathbf{X}) + F(\mathbf{0}_m)$. Then the tableau for F is the tableau for L with a row added on. The row has a 1 at the left and its components are those of $F(\mathbf{0}_m)$.

In matrix language let $\mathbf{A}$ be the matrix of L so that L is given by

$$\mathbf{XA} = \mathbf{Y}$$

Then F is given by

$$\mathbf{XA} + F(\mathbf{0}_m) = \mathbf{Y}$$

EXAMPLE 2

Let $F:R^2 \to R^3$ have the rule

$$(x_1, x_2) \mapsto (x_1 + 3x_2 + 2, 2x_1 - 2x_2 - 1, -x_1 + x_2 + 4)$$

or

$$(x_1, x_2) \mapsto (x_1 + 3x_2, 2x_1 - 2x_2, -x_1 + x_2) + (2, -1, 4)$$

Then

$$F(\mathbf{0}_2) = (2, -1, 4)$$

$L(\mathbf{X})$ has the rule

$$(x_1, x_2) \mapsto (x_1 + 3x_2, 2x_1 - 2x_2, -x_1 + x_2)$$

The tableau for $L(\mathbf{X})$ is

(1)

x_1	1	2	-1
x_2	3	-2	1
	$= y_1$	$= y_2$	$= y_3$

The tableau for $F(\mathbf{X})$ is obtained by adding $F(\mathbf{0}_2)$ to (1):

(2)

x_1	1*	2	-1
x_2	3	-2	1
1	2	-1	4
	$= y_1$	$= y_2$	$= y_3$

The matrix equation for F is

$$(3) \quad \begin{bmatrix} x_1 & x_2 \end{bmatrix} \begin{bmatrix} 1 & 2 & -1 \\ 1 & -2 & 1 \end{bmatrix} + \begin{bmatrix} 2 & -1 & 4 \end{bmatrix} = \begin{bmatrix} y_1 & y_2 & y_3 \end{bmatrix}$$

To find the image set of F use the parametric elimination algorithm on (2), yielding

(4)

y_1	2	-1
x_2	-8	4*
1	-5	6
	$= y_2$	$= y_3$

(5)

y_1	0
y_3	-2
1	7
	$= y_2$

Tableau (5) represents an affine plane in R^3. Adding unit columns gives

(6)

y_1	1	0	0
y_3	0	-2	1
1	0	7	0
	$= y_1$	$= y_2$	$= y_3$

which means that (1, 0, 0) and (0, −2, 1) form an elementary basis for the image set of F.

EXAMPLE 3

The tableau for $T: R^3 \to R^3$ with rule

$$(x_1, x_2, x_3) \mapsto (x_1, x_2, x_3) + (a_1, a_2, a_3)$$

is

x_1	1	0	0
x_2	0	1	0
x_3	0	0	1
1	a_1	a_2	a_3
	$= y_1$	$= y_2$	$= y_3$

The associated linear mapping is the identity mapping which takes each element of R^3 to itself. T is called a *translation.*

EXAMPLE 4

Let $\mathbf{A} = (3, 1, 4)$ be a vector (point) in R^3 and S the mapping $R^3 \to R^3$ which associates with each vector $\mathbf{X}$ the vector $\mathbf{Y}$ such that $\mathbf{A}$ is the midpoint of the segment $\mathbf{XY}$. (See Figure 7.2.)

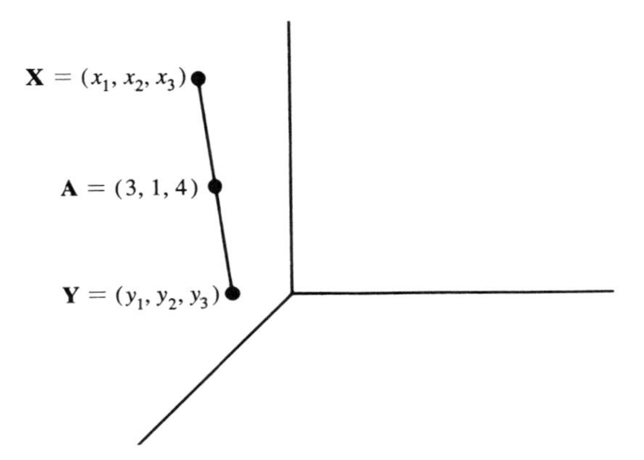

Figure 7.2

Then $(\mathbf{X} + \mathbf{Y})/2 = \mathbf{A}$, or $\mathbf{Y} = 2\mathbf{A} - \mathbf{X}$. Let $\mathbf{Y} = (y_1, y_2, y_3)$ and $\mathbf{x} = (x_1, x_2, x_3)$. Then

$$(y_1, y_2, y_3) = 2(3, 1, 4) - (x_1, x_2, x_3) = (6 - x_1, 2 - x_2, 8 - x_3)$$

In tableau form,

x_1	-1	0	0
x_2	0	-1	0
x_3	0	0	-1
1	6	2	8
	$= y_1$	$= y_2$	$= y_3$

This is an affine mapping. Had we started with $\mathbf{A} = (0, 0, 0)$ it would be a linear mapping. This kind of mapping, for any $\mathbf{A}$, is called a *symmetry in a point.*

EXAMPLE 5

Find the category of the affine mapping $A : R^2 \rightarrow R^3$ with tableau

(1)

x_1	1^*	2	-1
x_2	3	-2	1
1	2	-1	4
	$= y_1$	$= y_2$	$= y_3$

Pivoting as indicated,

(2)

y_1	1	2	-1
x_2	-3	-8	4*
1	-2	-5	6
	$= x_1$	$= y_2$	$= y_3$

(3)

y_1	$7/4$	0	$1/4$
y_3	$-3/4$	-2	$1/4$
1	$5/2$	7	$-3/2$
	$= x_1$	$= y_2$	$= x_2$

All x's in the fully exchanged tableau (3) are nonbasic variables. Therefore, any (y_1, y_2, y_3) in the image set of A has exactly one preimage. Hence A is injective. (The definitions of surjective, injective, and bijective for affine mappings are obtained from Definitions 7.2, 7.3, and 7.4 by replacing the word "linear" with the word "affine.") On the other hand, not all y's are nonbasic variables. In fact, a triple (y_1, y_2, y_3) is in the image of A if and only if $y_2 = -2y_3 + 7$. This equation defines a vector plane in R^3, not all of R^3. Hence A is not surjective.

When evaluating linear mappings $L: R^m \rightarrow R^n$, the kernel of the mapping is the preimage set of $\mathbf{0}_n$. The kernel of an affine mapping $F: R^m \rightarrow R^n$ is not the preimage set of $\mathbf{0}_n$. It is the preimage set of $F(\mathbf{0}_m)$. But $F(\mathbf{0}_m)$ is the vector of the row of the 1. Thus the preimage set of this vector is the kernel of F. In Example 5, $A(0, 0) = (2, -1, 4)$. Therefore, the kernel is the preimage set of $(2, -1, 4)$.

Continuing Example 5, to find the kernel of A, set $(y_1, y_2, y_3) = (2, -1, 4)$. Then (3) becomes

(4)

2	$7/4$	0	$1/4$
4	$-3/4$	-2	$1/4$
1	$5/2$	7	$-3/2$
	$= x_1$	$= -1$	$= x_2$

From columns 1 and 3 we deduce that $(x_1, x_2) = (0, 0)$ is the only possible preimage of $(2, -1, 4)$. Therefore, the kernel is $\{(0, 0)\}$. This agrees with the fact that A is injective.

EXAMPLE 6

Let $S: R^3 \rightarrow R^2$ be the affine mapping with tableau

(1)

x_1	1*	2
x_2	3	4
x_3	5	6
1	7	8
	$= y_1$	$= y_2$

Pivot-exchange, as indicated, yielding

(2)

y_1	1	2
x_2	−3	−2*
x_3	−5	−4
1	−7	−6
	$= x_1$	$= y_2$

(3)

y_1	−2	1
y_2	3/2	−1/2
x_3	1	−2
1	2	−3
	$= x_1$	$= x_2$

S is not injective because x_3 continues to be a nonbasic variable in the fully exchanged tableau (3). However, S is surjective because all y's have become nonbasic variables in (3). The image of S is R^2. The kernel of S is the set of preimages of $(y_1, y_2) = (7, 8)$. Substituting (7, 8) in (3) gives

(4)

7	−2	1
8	3/2	−1/2
x_3	1	−2
1	2	−3
	$= x_1$	$= x_2$

which reduces to

(5)

x_3	1	-2
	$= x_1$	$= x_2$

The kernel is read directly from the tableau (3) by reading only the row of x_3. This is because the kernel must contain $(x_1, x_2, x_3) = (0, 0, 0)$. Substituting these values in (3) gives

(6)

y_1	-2	1
y_2	$3/2$	$-1/2$
0	1	-2
1	2	-3
	$= 0$	$= 0$

Clearly from (6) the values of (y_1, y_2) must be selected so that each equation adds to zero without the row of x_3 being considered.

The schema for the fully exchanged tableau for linear mappings holds for affine mappings with the row of the 1 adjoined to the image and inverse blocks.

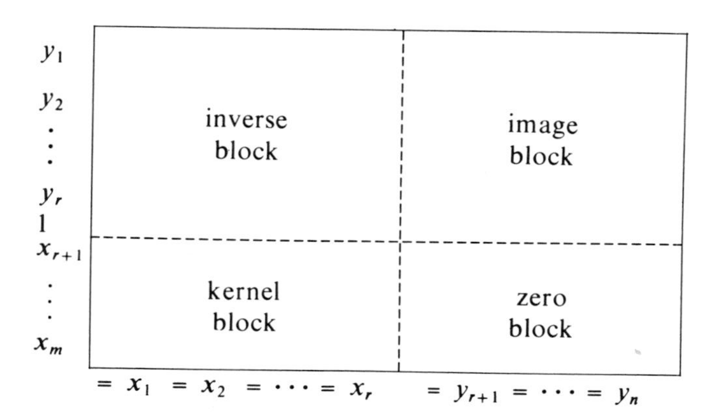

Thus for an affine mapping $T: R^m \rightarrow R^n$ with pivot rank r the image set is an affine subspace of dimension r and the kernel a vector subspace of dimension $m - r$.

As a point of interest, recall from our discussion of tableaus in Chapter 2 that a linear mapping may be regarded as an affine mapping with translate the zero vector; that is, there are all zeros in the row of the "1." Viewing a linear mapping in this light we see that the definition of kernel for affine mappings as the preimage set of $F(\mathbf{0}_m)$ and the definition of kernel for linear mappings as the preimage set of $\mathbf{0}_n$ are consistent with each other.

Exercises

1. Show that each mapping defined below is an affine mapping by showing that it satisfies Definition 7.5 $[F((1-t)\mathbf{X} + t\mathbf{Y}) = (1-t)F(\mathbf{X}) + tF(\mathbf{Y})]$.

(a)

x_1	1	2	3
x_2	2	−1	0
1	1	4	−2
	$= y_1$	$= y_2$	$= y_3$

(b)

x	1	2	3	0
	$= y_1$	$= y_2$	$= y_3$	$= y_4$

(c)

x_1	3
x_2	1
x_3	0
1	2
	$= y$

(d)

x_1	1	2	3
x_2	2	4	−1
x_3	0	6	2
1	4	8	0
	$= y_1$	$= y_2$	$= y_3$

(e)

x_1	1	2	0	4
x_2	0	−1	−1	1
x_3	−1	6	3	3
1	3	2	4	−1
	$= y_1$	$= y_2$	$= y_3$	$= y_4$

(f)

x_1	2	−1	0
x_2	1	2	3
x_3	−2	3	2
x_4	0	4	4
1	3	2	−1
	$= y_1$	$= y_2$	$= y_3$

2. Express as a matrix equation the rule of the mapping in:
   (a) Exercise 1 (a).
   (b) Exercise 1 (c).
   (c) Exercise 1 (e).

3. Find the kernel and image of the mapping in:
   (a) Exercise 1 (a).
   (b) Exercise 1 (b).
   (c) Exercise 1 (d).
   (d) Exercise 1 (e).
   (e) Exercise 1 (f).
   (Keep your results for Exercises 4 and 5).

4. What are the dimensions of the kernel and the image of the mapping in:
   (a) Exercise 1 (a).
   (b) Exercise 1 (b).
   (c) Exercise 1 (d).
   (d) Exercise 1 (e).
   (e) Exercise 1 (f).

5. Determine for each mapping whether it is surjective, injective, bijective, or none of these. The mappings are those in:
   (a) Exercise 1 (a).
   (b) Exercise 1 (b).
   (c) Exercise 1 (d).
   (d) Exercise 1 (e).
   (e) Exercise 1 (f).

6. (a) Explain why a translation is necessarily affine but not necessarily linear.
   (b) Find the rule of the translation $R^4 \rightarrow R^4$ that moves (2, 1, 3, 2) to (−1, 3, 0, 1).

7. Find the rule of the symmetry $R^3 \rightarrow R^3$ in:
   (a) (0, 0, 0).
   (b) (3, −1, 2).

8. Let M be an affine mapping $R^m \rightarrow R^n$ whose tableau has rank r. Prove that its image is an affine subspace with dimension r and that its kernel is a vector subspace with dimension $m - r$.

9. Let **A** and **B** be vectors in R^n with **M** the midpoint of the segment between **A** and **B**. That is, $\mathbf{M} = \frac{1}{2}(\mathbf{A} + \mathbf{B})$. Let T be an affine mapping. Prove that $T(\mathbf{M})$ is the midpoint of the segment between $T(\mathbf{A})$ and $T(\mathbf{B})$.

7.4 Euclidean Mappings

Recall from Chapter 6 that when an inner product is imposed on the vectors of R^n, then R^n is called E^n, Euclidean n-dimensional space. Just as E^n derives from R^n, so Euclidean mappings derive from linear and affine mappings. Linear and affine mappings are defined in terms of their intrinsic properties. We define a *Euclidean mapping* to be a linear or affine mapping $F: E^n \rightarrow E^n$.

EXAMPLE 1

Let $\mathbf{A} = (a_1, a_2)$, $\mathbf{B} = (b_1, b_2)$ be vectors in E^2. Let $S: E^2 \rightarrow E^2$ be a linear mapping with $(x_1, x_2) \mapsto (x_1, -x_2)$. Then

$$S(\mathbf{A}) = (a_1, -a_2)$$
$$S(\mathbf{B}) = (b_1, -b_2)$$

(See Figure 7.3.)

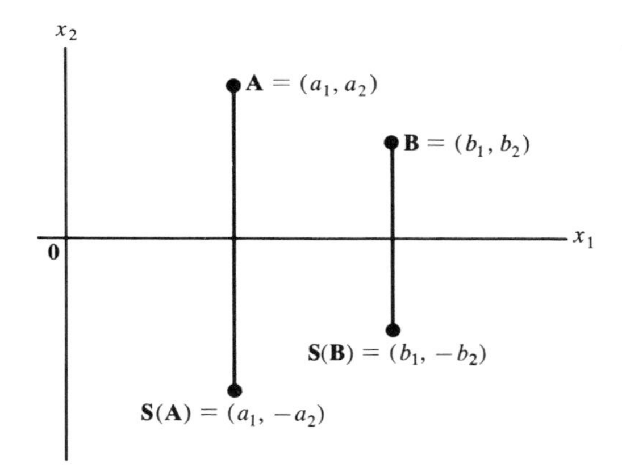

Figure 7.3

Let us consider how S affects the inner product $\mathbf{A} \cdot \mathbf{B}$; that is, what relationship exists between $\mathbf{A} \cdot \mathbf{B}$ and $S(\mathbf{A}) \cdot S(\mathbf{B})$?

$$\mathbf{A} \cdot \mathbf{B} = (a_1, a_2) \cdot (b_1, b_2) = a_1 b_1 + a_2 b_2$$
$$S(\mathbf{A}) \cdot S(\mathbf{B}) = (a_1, -a_2) \cdot (b_1, -b_2) = a_1 b_1 + a_2 b_2$$

Thus S preserves inner products. Since S preserves inner products it preserves norms, because norms are the square roots of inner products. Thus S preserves distance between points, because $\| \mathbf{A} - \mathbf{B} \|$, the distance from $\mathbf{A}$ to $\mathbf{B}$, is the square root of $(\mathbf{A} - \mathbf{B}) \cdot (\mathbf{A} - \mathbf{B})$. For a similar reason, S preserves angle measure. A tableau for S is

x_1	1	0
x_2	0	-1
	$= y_1$	$= y_2$

The mapping S is a *Euclidean mapping,* which preserves distance. Distance-preserving Euclidean mappings are called *isometries.*

EXAMPLE 2

Let $T : E^2 \to E^2$ have rule

$$(x_1, x_2) \mapsto (x_1 + p, x_2 + q)$$

We see that T is an affine mapping. It is a translation by (p, q). (See Figure 7.4.) Its tableau is

x_1	1	0
x_2	0	1
1	p	q
	$= y_1$	$= y_2$

Let $(p, q) = \mathbf{C}$. Then

$$\| T(\mathbf{A}) - T(\mathbf{B}) \| = \| (\mathbf{A} + \mathbf{C}) - (\mathbf{B} + \mathbf{C}) \| = \| \mathbf{A} - \mathbf{B} \|$$

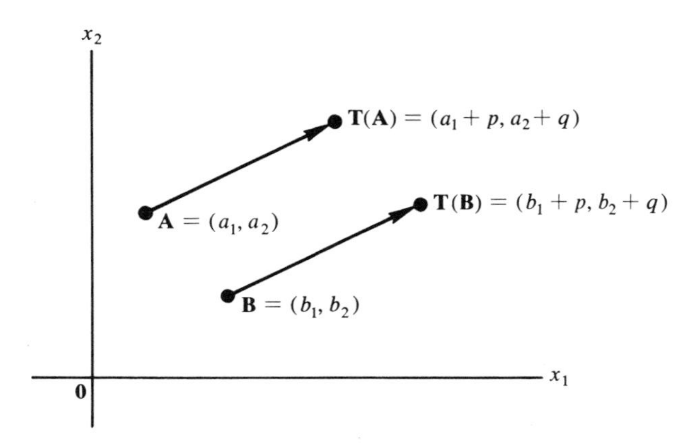

Figure 7.4

Therefore, the distance from **A** to **B** is equal to the distance from $T(\mathbf{A})$ to $T(\mathbf{B})$; that is, T preserves distance. $\mathbf{A}\cdot\mathbf{B} = a_1b_1 + a_2b_2$ while

$$T(\mathbf{A})\cdot T(\mathbf{B}) = (a_1 + p)(b_1 + p) + (a_2 + q)(b_2 + q)$$

Thus T does not preserve inner products (and therefore does not preserve norms).

Thus the fact that a Euclidean mapping preserves distances does not imply that it preserves inner products, although clearly if it preserves inner products of necessity it preserves distance. We have that T is an isometry since it does preserve distance.

Let α be any vector line in E^n and **X** any vector in E^n. Let the vector line α be generated by the unit vector **A**. (Thus $\mathbf{A}\cdot\mathbf{A} = 1$.) Let **B** be the vector in α such that $\mathbf{B} - \mathbf{X}$ and **A** are orthogonal. (See Figure 7.5.) (Pictorially, in two dimensions, the line segment $\overline{\mathbf{BX}}$ is perpendicular to the line α.) Since **B** is in α and **A** generates α, there exists a number t such that

$$\mathbf{B} = t\mathbf{A} \tag{1}$$

To find t we solve

$$\mathbf{A}\cdot(\mathbf{B} - \mathbf{X}) = 0 \tag{2}$$

$$\mathbf{A}\cdot(t\mathbf{A} - \mathbf{X}) = 0 \tag{3}$$

$$\mathbf{A}\cdot t\mathbf{A} - \mathbf{A}\cdot\mathbf{X} = 0 \tag{4}$$

$$t(\mathbf{A}\cdot\mathbf{A}) - \mathbf{A}\cdot\mathbf{X} = 0 \tag{5}$$

$$t - \mathbf{A}\cdot\mathbf{X} = 0 \qquad (\text{since } \mathbf{A}\cdot\mathbf{A} = 1) \tag{6}$$

$$t = \mathbf{A}\cdot\mathbf{X} \tag{7}$$

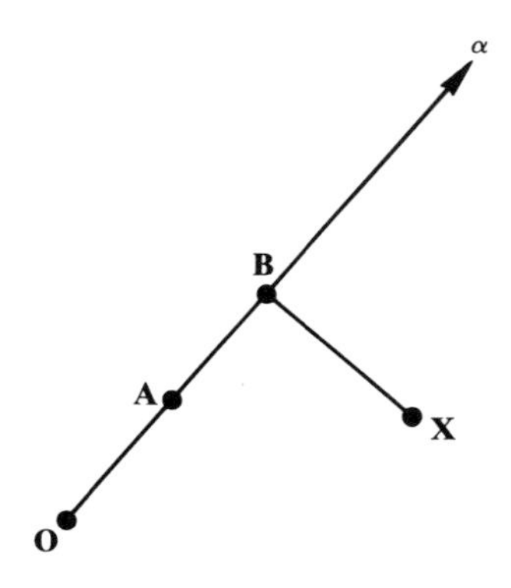

Figure 7.5

Therefore substituting (7) into (1) yields

$$\mathbf{B} = (\mathbf{A} \cdot \mathbf{X})\mathbf{A} \tag{8}$$

B is called the *orthogonal projection* of **X** on α. It corresponds to the foot of the perpendicular from a point to a line. The mapping under which each vector **X** in E^n is assigned its projection on α is denoted P_α. We write $P_\alpha: E^n \to E^n$ or $\mathbf{X} \mapsto (\mathbf{A} \cdot \mathbf{X})\mathbf{A}$, where **A** is the unit vector that generates α. $\mathbf{P}_\alpha$ is called a *projection* on α.

EXAMPLE 3

Let $\mathbf{A} = (2/3, 1/3, -2/3)$. (Verify that **A** is a unit vector.) Let α be the vector line generated by **A**.

If $\mathbf{X} = (x_1, x_2, x_3)$, then

$$\mathbf{A} \cdot \mathbf{X} = (2/3\,x_1 + 1/3\,x_2 - 2/3\,x_3)$$

$$\begin{aligned}(\mathbf{A} \cdot \mathbf{X})\mathbf{A} &= (2/3\,x_1 + 1/3\,x_2 - 2/3\,x_3)(2/3, 1/3, -2/3)\\ &= (4/9\,x_1 + 2/9\,x_2 - 4/9\,x_3, 2/9\,x_1 + 1/9\,x_2 - 2/9\,x_3,\\ &\quad -4/9\,x_1 - 2/9\,x_2 + 4/9\,x_3)\end{aligned}$$

Thus a tableau for P_α is

x_1	$4/9$	$2/9$	$-4/9$
x_2	$2/9$	$1/9$	$-2/9$
x_3	$-4/9$	$-2/9$	$4/9$
	$= b_1$	$= b_2$	$= b_3$

This shows that P_α is a linear mapping. However, P_α does not preserve distance, so it is a Euclidean mapping that is not an isometry. (Show that it does not preserve distance.)

Let α be a vector line generated by the unit vector **A**. Let $\mathbf{B} = (\mathbf{A} \cdot \mathbf{X})\mathbf{A}$ be the orthogonal projection of **X** on α. Let **Y** be such that

$$\frac{\mathbf{X} + \mathbf{Y}}{2} = \mathbf{B} \tag{1}$$

(See Figure 7.6. Pictorially, in two dimensions, **B** is the midpoint of the line segment $\overline{\mathbf{XY}}$ with α the perpendicular bisector of this line segment.) We get

$$\mathbf{Y} = 2\mathbf{B} - \mathbf{X} \tag{2}$$

$$\mathbf{Y} = 2(\mathbf{A} \cdot \mathbf{X})\mathbf{A} - \mathbf{X} \tag{3}$$

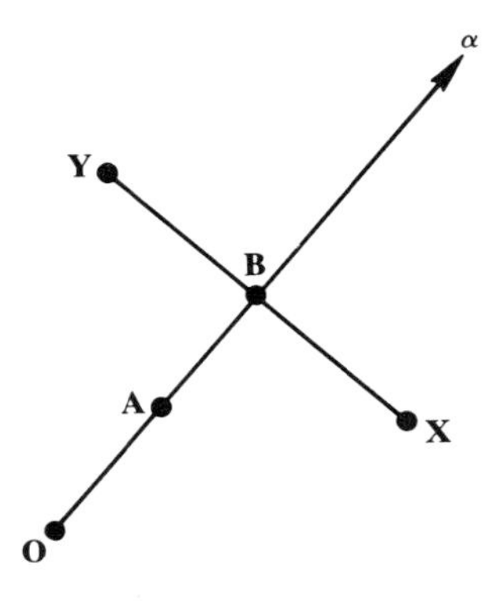

Figure 7.6

Figure 7.6 shows why the mapping $R_\alpha: E^n \to E^n$ given by $\mathbf{X} \mapsto 2(\mathbf{A}\cdot\mathbf{X})\mathbf{A} - \mathbf{X}$ is called a *reflection* in α.

EXAMPLE 4

We use the data in Example 3 to illustrate R_α, a reflection in α.

$$\begin{aligned} 2(\mathbf{A}\cdot\mathbf{X})\mathbf{A} - \mathbf{X} &= 2(\tfrac{4}{9}x_1 + \tfrac{2}{9}x_2 - \tfrac{4}{9}x_3, \tfrac{2}{9}x_1 + \tfrac{1}{9}x_2 - \tfrac{2}{9}x_3, \\ &\qquad -\tfrac{4}{9}x_1 - \tfrac{2}{9}x_2 + \tfrac{4}{9}x_3) - (x_1, x_2. x_3) \\ &= (-\tfrac{1}{9}x_1 + \tfrac{4}{9}x_2 - \tfrac{8}{9}x_3, \tfrac{4}{9}x_1 - \tfrac{7}{9}x_2 - \tfrac{4}{9}x_3, \\ &\qquad -\tfrac{1}{9}x_1 + \tfrac{4}{9}x_2 - \tfrac{8}{9}x_3) \end{aligned}$$

Thus the reflection in α has tableau

x_1	$-\tfrac{1}{9}$	$\tfrac{4}{9}$	$-\tfrac{1}{9}$
x_2	$\tfrac{4}{9}$	$-\tfrac{7}{9}$	$\tfrac{4}{9}$
x_3	$-\tfrac{8}{9}$	$-\tfrac{4}{9}$	$-\tfrac{8}{9}$
	$= y_1$	$= y_2$	$= y_3$

Example 1 is also a reflection $R_\alpha: E^2 \to E^2$.

Any reflection R_α is linear since its tableau has no constant row. That R_α preserves inner products is clear from the following. Let

$$\mathbf{Y}_1 = 2(\mathbf{A}\cdot\mathbf{X}_1)\mathbf{A} - \mathbf{X}_1 \qquad \text{and} \qquad \mathbf{Y}_2 = 2(\mathbf{A}\cdot\mathbf{X}_2)\mathbf{A} - \mathbf{X}_2$$

Recalling that $\mathbf{A}\cdot\mathbf{A} = 1$,

$$\mathbf{Y}_1 \cdot \mathbf{Y}_2 = 4(\mathbf{A}\cdot\mathbf{X}_1)(\mathbf{A}\cdot\mathbf{X}_2) - 2(\mathbf{A}\cdot\mathbf{X}_1)(\mathbf{A}\cdot\mathbf{X}_2) - 2(\mathbf{A}\cdot\mathbf{X}_2)(\mathbf{A}\cdot\mathbf{X}_1) + \mathbf{X}_1\cdot\mathbf{X}_2$$
$$= \mathbf{X}_1\cdot\mathbf{X}_2$$

Since R_α preserves inner products it preserves distances and is an isometry.

Since α is a line, $\alpha^\perp$, its orthogonal complement, is a hyperplane. Corresponding to the reflection

$$R_\alpha : E^n \to E^n \qquad \text{given by} \qquad \mathbf{X} \mapsto 2(\mathbf{A}\cdot\mathbf{X})\mathbf{A} - \mathbf{X}$$

is

$$R_{\alpha\perp} : E^n \to E^n \qquad \text{given by} \qquad \mathbf{X} \mapsto \mathbf{X} - 2(\mathbf{A}\cdot\mathbf{X})\mathbf{A}$$

The tableau of $R_{\alpha\perp}$ is obtained by changing the sign of each entry in the tableau of R_α. The mapping $R_{\alpha\perp}$ is also an isometry.

EXAMPLE 5

For R_α as in Example 4, $R_{\alpha\perp}$ has the tableau

x_1	$1/9$	$-4/9$	$1/9$
x_2	$-4/9$	$7/9$	$-4/9$
x_3	$8/9$	$4/9$	$8/9$
	$= y_1$	$= y_2$	$= y_3$

Exercises

1. Let $\mathbf{A} = (3/5, 4/5)$ generate the vector line α in E^2.
   (a) Find the tableau for the projection P_α.
   (b) Show that the P_α is neither surjective nor injective.
   (c) Find the image set and kernel of P_α.

2. Let $\mathbf{A} = (3/5, 4/5)$ generate vector line α in E^2.
   (a) Find the tableau of the reflection R_α.
   (b) Show that R_α is bijective and find its inverse.
   (c) Find the image set and kernel of R_α.

3. Using $\mathbf{A}$ and α as defined in Exercises 1 and 2:
   (a) Find the tableau of $R_{\alpha\perp}$.
   (b) Show that $R_{\alpha\perp}$ is its own inverse.

4. Let $\mathbf{A} = (3/5, 4/5, 0)$ generate the vector line α in E^3. Carry out the instructions in Exercise 1.

5. Using the data in Exercise 4 carry out the instruction in Exercise 2.

6. Using the data in Exercise 4 carry out the instructions in Exercise 3.

7. Show that a projection on a vector line in E^n is linear but not an isometry.

8. Prove that a projection on a vector hyperplane in E^n is linear but not an isometry.

9. Prove that reflections in a vector line in E^n are bijections; also that reflections in a vector hyperplane are bijections.

10. Let $\mathbf{A} = (6, 8, 0)$ generate a vector line α in R^3. Find the rules for R_α and $R_{\alpha\perp}$. (*Hint:* Use the results of Exercises 4 and 5.)

7.5 Composition of Mappings

In Chapter 3 we saw that for two linear mappings $T_1: R^m \to R^p$ and $T_2: R^p \to R^n$, there is a composite mapping $(T_2 \circ T_1): R^m \to R^n$. Also if $\mathbf{A}_1$ is the matrix of T_1 and $\mathbf{A}_2$ the matrix of T_2, then $\mathbf{A}_2\mathbf{A}_1$ is the matrix of $T_2 \circ T_1$. Now that we are writing linear mappings with the nonbasic variables at the left instead of at the top, it turns out, as illustrated in Example 1, that the matrix of $T_2 \circ T_1$ is $\mathbf{A}_1\mathbf{A}_2$, not $\mathbf{A}_2\mathbf{A}_1$.

EXAMPLE 1

Let $T_1: R^3 \to R^2$ with tableau

(1)

x_1	a_1	a_2
x_2	b_1	b_2
x_3	c_1	c_2
	$= y_1$	$= y_2$

and $T_2: R^2 \to R^4$ with tableau

(2)

y_1	d_1	d_2	d_3	d_4
y_2	e_1	e_2	e_3	e_4
	$= z_1$	$= z_2$	$= z_3$	$= z_4$

Then $T_2 \circ T_1$ has as its matrix

(3)
$$\mathbf{A}_1\mathbf{A}_2 = \begin{bmatrix} a_1 & a_2 \\ b_1 & b_2 \\ c_1 & c_2 \end{bmatrix} \begin{bmatrix} d_1 & d_2 & d_3 & d_4 \\ e_1 & e_2 & e_3 & e_4 \end{bmatrix}$$

Again $\mathbf{A}_1\mathbf{A}_2$, rather than $\mathbf{A}_2\mathbf{A}_1$, is the matrix for $T_2 \circ T_1$. The reason for this is that in Chapter 3 we write tableaus with variables at the top and at the right. [You should rewrite (1) and (2) in that format and satisfy yourself that this commutes the order of matrix multiplication.]

From (3) we get

(4)
$$\mathbf{A}_1\mathbf{A}_2 = \begin{bmatrix} a_1d_1 + a_2e_1 & a_1d_2 + a_2e_2 & a_1d_3 + a_2e_3 & a_1d_4 + a_2e_4 \\ b_1d_1 + b_2e_1 & b_1d_2 + b_2e_2 & b_1d_3 + b_2e_3 & b_1d_4 + b_2e_4 \\ c_1d_1 + c_2e_1 & c_1d_2 + c_2e_2 & c_1d_3 + c_2e_3 & c_1d_4 + c_2e_4 \end{bmatrix}$$

Thus a tableau for $T_2 \circ T_1$ is

(5)

x_1	$a_1d_1 + a_2e_1$	$a_1d_2 + a_2e_2$	$a_1d_3 + a_2e_3$	$a_1d_4 + a_2e_4$
x_2	$b_1d_1 + b_2e_1$	$b_1d_2 + b_2e_2$	$b_1d_3 + b_2e_3$	$b_1d_4 + b_2e_4$
x_3	$c_1d_1 + c_2e_1$	$c_1d_2 + c_2e_2$	$c_1d_3 + c_2e_3$	$c_1d_4 + c_2e_4$
	$= z_1$	$= z_2$	$= z_3$	$= z_4$

Tableaus (1) and (2) in matrix language are

(1′) $\mathbf{X}\mathbf{A}_1 = \mathbf{Y} \quad (\mathbf{X} = [x_1 \quad x_2 \quad x_3],\ \mathbf{Y} = [y_1 \quad y_2])$

(2′) $\mathbf{Y}\mathbf{A}_2 = \mathbf{Z} \quad (\mathbf{Z} = [z_1 \quad z_2 \quad z_3 \quad z_4])$

Substituting for $\mathbf{Y}$ in (2′) from (1′) gives

$$(\mathbf{X}\mathbf{A}_1)\mathbf{A}_2 = \mathbf{Z}$$

or, since matrix multiplication is associative,

(5′)
$$\mathbf{X}(\mathbf{A}_1\mathbf{A}_2) = \mathbf{Z}$$

This is the matrix form for (5).

The results of Example 1 can be generalized. Thus the matrix of the composite $T_2 \circ T_1$ of two linear mappings T_1 and T_2 with respective matrices $\mathbf{A}_1$ and $\mathbf{A}_2$ is $\mathbf{A}_1\mathbf{A}_2$. We deduce that $T_2 \circ T_1$ is a linear mapping.

Let us examine the notion of composite mapping for Euclidean mappings. We leave as an exercise that if two Euclidean mappings are isometries, then so is their composite. (Just check that inner products are preserved.)

EXAMPLE 2

Let $R_1 : E^2 \rightarrow E^2$ with

x_1	1	0
x_2	0	-1
	$= y_1$	$= y_2$

This is a reflection in the x_1 axis. Let $R_2 : E^2 \rightarrow E^2$ with

y_1	-1	0
y_2	0	1
	$= z_1$	$= z_2$

This is a reflection in the x_2 axis. The composite $R_2 \circ R_1$ has the matrix

$$\begin{bmatrix} 1 & 0 \\ 0 & -1 \end{bmatrix} \begin{bmatrix} -1 & 0 \\ 0 & 1 \end{bmatrix} = \begin{bmatrix} -1 & 0 \\ 0 & -1 \end{bmatrix}$$

Thus the rule of the composite is

$$\begin{bmatrix} x_1 & x_2 \end{bmatrix} \begin{bmatrix} -1 & 0 \\ 0 & -1 \end{bmatrix} = \begin{bmatrix} z_1 & z_2 \end{bmatrix}$$

The tableau is

x_1	-1	0
x_2	0	-1
	$= z_1$	$= z_2$

This composite is called a *central symmetry* in the origin, so called because for all $\mathbf{X}$ and its image $\mathbf{X}'$, the origin is the midpoint of the segment $\mathbf{XX}'$.

EXAMPLE 3

This example is a generalization of Example 1, using vector lines, not necessarily axes, for reflections. Let α be the vector line E^n generated by the unit vector $\mathbf{A}$, and β the vector line generated by the unit vector $\mathbf{B}$. Then

$$R_\alpha(\mathbf{X}) = 2(\mathbf{X}\cdot\mathbf{A})\mathbf{A} - \mathbf{X} \tag{1}$$

$$R_\beta(\mathbf{X}) = 2(\mathbf{X}\cdot\mathbf{B})\mathbf{B} - \mathbf{X} \tag{2}$$

To find the rule of $R_\beta \circ R_\alpha$ replace each $\mathbf{X}$ in (2) by $R_\alpha(\mathbf{X})$. The rule of $R_\beta \circ R_\alpha$ is

$$R_\beta \circ R_\alpha(\mathbf{X}) = 2(R_\alpha(\mathbf{X})\cdot\mathbf{B})\mathbf{B} - R_\alpha(\mathbf{X}) \tag{3}$$

$$= 2([2(\mathbf{X}\cdot\mathbf{A})\mathbf{A} - \mathbf{X}]\cdot\mathbf{B})\mathbf{B} - (2(\mathbf{X}\cdot\mathbf{A})\mathbf{A} - \mathbf{X}) \tag{4}$$

$$= 2(2(\mathbf{X}\cdot\mathbf{A})\mathbf{A}\cdot\mathbf{B} - \mathbf{X}\cdot\mathbf{B})\mathbf{B} - 2(\mathbf{X}\cdot\mathbf{A})\mathbf{A} + \mathbf{X} \tag{5}$$

$$= 4(\mathbf{X}\cdot\mathbf{A})(\mathbf{A}\cdot\mathbf{B})\mathbf{B} - 2(\mathbf{X}\cdot\mathbf{B})\mathbf{B} - 2(\mathbf{X}\cdot\mathbf{A})\mathbf{A} + \mathbf{X} \tag{6}$$

Consider the special case in which $\mathbf{A} = \mathbf{B}$. Then $\alpha = \beta$ and the rule of the composite, from (6), becomes (remember $\mathbf{A}\cdot\mathbf{A} = 1$)

$$R_\alpha \circ R_\alpha(\mathbf{X}) = 4(\mathbf{X}\cdot\mathbf{A})\mathbf{A} - 2(\mathbf{X}\cdot\mathbf{A})\mathbf{A} - 2(\mathbf{X}\cdot\mathbf{A})\mathbf{A} + \mathbf{X} \tag{7}$$

$$= \mathbf{X} \tag{8}$$

Thus $R_\alpha \circ R_\alpha$ leaves each vector unaltered. It is the *identity* mapping. The mapping R_α is its own inverse.

Now consider the case in which $\mathbf{A}$ and $\mathbf{B}$ are orthogonal, whence $\mathbf{A}\cdot\mathbf{B} = 0$. Then the rule of the composite from (6) becomes

$$\mathbf{X} \mapsto -2(\mathbf{X}\cdot\mathbf{B})\mathbf{B} - 2(\mathbf{X}\cdot\mathbf{A})\mathbf{A} + \mathbf{X} \tag{9}$$

As a special instance of this case, let E^n be E^3 with $\mathbf{A} = (1, 0, 0)$, $\mathbf{B} = (0, 1, 0)$. With $\mathbf{X} = (x_1, x_2, x_3)$, $\mathbf{X}\cdot\mathbf{A} = x_1$ and $\mathbf{X}\cdot\mathbf{B} = x_2$. Then (9) becomes

$$(0, -2_{x_2}, 0) + (-2_{x_1}, 0, 0) + (x_1, x_2, x_3) = (-x_1, -x_2, x_3) \tag{10}$$

For our special case with $\mathbf{A} = (1, 0, 0)$ the matrix associated with R_α is

$$(11) \qquad \begin{bmatrix} 1 & 0 & 0 \\ 0 & -1 & 0 \\ 0 & 0 & -1 \end{bmatrix} \qquad \text{(check this)}$$

Similarly, with $\mathbf{B} = (0, 1, 0)$ the matrix associated with R_β is

$$(12) \qquad \begin{bmatrix} -1 & 0 & 0 \\ 0 & 1 & 0 \\ 0 & 0 & -1 \end{bmatrix}$$

Therefore, the matrix associated with $R_\beta \circ R_\alpha$ is

$$(13) \qquad \begin{bmatrix} 1 & 0 & 0 \\ 0 & -1 & 0 \\ 0 & 0 & -1 \end{bmatrix} \begin{bmatrix} -1 & 0 & 0 \\ 0 & 1 & 0 \\ 0 & 0 & -1 \end{bmatrix} = \begin{bmatrix} -1 & 0 & 0 \\ 0 & -1 & 0 \\ 0 & 0 & 1 \end{bmatrix}$$

Let δ be the x_3 axis generated by $(0, 0, 1)$. Then the rule of $R_\delta \circ R_\beta \circ R_\alpha$ has the matrix

$$(14) \qquad \begin{bmatrix} 1 & 0 & 0 \\ 0 & -1 & 0 \\ 0 & 0 & -1 \end{bmatrix} \begin{bmatrix} -1 & 0 & 0 \\ 0 & 1 & 0 \\ 0 & 0 & -1 \end{bmatrix} \begin{bmatrix} -1 & 0 & 0 \\ 0 & -1 & 0 \\ 0 & 0 & 1 \end{bmatrix}$$

$$= \begin{bmatrix} 1 & 0 & 0 \\ 0 & 1 & 0 \\ 0 & 0 & 1 \end{bmatrix} = \mathbf{I}_3$$

Thus the composition of all reflections in the three axes (in any order it happens) is the identity.

EXAMPLE 4

Let α be the vector generated by a unit vector $\mathbf{A}$ in E^n. Recall that the rules of R_α and $R_{\alpha\perp}$, respectively, are (see Section 7.4)
for R_α,

$$\mathbf{X} \mapsto 2(\mathbf{X} \cdot \mathbf{A})\mathbf{A} - \mathbf{X}$$

for $R_{\alpha\perp}$,

$$\mathbf{X} \mapsto \mathbf{X} - 2(\mathbf{X} \cdot \mathbf{A})\mathbf{A}$$

under $R_{\alpha\perp} \circ R_\alpha$,

$$\mathbf{X} \mapsto -\mathbf{X} \qquad \text{(check this; remember that } \mathbf{A}\cdot\mathbf{A} = 1\text{)}$$

We conclude that $R_{\alpha\perp} \circ R_\alpha$ is a central symmetry whose matrix is $-\mathbf{I}$.

Exercises

1. Find the rule of the composite $T_2 \circ T_1$ where:
   (a) $T_1 : R^2 \to R^2, (x_1, x_2) \mapsto (x_1, -x_2)$
   $T_2 : R^2 \to R^2, (x_1, x_2) \mapsto (-x_1, x_2)$.
   (b) $T_1 : R^2 \to R^2, (x_1, x_2) \mapsto (x_1 + 3, x_2 - 1)$
   $T_2 : R^2 \to R^2, (x_1, x_2) \mapsto (-x_1, x_2)$.
   (c) $T_1 : R^3 \to R^3, (x_1, x_2, x_3) \mapsto (x_1, x_2, 0)$
   $T_2 : R^3 \to R^3, (x_1, x_2, x_3) \mapsto (-x_1, x_2, -x_3)$.
   (d) $T_1 : R^2 \to R^2, (x_1, x_2) \mapsto (x_1 + 2x_2, x_2)$
   $T_2 : R^2 \to R^2, (x_1, x_2) \mapsto (2x_1 + 1, x_2)$.

2. For which of T_1 and T_2 in Exercises 1(a)–1(d) is it true that

$$T_2 \circ T_1 = T_1 \circ T_2?$$

3. Prove that $R_\alpha \circ R_{\alpha\perp} = -\mathbf{I}$.

4. Let α be a vector line in R^n. Prove that the inverse of a reflection in $\alpha^\perp$ is itself.

5. Let $\mathbf{A} = (.6, .8, 0)$ generate vector line α and $\mathbf{B} = (-.8, 0, .6)$ generate vector line β. (*Note:* **A** and **B** are unit vectors.) Find the rules of:
   (a) $R_\beta \circ R_\alpha$ and $R_\alpha \circ R_\beta$.
   (b) $R_{\beta\perp} \circ R_{\alpha\perp}$ and $R_{\alpha\perp} \circ R_{\beta\perp}$.
   (c) $R_\alpha \circ R_{\alpha\perp}$ and $R_{\alpha\perp} \circ R_\alpha$.

6. Would any of the results in Exercise 5 change if $\mathbf{A} = (6, 8, 0)$ and $\mathbf{B} = (-8, 0, 6)$? Explain.

7. Let $\mathbf{A} = (1, 1, 1)$ and α the vector line generated by **A**. Verify that $R_\alpha \circ R_{\alpha\perp} = R_{\alpha\perp} \circ R_\alpha$ and both are central symmetries.

7.6 Decomposition of Linear Mappings

Decomposition of a linear mapping seeks to express a given linear mapping as the composite of more elementary linear mappings. The method of decomposition explained here (there are others) is used for a linear mapping that is neither surjective nor injective and yields three components: a surjection, a bijection, and an injection. It is based on the method of finding the image set and kernel of the mapping.

In Chapter 2 and again in Section 7.2 we saw that by using the fundamental pivot-exchange algorithm for a linear mapping $M: R^m \rightarrow R^n$ we obtain

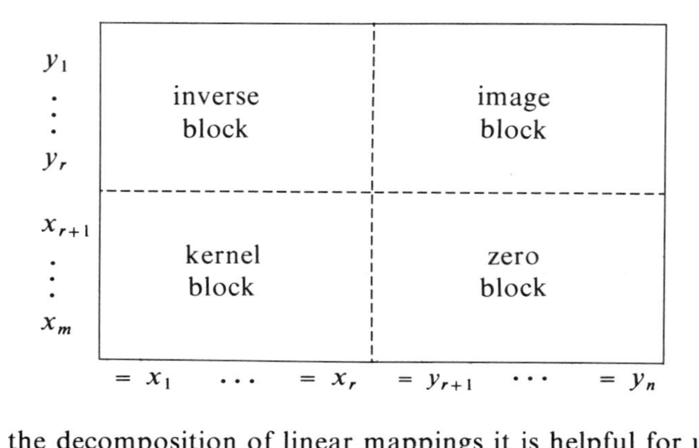

To illustrate the decomposition of linear mappings it is helpful for us to let $m = 5, r = 3$, and $n = 7$. The reader can get the general case by substituting m for 5, r for 3, and n for 7 at the appropriate places in the following development. If

$$\mathbf{X} = \begin{bmatrix} x_1 & x_2 \\ x_3 & x_4 \end{bmatrix} \quad \text{and} \quad \mathbf{Y} = \begin{bmatrix} y_1 & y_2 \\ y_3 & y_4 \end{bmatrix}$$

then

$$\begin{bmatrix} \mathbf{X} \\ \hline \mathbf{Y} \end{bmatrix} \quad \text{means the matrix} \quad \begin{bmatrix} x_1 & x_2 \\ x_3 & x_4 \\ y_1 & y_2 \\ y_3 & y_4 \end{bmatrix}$$

and

$$[\mathbf{X} \mid \mathbf{Y}] \quad \text{means the matrix} \quad \begin{bmatrix} x_1 & x_2 & y_1 & y_2 \\ x_3 & x_4 & y_3 & y_4 \end{bmatrix}$$

This notation is used for matrices of any size. Of course,

$$\begin{bmatrix} \mathbf{X} \\ \hline \mathbf{Y} \end{bmatrix}$$

implies that $\mathbf{X}$ and $\mathbf{Y}$ have the same number of columns and $[\mathbf{X} \mid \mathbf{Y}]$ implies that $\mathbf{X}$ and $\mathbf{Y}$ have the same number of rows. Thus if

$$\mathbf{X} = \begin{bmatrix} 2 & 1 \\ 3 & 5 \end{bmatrix} \quad \text{and} \quad \mathbf{Y} = \begin{bmatrix} 1 & 1 & 1 \\ 2 & 3 & 4 \end{bmatrix}$$

$[\mathbf{X} \mid \mathbf{Y}]$ makes sense but

$$\begin{bmatrix} \mathbf{X} \\ \hline \mathbf{Y} \end{bmatrix}$$

does not.

Let $M : R^5 \rightarrow R^7$ have tableau

$$(1) \qquad \begin{array}{c|ccc:cccc|} \cline{2-8} x_1 & & & & & & & \\ x_2 & & \mathbf{A} & & & & \mathbf{B} & \\ x_3 & & & & & & & \\ \hdashline x_4 & & & & & & & \\ x_5 & & \mathbf{C} & & & & \mathbf{D} & \\ \cline{2-8} & = y_1 & = y_2 & = y_3 & = y_4 & = y_5 & = y_6 & = y_7 \end{array}$$

where all x's and y's have been rearranged so that (since $r = 3$) the three pivots are the main diagonal elements of $\mathbf{A}$. Then, after pivoting, the resulting fully exchanged tableau is

$$(2) \qquad \begin{array}{c|ccc:cccc|} \cline{2-8} y_1 & & & & & & & \\ y_2 & & \mathbf{A}^{-1} & & & & \overline{\mathbf{B}} & \\ y_3 & & & & & & & \\ \hdashline x_4 & & & & & & & \\ x_5 & & \overline{\mathbf{C}} & & & & \mathbf{0} & \\ \cline{2-8} & = x_1 & = x_2 & = x_3 & = y_4 & = y_5 & = y_6 & = y_7 \end{array}$$

where, as we know, $\overline{\mathbf{C}}$ determines the kernel of M and $\overline{\mathbf{B}}$ the image of M and

$$\mathbf{0} = \begin{bmatrix} 0 & 0 & 0 & 0 \\ 0 & 0 & 0 & 0 \end{bmatrix}$$

THEOREM 7.6. The linear mapping M is the composite of

(3)

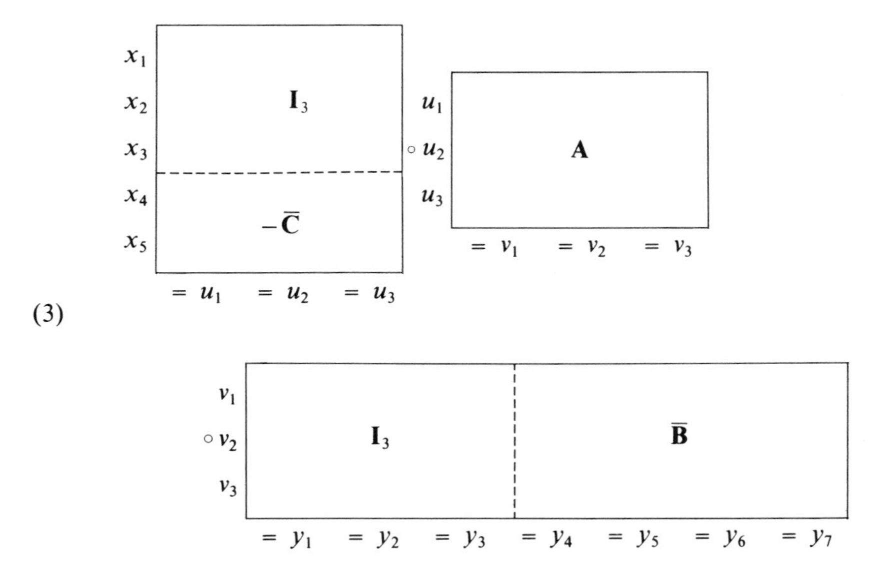

($\mathbf{I}_3$ is the identity matrix of order 3.) The left-hand mapping in (3) is a surjection, the middle one a bijection, and the right-hand one an injection.

(This theorem states that a linear mapping neither injective nor surjective is decomposable into a composition of a surjection, bijection, and injection.)

Proof: Let

$$\mathbf{X} = [x_1 \quad x_2 \quad x_3] \tag{4}$$

$$\overline{\mathbf{X}} = [x_4 \quad x_5] \tag{5}$$

$$\mathbf{Y} = [y_1 \quad y_2 \quad y_3] \tag{6}$$

$$\overline{\mathbf{Y}} = [y_4 \quad y_5 \quad y_6 \quad y_7] \tag{7}$$

From (1) and (4)–(7) we get

$$\mathbf{XA} + \overline{\mathbf{X}}\mathbf{C} = \mathbf{Y} \tag{8}$$

$$\mathbf{XB} + \overline{\mathbf{X}}\mathbf{D} = \overline{\mathbf{Y}} \tag{9}$$

From (2) and (4)–(7) we get

$$\mathbf{YA}^{-1} + \overline{\mathbf{X}}\overline{\mathbf{C}} = \mathbf{X} \tag{10}$$

(11) $$\mathbf{Y\overline{B}} + \mathbf{\overline{X}0} = \mathbf{\overline{Y}}$$

multiplying (10) on the right by **A** gives

(12) $$\mathbf{Y} + \mathbf{\overline{X}\overline{C}A} = \mathbf{XA}$$

or

(13) $$\mathbf{Y} = \mathbf{XA} - \mathbf{\overline{X}\overline{C}A}$$

From (8) and (13) we obtain

(14) $$\mathbf{XA} + \mathbf{\overline{X}C} = \mathbf{XA} - \mathbf{\overline{X}\overline{C}A}$$

From which we deduce

(15) $$\mathbf{C} = -\mathbf{\overline{C}A} \qquad \text{(How?)}$$

Now multiplying (8) on the right by $\mathbf{A}^{-1}$ gives

(16) $$\mathbf{X} = \mathbf{YA}^{-1} - \mathbf{\overline{X}CA}^{-1}$$

which when substituted for **X** in (9) yields

(17) $$(\mathbf{YA}^{-1} - \mathbf{\overline{X}CA}^{-1})\mathbf{B} + \mathbf{\overline{X}D} = \mathbf{\overline{Y}}$$

(17) and (11) yield

(18) $$(\mathbf{YA}^{-1} - \mathbf{\overline{X}CA}^{-1})\mathbf{B} + \mathbf{\overline{X}D} = \mathbf{Y\overline{B}} + \mathbf{\overline{X}0}$$

or

(19) $$\mathbf{Y}(\mathbf{A}^{-1}\mathbf{B}) + \mathbf{\overline{X}}(\mathbf{D} - \mathbf{CA}^{-1}\mathbf{B}) = \mathbf{Y\overline{B}} + \mathbf{\overline{X}0}$$

Comparing coefficients of **Y** in (19) gives

(20) $$\mathbf{A}^{-1}\mathbf{B} = \mathbf{\overline{B}}$$

or multiplying on the left by **A**

(21) $$\mathbf{B} = \mathbf{A\overline{B}}$$

Comparing coefficients of $\mathbf{\overline{X}}$ in (19) gives

(22) $$\mathbf{D} - \mathbf{CA}^{-1}\mathbf{B} = \mathbf{0}$$

or

$$\mathbf{D} = \mathbf{C}\mathbf{A}^{-1}\mathbf{B} \tag{23}$$

Now substituting (15) and (21) in (23) gives

$$\mathbf{D} = (-\bar{\mathbf{C}}\mathbf{A})\,\mathbf{A}^{-1}(\mathbf{A}\bar{\mathbf{B}}) \tag{24}$$

Thus

$$\mathbf{D} = -\bar{\mathbf{C}}\mathbf{A}\bar{\mathbf{B}} \tag{25}$$

We now show, by multiplying the matrices of (3) (which is the equivalent of composing the mapping), that the theorem is true.

First,

$$\begin{bmatrix} \mathbf{I}_3 \\ \hline -\bar{\mathbf{C}} \end{bmatrix} \begin{bmatrix} \mathbf{A} \end{bmatrix} = \begin{bmatrix} \mathbf{A} \\ \hline -\bar{\mathbf{C}}\mathbf{A} \end{bmatrix} \tag{26}$$

Now

$$\begin{bmatrix} \mathbf{A} \\ \hline -\bar{\mathbf{C}}\mathbf{A} \end{bmatrix} [\mathbf{I}_3 \mid \bar{\mathbf{B}}] = \left[\begin{array}{c|c} \mathbf{A} & \mathbf{A}\bar{\mathbf{B}} \\ \hline -\bar{\mathbf{C}}\mathbf{A} & -\bar{\mathbf{C}}\mathbf{A}\bar{\mathbf{B}} \end{array}\right] \tag{27}$$

Thus

$$\begin{bmatrix} \mathbf{I}_3 \\ \hline -\bar{\mathbf{C}} \end{bmatrix} \begin{bmatrix} \mathbf{A} \end{bmatrix} \left[\begin{array}{c|c} \mathbf{I}_3 & \bar{\mathbf{B}} \end{array}\right] = \left[\begin{array}{c|c} \mathbf{A} & \mathbf{A}\bar{\mathbf{B}} \\ \hline -\bar{\mathbf{C}}\mathbf{A} & -\bar{\mathbf{C}}\mathbf{A}\bar{\mathbf{B}} \end{array}\right] \tag{28}$$

From (15), (21), and (25)

$$\left[\begin{array}{c|c} \mathbf{A} & \mathbf{A}\bar{\mathbf{B}} \\ \hline -\bar{\mathbf{C}}\mathbf{A} & -\bar{\mathbf{C}}\mathbf{A}\bar{\mathbf{B}} \end{array}\right] = \left[\begin{array}{c|c} \mathbf{A} & \mathbf{B} \\ \hline \mathbf{C} & \mathbf{D} \end{array}\right] \tag{29}$$

which completes the theorem.

Also notice that the matrix of the fully exchanged tableau (2) is, from (20), (22), and (15),

$$\left[\begin{array}{c|c} \mathbf{A}^{-1} & \mathbf{A}^{-1}\mathbf{B} \\ \hline -\mathbf{C}\mathbf{A}^{-1} & \mathbf{D} - \mathbf{C}\mathbf{A}^{-1}\mathbf{B} \end{array}\right] \tag{30}$$

Also from (15) and (20) the theorem can be restated as: M is the composite of

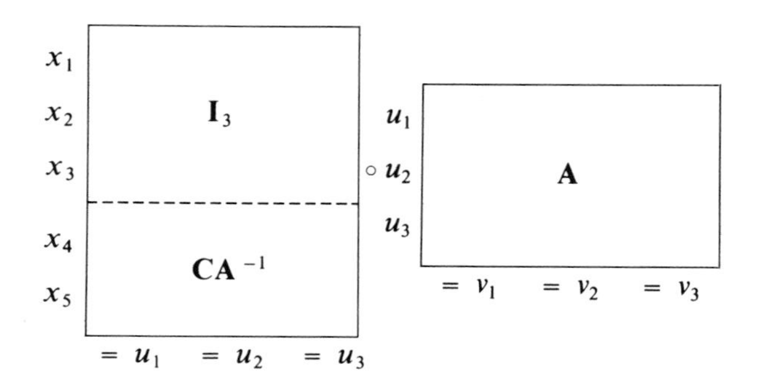

v_1							
$\circ\, v_2$		$\mathbf{I}_3$			$\mathbf{A}^{-1}\mathbf{B}$		
v_3							
	$= y_1$	$= y_2$	$= y_3$	$= y_4$	$= y_5$	$= y_6$	$= y_7$

Exercises

1. (a) Show that $T: R^3 \rightarrow R^3$, $(x_1, x_2, x_3) \mapsto (x_1 + 2x_2, -x_1 + 2x_3, x_1 + 3x_2 + x_3)$ is neither surjective nor injective.

(b) Decompose T into a surjection, followed by a bijection, followed by an injection.

2. Repeat the instructions of Exercise 1 for:

(a) $L: R^3 \rightarrow R^3$, $(x_1, x_2, x_3) \mapsto (x_1 - 2x_2 - x_3, 2x_1 + 4x_2 - 2x_3, 3x_1 + 6x_2 - 3x_3)$.

(b) $L: R^3 \rightarrow R^4$,

x_1	1	4	2	5
x_2	0	4	12	12
x_3	1	3	-1	2
	$= y_1$	$= y_2$	$= y_3$	$= y_4$

(See Exercise 10, Section 7.2.)

3. The fully exchanged tableau of the mapping $M: R^5 \rightarrow R^4$ with tableau

x_1	2	−4	1	5
x_2	−5	7	−3	−9
x_3	8	0	6	−2
x_4	5	2	4	−4
x_5	−9	8	−6	−10
	$= y_1$	$= y_2$	$= y_3$	$= y_4$

is

y_1	$21/2$	6	$5/4$	−4
y_2	$3/2$	1	$1/4$	−2
y_3	−14	−8	$-3/2$	5
x_4	$1/2$	0	$-3/4$	0
x_5	$-3/2$	−2	$1/4$	0
	$= x_1$	$= x_2$	$= x_3$	$= y_4$

Using this information, express M as the composite of a surjection, bijection, and an injection, and check the decomposition by multiplying matrices.

4. Decompose into a surjection, bijection, and injection the mapping $T: R^4 \rightarrow R^5$ with tableau

x_1	2	−2	1	−1	2
x_2	0	1	−2	1	−1
x_3	−3	0	1	0	1
x_4	1	−4	1	0	2
	$= y_1$	$= y_2$	$= y_3$	$= y_4$	$= y_5$

7.7 Chapter Review

In this chapter linear, affine, and Euclidean mappings are studied. They are shown to be closely related to vector spaces (linear spaces), affine spaces, and Euclidean spaces.

These mappings are formally defined and among the theorems proved are

1. There is a one-to-one correspondence between the set of linear mappings $R^m \to R^n$ and the set of $m \times n$ matrices in which the rows of the matrices are the images of the natural basis vectors.
2. Let $L: R^m \to R^n$ be a linear mapping whose initial tableau has pivot rank r; then
   (a) The image of L is a vector subspace of R^n.
   (b) The kernel of L is a vector subspace of R^m.
   (c) The dimension of the image is r.
   (d) The dimension of the kernel is $m - r$.
3. Let F and L be mappings $R^m \to R^n$ such that for all $\mathbf{X}$ in R^m,

$$F(\mathbf{X}) = L(\mathbf{X}) + F(\mathbf{0}_m)$$

Then if L is linear, F is affine; conversely, if F is affine, L is linear.

Isometries, projections, reflections, and symmetries are defined and discussed.

The decomposition of a nonsurjective and noninjective linear mapping into a composition of a surjection, bijection, and an injection is shown.

Exercises

1. Which, if any, of the following mappings are linear?
(a) $(x_1, x_2) \mapsto (3x_1 + x_2, 4)$.
(b) $(x_1, x_2, x_3) \mapsto \left(\frac{1}{x_1}, \frac{1}{x_2}, \frac{1}{x_3}\right)$.
(c) $(x_1, x_2) \mapsto (x_1^2 + x_2^2)$.

2. Given the linear mapping $L: R^4 \to R^2$ such that

$$L(1, 0, 0, 0) = (1, 2)$$

$$L(0, 1, 0, 0) = (3, 4)$$

$$L(0, 0, 1, 0) = (5, 6)$$

$$L(0, 0, 0, 1) = (7, 8)$$

(a) Find the tableau that expresses the rule of L.
(b) Find $L(1, 1, -1, 2)$.

3. Find the image and kernel of each of the following:

(a)

x_1	3	2
x_2	5	4
	$= y_1$	$= y_2$

(b)

x_1	1	2	3
x_2	2	4	8
	$= y_1$	$= y_2$	$= y_3$

(c)

x_1	1	2
x_2	3	6
x_3	−1	2
	$= y_1$	$= y_2$

(d)

x_1	1	0	1
x_2	2	0	2
x_3	3	0	−3
	$= y_1$	$= y_2$	$= y_3$

4. Find a linear mapping $R^2 \to R^2$ whose image is determined by

$$y_2 = {}^2\!/_3 y_1$$

and whose kernel is determined by

$$x_1 = -2x_2$$

5. Given the mapping with tableau

x_1	−1	2	3
x_2	2	−1	0
1	1	4	−3
	$= y_1$	$= y_2$	$= y_3$

(a) Explain why it is affine but not linear.
(b) Find its image and kernel.

6. (a) Find the rule of the translation that moves $(2, -1, 4)$ to $(0, 0, 0)$.
(b) Find the rule of the symmetry in $(2, -1, 4)$.

7. Let $\mathbf{A} = (1, 1, 1)$ generate the vector line α. Find the rule of:
(a) The projection P_α.
(b) The reflection R_α.
(c) The reflection $R_{\alpha\perp}$.
(d) Show that P_α is neither surjective nor injective.
(e) Decompose P_α as the composite of a surjection, bijection, and an injection.
(f) Find the rule for the composite $R_\alpha \circ R_{\alpha\perp}$.
(g) Show that R_α is a bijection and that its inverse is itself.

CLA CHAPTER 8

Determinants

A determinant is a number that is associated with a square matrix. Historically determinants were found useful in solving a system of n linear equations in n variables, in evaluating areas and volumes, and in stating some theorems in linear algebra. These uses of determinants are discussed in Section 8.4. In Section 8.6 some other applications of determinants are discussed. The notions of eigenvalue, eigenvector, and characteristic equations are introduced. A complete study of these notions, which are of great importance in many areas, is a course in itself.

As this chapter progresses certain properties of determinants (theorems) are proved or asserted. The purpose of these determinant properties is to aid in the calculation of determinants, a process that is sometimes tedious.

8.1 *Definition of a Determinant*

A determinant is a number associated with a square matrix. From a given square matrix M there is a precise way to determine its determinant. In this section we define, for any square matrix, what its determinant is.

To start, if **A** is a 1×1 matrix whose entry is a_{11}, its determinant is defined to be that entry, a_{11}. The determinant of **A** is denoted $|\mathbf{A}|$ or det **A**. Thus if $\mathbf{A} = [-4]$, $\det[-4] = |[-4]| = -4$. If **A** is a 2×2 matrix,

$$\begin{bmatrix} a_{11} & a_{12} \\ a_{21} & a_{22} \end{bmatrix}$$

its determinant is $a_{11}a_{22} - a_{12}a_{21}$. Thus, if

$$\mathbf{A} = \begin{bmatrix} 3 & 6 \\ 2 & 5 \end{bmatrix}$$

then $|\mathbf{A}| = 3 \cdot 5 - 6 \cdot 2 = 3$. It is convenient to write this determinant as

$$\begin{vmatrix} 3 & 6 \\ 2 & 5 \end{vmatrix} \quad \text{instead of} \quad \left|\begin{bmatrix} 3 & 6 \\ 2 & 5 \end{bmatrix}\right|$$

If **M** is an $n \times n$ matrix, its determinant is denoted $|\mathbf{M}|$.

The determinant of the 3×3 matrix

$$\mathbf{A} = \begin{bmatrix} a_{11} & a_{12} & a_{13} \\ a_{21} & a_{22} & a_{23} \\ a_{31} & a_{32} & a_{33} \end{bmatrix}$$

is

$$(1) \qquad |\mathbf{A}| = a_{11}|\mathbf{A}_{11}| - a_{21}|\mathbf{A}_{21}| + a_{31}|\mathbf{A}_{31}|$$

where the matrix $\mathbf{A}_{11}$ is obtained from **A** by deleting the row and column of a_{11}. That is,

$$\mathbf{A}_{11} = \begin{bmatrix} a_{22} & a_{23} \\ a_{32} & a_{33} \end{bmatrix}$$

The matrix $\mathbf{A}_{11}$ is called the *minor* of a_{11}. The minor of a_{21} is

$$\mathbf{A}_{21} = \begin{bmatrix} a_{12} & a_{13} \\ a_{32} & a_{33} \end{bmatrix}$$

The matrix $\mathbf{A}_{21}$ is obtained from $\mathbf{A}$ by deleting the row and column of a_{21}. The minor of a_{31} is

$$\mathbf{A}_{31} = \begin{bmatrix} a_{12} & a_{13} \\ a_{22} & a_{23} \end{bmatrix}$$

EXAMPLE 1

$$\begin{vmatrix} 2 & 1 & 1 \\ 3 & 2 & -1 \\ -1 & 3 & -2 \end{vmatrix} = (2)\begin{vmatrix} 2 & -1 \\ 3 & -2 \end{vmatrix} - (3)\begin{vmatrix} 1 & 1 \\ 3 & -2 \end{vmatrix} + (-1)\begin{vmatrix} 1 & 1 \\ 2 & -1 \end{vmatrix}$$

$$= 2(-4 + 3) - 3(-2 - 3) - 1(-1 - 2)$$

$$= -2 + 15 + 3$$

$$= 16$$

Let $\mathbf{A}$ be the 4×4 matrix

$$\begin{bmatrix} a_{11} & a_{12} & a_{13} & a_{14} \\ a_{21} & a_{22} & a_{23} & a_{24} \\ a_{31} & a_{32} & a_{33} & a_{34} \\ a_{41} & a_{42} & a_{43} & a_{44} \end{bmatrix}$$

Its determinant, $|\mathbf{A}|$, is defined by

$$|\mathbf{A}| = a_{11}|\mathbf{A}_{11}| - a_{21}|\mathbf{A}_{21}| + a_{31}|\mathbf{A}_{31}| - a_{41}|\mathbf{A}_{41}|$$

EXAMPLE 2

Let

$$\mathbf{A} = \begin{bmatrix} 1 & 2 & 3 & 4 \\ 3 & 7 & -1 & 0 \\ 2 & 1 & 4 & -2 \\ 1 & 3 & 2 & 4 \end{bmatrix}$$

Then

$$(1)\qquad |\mathbf{A}| = 1\begin{vmatrix} 7 & -1 & 0 \\ 1 & 4 & -2 \\ 3 & 2 & 4 \end{vmatrix} - 3\begin{vmatrix} 2 & 3 & 4 \\ 1 & 4 & -2 \\ 3 & 2 & 4 \end{vmatrix} + 2\begin{vmatrix} 2 & 3 & 4 \\ 7 & -1 & 0 \\ 3 & 2 & 4 \end{vmatrix} - 1\begin{vmatrix} 2 & 3 & 4 \\ 7 & -1 & 0 \\ 1 & 4 & -2 \end{vmatrix}$$

Each of the 3×3 matrices of (1) is evaluated as in Example 1. Thus

$$\begin{vmatrix} 7 & -1 & 0 \\ 1 & 4 & -2 \\ 3 & 2 & 4 \end{vmatrix} = 7\begin{vmatrix} 4 & -2 \\ 2 & 4 \end{vmatrix} - 1\begin{vmatrix} -1 & 0 \\ 2 & 4 \end{vmatrix} + 3\begin{vmatrix} -1 & 0 \\ 4 & -2 \end{vmatrix}$$
$$= 7(20) - 1(-4) + 3(2)$$
$$= 150$$

$$\begin{vmatrix} 2 & 3 & 4 \\ 1 & 4 & -2 \\ 3 & 2 & 4 \end{vmatrix} = 2\begin{vmatrix} 4 & -2 \\ 2 & 4 \end{vmatrix} - 1\begin{vmatrix} 3 & 4 \\ 2 & 4 \end{vmatrix} + 3\begin{vmatrix} 3 & 4 \\ 4 & -2 \end{vmatrix}$$
$$= 2(20) - 1(4) + 3(-22)$$
$$= -30$$

$$\begin{vmatrix} 2 & 3 & 4 \\ 7 & -1 & 0 \\ 3 & 2 & 4 \end{vmatrix} = 2\begin{vmatrix} 4 & 0 \\ 2 & 4 \end{vmatrix} - 7\begin{vmatrix} 3 & 4 \\ 2 & 4 \end{vmatrix} + 3\begin{vmatrix} 3 & 4 \\ -1 & 0 \end{vmatrix}$$
$$= 2(-4) - 7(4) + 3(4)$$
$$= -24$$

$$\begin{vmatrix} 2 & 3 & 4 \\ 7 & -1 & 0 \\ 1 & 4 & -2 \end{vmatrix} = 2\begin{vmatrix} -1 & 0 \\ 4 & -2 \end{vmatrix} - 7\begin{vmatrix} 3 & 4 \\ 4 & -2 \end{vmatrix} + 1\begin{vmatrix} 3 & 4 \\ -1 & 0 \end{vmatrix}$$
$$= 2(2) - 7(-22) + 1(4)$$
$$= 162$$

Thus

$$|\mathbf{A}| = 1(150) - 3(-30) + 2(-24) - 1(162)$$
$$= 30$$

The definition of the determinant of any square matrix $\mathbf{A}$ of any size n is now given. Multiply each element of the first column of $\mathbf{A}$ by the determinant of its minor. Then alternately add and subtract the products.

If $\mathbf{A}$ is 5×5, the definition yields

$$(1) \quad |\mathbf{A}| = a_{11}|\mathbf{A}_{11}| - a_{21}|\mathbf{A}_{21}| + a_{31}|\mathbf{A}_{31}| - a_{41}|\mathbf{A}_{41}| + a_{51}|\mathbf{A}_{51}|$$

Each minor of (1) is itself a 4×4 matrix. Therefore, in order to compute $|\mathbf{A}|$ it is necessary to first compute the determinants of the five 4×4 minors $\mathbf{A}_{11}$, $\mathbf{A}_{21}$, $\mathbf{A}_{31}$, $\mathbf{A}_{41}$, and $\mathbf{A}_{51}$. We know from the previous examples that to evaluate the determinant of each of the 4×4's requires expressing each of them in terms of the determinants of 3×3 minors. Each of these 3×3's is then expressed in terms of three 2×2 minors. This means that to evaluate the determinant of a single 5×5 matrix we must evaluate the determinants of sixty 2×2 matrices (a 6×6 requires evaluating three hundred and sixty 2×2 determinants). In the following sections we investigate methods to reduce this number of calculations.

Exercises

Find the determinant of each matrix in Exercises 1–15.

1. $\begin{bmatrix} 6 \end{bmatrix}$ **2.** $\begin{bmatrix} 6 & 3 \\ 5 & 2 \end{bmatrix}$ **3.** $\begin{bmatrix} -1 & 0 \\ 4 & -1 \end{bmatrix}$

4. $\begin{bmatrix} a & c \\ 0 & 1 \end{bmatrix}$ **5.** $\begin{bmatrix} 1 & -1 & 1 \\ 2 & 1 & 1 \\ 1 & 0 & 1 \end{bmatrix}$ **6.** $\begin{bmatrix} 1 & 2 & 3 \\ 2 & 3 & 1 \\ 3 & 1 & 2 \end{bmatrix}$

7. $\begin{bmatrix} a & 1 & 0 \\ b & 1 & 0 \\ c & 0 & 1 \end{bmatrix}$ **8.** $\begin{bmatrix} -1 & 2 & -1 \\ 1 & -1 & 2 \\ 2 & 4 & 5 \end{bmatrix}$

9. $\begin{bmatrix} 1+a & 1 \\ 1 & 1+b \end{bmatrix}$ **10.** $\begin{bmatrix} a & 0 & 0 \\ 0 & a & 0 \\ 0 & 0 & a \end{bmatrix}$

11. $\begin{bmatrix} 1 & a & a \\ a & a^2 & 1 \\ a^2 & a & 1 \end{bmatrix}$

12. $\begin{bmatrix} 3 & 2 & 4 \\ 0 & 2 & 5 \\ 0 & 0 & 1 \end{bmatrix}$

13. $\begin{bmatrix} -2 & 8 & 2 & 1 \\ 0 & 1 & 0 & 0 \\ 0 & 0 & 1 & 0 \\ 0 & 0 & 0 & 1 \end{bmatrix}$

14. $\begin{bmatrix} a & 0 & 0 & 0 \\ 0 & b & 0 & 0 \\ 0 & 0 & c & 0 \\ 0 & 0 & 0 & d \end{bmatrix}$

15. $\begin{bmatrix} 1 & -1 & 2 & 1 \\ 0 & 1 & 1 & 0 \\ 0 & 0 & 1 & 0 \\ 2 & 0 & 1 & 0 \end{bmatrix}$

16. A column of an $n \times n$ matrix consists only of zeros. Prove that its determinant is zero if:
(a) $n = 2$.
(b) $n = 3$.
(c) $n = 4$.

17. Prove that the determinant of an identity matrix of order n is 1 if:
(a) $n = 1$.
(b) $n = 2$.
(c) $n = 3$.
(d) $n = p$.

18. Solve for x if:

(a) $\begin{vmatrix} x & 1 \\ 2 & 3 \end{vmatrix} = 2$ (b) $\begin{vmatrix} 1 & 3 \\ x & 2 \end{vmatrix} = 0$

(c) $\begin{vmatrix} x-1 & 1 \\ 1 & x-1 \end{vmatrix} = 8$ (d) $\begin{vmatrix} x & 1 & 1 \\ 0 & x & 1 \\ 0 & 0 & x \end{vmatrix} = 8$

19. Prove that

$$\begin{vmatrix} a & b \\ c & d \end{vmatrix} = -\begin{vmatrix} c & d \\ a & b \end{vmatrix} = -\begin{vmatrix} b & a \\ d & c \end{vmatrix}$$

20. $$k\begin{vmatrix} a & b \\ c & d \end{vmatrix} = \begin{vmatrix} ka & b \\ kc & d \end{vmatrix} = \begin{vmatrix} ka & kb \\ c & d \end{vmatrix} = \begin{vmatrix} a & kb \\ c & kd \end{vmatrix} = \begin{vmatrix} a & b \\ kc & kd \end{vmatrix}$$

21. Prove that:

(a) $$\begin{vmatrix} a_1 & b \\ c_1 & d \end{vmatrix} + \begin{vmatrix} a_2 & b \\ c_2 & d \end{vmatrix} = \begin{vmatrix} a_1 + a_2 & b \\ c_1 + c_2 & d \end{vmatrix}$$

(b) $$\begin{vmatrix} a & b_1 \\ c & d_1 \end{vmatrix} + \begin{vmatrix} a & b_2 \\ c & d_2 \end{vmatrix} = \begin{vmatrix} a & b_1 + b_2 \\ c & d_1 + d_2 \end{vmatrix}$$

(c) $$\begin{vmatrix} a_1 & b_1 \\ c & d \end{vmatrix} + \begin{vmatrix} a_2 & b_2 \\ c & d \end{vmatrix} = \begin{vmatrix} a_1 + a_2 & b_1 + b_2 \\ c & d \end{vmatrix}$$

(d) $$\begin{vmatrix} a & b \\ c_1 & d_1 \end{vmatrix} + \begin{vmatrix} a & b \\ c_2 & d_2 \end{vmatrix} = \begin{vmatrix} a & b \\ c_1 + c_2 & d_1 + d_2 \end{vmatrix}$$

22. Prove that

$$\begin{vmatrix} a & b \\ c & d \end{vmatrix} = \begin{vmatrix} a & b \\ c + ka & d + kb \end{vmatrix} = \begin{vmatrix} a & b + ka \\ c & d + kc \end{vmatrix}$$

23. Prove that

$$\begin{vmatrix} a_1 & a_2 & a_3 \\ b_1 & b_2 & b_3 \\ c_1 & c_2 & c_3 \end{vmatrix} = -\begin{vmatrix} b_1 & b_2 & b_3 \\ a_1 & a_2 & a_3 \\ c_1 & c_2 & c_3 \end{vmatrix} = \begin{vmatrix} b_1 & b_2 & b_3 \\ c_1 & c_2 & c_3 \\ a_1 & a_2 & a_3 \end{vmatrix}$$

24. Prove that if one row of

$$\mathbf{A} = \begin{vmatrix} a_1 & a_2 & a_3 \\ b_1 & b_2 & b_3 \\ c_1 & c_2 & c_3 \end{vmatrix}$$

is multiplied by k, then $|\mathbf{A}|$ is multiplied by k.

25. Prove that if two rows of a 3×3 matrix are identical, then its determinant is zero.

8.2 Properties of Determinants

In Exercise 20 of Section 8.1 we show (among other things) that

$$k\begin{vmatrix} a_{11} & a_{12} \\ a_{21} & a_{22} \end{vmatrix} = \begin{vmatrix} ka_{11} & ka_{12} \\ a_{21} & a_{22} \end{vmatrix} = \begin{vmatrix} a_{11} & a_{12} \\ ka_{21} & ka_{22} \end{vmatrix}$$

In words, if a row of a 2×2 matrix $\mathbf{A}$ is multiplied by a constant k, then the determinant of the new matrix is equal to k times the determinant of $\mathbf{A}$.

EXAMPLE 1

$$\begin{vmatrix} 2 & 3 \\ 4 & 7 \end{vmatrix} = 2$$

$$\begin{vmatrix} 10 & 15 \\ 4 & 7 \end{vmatrix} = 10 = 5\begin{vmatrix} 2 & 3 \\ 4 & 7 \end{vmatrix}$$

$$\begin{vmatrix} 2 & 3 \\ 20 & 35 \end{vmatrix} = 10 = 5\begin{vmatrix} 2 & 3 \\ 4 & 7 \end{vmatrix}$$

Does this property extend to 3×3 matrices? That is, is it true that

$$k\begin{vmatrix} a_{11} & a_{12} & a_{13} \\ a_{21} & a_{22} & a_{23} \\ a_{31} & a_{32} & a_{33} \end{vmatrix} = \begin{vmatrix} ka_{11} & ka_{12} & ka_{13} \\ a_{21} & a_{22} & a_{23} \\ a_{31} & a_{32} & a_{33} \end{vmatrix} = \begin{vmatrix} a_{11} & a_{12} & a_{13} \\ ka_{21} & ka_{22} & ka_{23} \\ a_{31} & a_{32} & a_{33} \end{vmatrix}$$

$$= \begin{vmatrix} a_{11} & a_{12} & a_{13} \\ a_{21} & a_{22} & a_{23} \\ ka_{31} & ka_{32} & ka_{33} \end{vmatrix} ?$$

The answer is "yes." To see this observe the following:

(1) $$k\begin{vmatrix} a_{11} & a_{12} & a_{13} \\ a_{21} & a_{22} & a_{23} \\ a_{31} & a_{32} & a_{33} \end{vmatrix} = k(a_{11}\,|\,\mathbf{A}_{11}\,| - a_{21}\,|\,\mathbf{A}_{21}\,| + a_{31}\,|\,\mathbf{A}_{31}\,|)$$

(2) $$\begin{vmatrix} ka_{11} & ka_{12} & ka_{13} \\ a_{21} & a_{22} & a_{23} \\ a_{31} & a_{32} & a_{33} \end{vmatrix} = k(a_{11}\,|\,\mathbf{A}_{11}\,| - a_{21}\,|\,\mathbf{A}_{21}\,| + a_{31}\,|\;|\,\mathbf{A}_{31}\,|)$$

(3) $$\begin{vmatrix} a_{11} & a_{21} & a_{13} \\ ka_{21} & ka_{22} & ka_{23} \\ a_{31} & a_{32} & a_{33} \end{vmatrix} = a_{11}k\begin{vmatrix} a_{22} & a_{23} \\ a_{32} & a_{33} \end{vmatrix} - a_{21}k\begin{vmatrix} a_{12} & a_{13} \\ a_{32} & a_{33} \end{vmatrix} + a_{31}k\begin{vmatrix} a_{12} & a_{13} \\ a_{22} & a_{23} \end{vmatrix}$$

(by Exercise 20 in Section 8.1)

As an exercise the reader may show that

$$\begin{vmatrix} a_{11} & a_{12} & a_{13} \\ a_{21} & a_{22} & a_{23} \\ ka_{31} & ka_{32} & ka_{33} \end{vmatrix} = k\,|\,\mathbf{A}\,| \tag{4}$$

EXAMPLE 2

$$\begin{vmatrix} 2 & -1 & 3 \\ 1 & 3 & -1 \\ 2 & 3 & 4 \end{vmatrix} = 27$$

$$\begin{vmatrix} 4 & -2 & 6 \\ 1 & 3 & -1 \\ 2 & 3 & 4 \end{vmatrix} = 54 = 2\begin{vmatrix} 2 & -1 & 3 \\ 1 & 3 & -1 \\ 2 & 3 & 4 \end{vmatrix}$$

$$\begin{vmatrix} 2 & -1 & 3 \\ 2 & 6 & -2 \\ 2 & 3 & 4 \end{vmatrix} = 54$$

$$\begin{vmatrix} 2 & -1 & 3 \\ 1 & 3 & -1 \\ 4 & 6 & 8 \end{vmatrix} = 54$$

Do 4×4 matrices have this property called the *homogeneity* property? That is, if $\mathbf{B}$ is obtained from $\mathbf{A}$ by multiplying any row of $\mathbf{A}$ by a constant k, is it true that

$$|\,\mathbf{B}\,| = k\,|\,\mathbf{A}\,|\,?$$

The answer is "yes." To see this, the reader can proceed as above in the 3×3 case. In the 3×3 case the proof depends on the 2×2 case, which we proved in Exercise 20 of Section 8.1. In the 4×4 case the proof depends on the 3×3 case, as shown above. The homogeneity property holds for any $n \times n$ matrix regardless of n. The proof for each n depends on the existence of the proof for the $n - 1$ case and eventually the proof for the case when $n = 2$. Such proofs are called *inductive.* The proofs for many of the properties of determinants are *inductive.* In this book we indicate for $n = 2, 3$, and sometimes 4, how the *induction* proceeds. The reader may care to try the case for the next n to help establish the extension to n of higher order. Meanwhile we have initiated the inductive proof for Determinant Property 1.

DP1 (HOMOGENEITY). Let $\mathbf{A}$ be an $n \times n$ matrix. If $\mathbf{B}$ is obtained from $\mathbf{A}$ by multiplying each entry in any row by k, then

$$|\mathbf{B}| = k\,|\mathbf{A}|$$

Since there are no restrictions on k, the result is true if $k = 0$. Thus

COROLLARY. If a matrix contains a row of zeros, its determinant is zero.

EXAMPLE 3

(a)
$$\begin{vmatrix} 3 & -3 & 3 \\ -2 & 0 & 3 \\ -1 & 2 & -4 \end{vmatrix} = 3\begin{vmatrix} 1 & -1 & 1 \\ -2 & 0 & 3 \\ -1 & 2 & -4 \end{vmatrix}$$

$$= 3\left(1\begin{vmatrix} 0 & 3 \\ 2 & -4 \end{vmatrix} + 2\begin{vmatrix} -1 & 1 \\ 2 & -4 \end{vmatrix} - 1\begin{vmatrix} -1 & 1 \\ 0 & 3 \end{vmatrix}\right)$$

$$= 3\,(-6 + 4 + 3)$$

$$= 3$$

(b)
$$\begin{vmatrix} 15 & 20 & 30 \\ 8 & 12 & 28 \\ 15 & 6 & 15 \end{vmatrix} = 5\begin{vmatrix} 3 & 4 & 6 \\ 8 & 12 & 28 \\ 15 & 6 & 15 \end{vmatrix}$$

$$= 5\cdot 4\begin{vmatrix} 3 & 4 & 6 \\ 2 & 3 & 7 \\ 15 & 6 & 15 \end{vmatrix} = 5\cdot 4\cdot 3\begin{vmatrix} 3 & 4 & 6 \\ 2 & 3 & 7 \\ 5 & 2 & 5 \end{vmatrix}$$

EXAMPLE 4

Let

$$\mathbf{A}_1 = \begin{bmatrix} 2 & 1 & 6 \\ 3 & -1 & 2 \\ 5 & 1 & 8 \end{bmatrix}$$

Then

$$|\mathbf{A}_1| = 14$$

Let

$$\mathbf{A}_2 = \begin{bmatrix} 1 & 3 & 4 \\ 3 & -1 & 2 \\ 5 & 1 & 8 \end{bmatrix}$$

Thus $\mathbf{A}_2$ is the same as $\mathbf{A}_1$ except in the first row.

$$|\mathbf{A}_2| = -20$$

Let

$$\mathbf{A}_3 = \begin{bmatrix} 3 & 4 & 10 \\ 3 & -1 & 2 \\ 5 & 1 & 8 \end{bmatrix}$$

Thus $\mathbf{A}_3$ is the same as $\mathbf{A}_1$ and $\mathbf{A}_2$ except that its first row is the sum of the first rows of $\mathbf{A}_1$ and $\mathbf{A}_2$.

$$|\mathbf{A}_3| = -6$$

In this case $|\mathbf{A}_3| = |\mathbf{A}_1| + |\mathbf{A}_2|$. The general question is now: If two matrices $\mathbf{A}_1$ and $\mathbf{A}_2$ are identical (except for one row), is it true that $|\mathbf{A}_3| = |\mathbf{A}_1| + |\mathbf{A}_2|$? (Here $\mathbf{A}_3$ is identical to $\mathbf{A}_1$ and $\mathbf{A}_2$ except that in the differing row the corresponding entries of $\mathbf{A}_3$ are the sum of those in $\mathbf{A}_1$ and $\mathbf{A}_2$.) The answer is "yes." To see this, the reader should proceed as follows.

First consider the 2×2 case with

$$\mathbf{A}_1 = \begin{bmatrix} a_{11} & a_{12} \\ a_{21} & a_{22} \end{bmatrix} \quad \text{and} \quad \mathbf{A}_2 = \begin{bmatrix} b_{11} & b_{12} \\ a_{21} & a_{22} \end{bmatrix}$$

and thus

$$\mathbf{A}_3 = \begin{bmatrix} a_{11} + b_{11} & a_{12} + b_{12} \\ a_{21} & a_{22} \end{bmatrix}$$

It is easy to see that $|\mathbf{A}_3| = |\mathbf{A}_1| + |\mathbf{A}_2|$.

Next consider the 2×2 case with

$$\mathbf{A}_1 = \begin{bmatrix} a_{11} & a_{12} \\ a_{21} & a_{22} \end{bmatrix} \quad \text{and} \quad \mathbf{A}_2 = \begin{bmatrix} a_{11} & a_{12} \\ b_{21} & b_{22} \end{bmatrix}$$

and thus

$$\mathbf{A}_3 = \begin{bmatrix} a_{11} & a_{12} \\ a_{21} + b_{21} & a_{22} + b_{22} \end{bmatrix}$$

Again, it is easy to see that

$$|\mathbf{A}_3| = |\mathbf{A}_1| + |\mathbf{A}_2|$$

Proceeding in a similar manner for the 3×3 case, first letting $\mathbf{A}_1$ and $\mathbf{A}_2$ differ in the first row, then the second row, and then the third will establish the *additive* result for 3×3 matrices. The reader will see that the 3×3 case depends on the 2×2 case and will also see the general inductive procedure required for $n = 4, 5, \ldots$. Thus we assert Determinant Property 2.

DP 2 (THE ADDITIVE PROPERTY). Let $\mathbf{A}_1$ and $\mathbf{A}_2$ be identical matrices except for the kth row. Then $|\mathbf{A}_3| = |\mathbf{A}_1| + |\mathbf{A}_2|$ (here $\mathbf{A}_3$ is identical to $\mathbf{A}_1$ and $\mathbf{A}_2$ except that its kth row is the sum of the kth rows of $\mathbf{A}_1$ and $\mathbf{A}_2$).

Using DP2 we show that a matrix with a row of zeros has determinant zero.

Let $\mathbf{A}_1$ have only zeros in its kth row. Let $\mathbf{A}_2$ be identical to $\mathbf{A}_1$ except in the kth row. Let $\mathbf{A}_3$ be identical to $\mathbf{A}_1$ and $\mathbf{A}_2$ except that its kth row is the sum of the kth rows of $\mathbf{A}_1$ and $\mathbf{A}_2$. Then by DP2,

$$(1) \qquad |\mathbf{A}_3| = |\mathbf{A}_1| + |\mathbf{A}_2|$$

But $\mathbf{A}_3$ is actually equal to $\mathbf{A}_2$ since the kth row of $\mathbf{A}_1$ contains only zeros. Thus $|\mathbf{A}_3| = |\mathbf{A}_2|$. Substituting this result in (1) yields $|\mathbf{A}_1| = 0$.

EXAMPLE 5

$$(1) \qquad \begin{vmatrix} 3 & 4 & 10 \\ 3 & -1 & 2 \\ 5 & 1 & 8 \end{vmatrix} = -6$$

Interchanging rows 2 and 3 gives

(2)
$$\begin{vmatrix} 3 & 4 & 10 \\ 5 & 1 & 8 \\ 3 & -1 & 2 \end{vmatrix} = 6$$

Determinant (3) is also obtained from (1) by interchanging two adjacent rows.

(3)
$$\begin{vmatrix} 3 & -1 & 2 \\ 3 & 4 & 10 \\ 5 & 1 & 8 \end{vmatrix} = 6$$

In both cases of Example 5 only, the sign of the determinant changes. Is this true in general? Does interchanging adjacent rows of a matrix change the sign of the determinant? The answer is "yes." The reader can easily establish this for 2×2 matrices by working with

$$\begin{vmatrix} a_{11} & a_{12} \\ a_{21} & a_{22} \end{vmatrix} \quad \text{and} \quad \begin{vmatrix} a_{21} & a_{22} \\ a_{11} & a_{12} \end{vmatrix}$$

For the 3×3 case, let

(1)
$$\mathbf{A} = \begin{bmatrix} a_{11} & a_{12} & a_{13} \\ a_{21} & a_{22} & a_{23} \\ a_{31} & a_{32} & a_{33} \end{bmatrix}$$

Then

(2)
$$|\mathbf{A}| = a_{11}|\mathbf{A}_{11}| - a_{21}|\mathbf{A}_{21}| + a_{31}|\mathbf{A}_{31}|$$

Let

(3)
$$\mathbf{B} = \begin{bmatrix} a_{21} & a_{22} & a_{23} \\ a_{11} & a_{12} & a_{13} \\ a_{31} & a_{32} & a_{33} \end{bmatrix}$$

Then

(4)
$$|\mathbf{B}| = a_{21}|\mathbf{A}_{21}| - a_{11}|\mathbf{A}_{11}| + a_{31}\begin{vmatrix} a_{22} & a_{23} \\ a_{12} & a_{13} \end{vmatrix}$$

By the 2 × 2 case,

$$\begin{vmatrix} a_{22} & a_{23} \\ a_{12} & a_{13} \end{vmatrix} = -\begin{vmatrix} a_{12} & a_{13} \\ a_{22} & a_{23} \end{vmatrix} = -|\mathbf{A}_{31}|$$

Substituting in (4) gives

$$|\mathbf{B}| = -a_{11}|\mathbf{A}_{11}| + a_{21}|\mathbf{A}_{21}| - a_{31}|\mathbf{A}_{31}| \tag{5}$$

From (2) and (5),

$$|\mathbf{B}| = -|\mathbf{A}| \tag{6}$$

A similar result holds when interchanging rows 2 and 3 of **A**.

The inductive procedure is clear. We now state

DP3. If any two adjacent rows of an $n \times n$ matrix **A** are interchanged to give a matrix **B**, then

$$|\mathbf{B}| = -|\mathbf{A}|$$

From DP3 we immediately deduce

DP4. If *any* two rows of **A** are interchanged yielding a matrix **B**, then

$$|\mathbf{B}| = -|\mathbf{A}|$$

Proof: Suppose that rows k and m of matrix **A** are to be interchanged and that there are r intermediary rows between them. Interchange row k with each of the r intermediary rows in turn, and finally with row m. This requires $r + 1$ interchanges. Now interchange row m (which is moved one row) with each of the r intermediary rows. This puts row m in the original place of row k and restores the intermediary rows to their original places. The total number of interchanges is $2r + 1$. Each interchange changes the determinant (by DP3) by multiplying it by (-1). Since $2r + 1$ is an odd number, the determinant of **A** is multiplied by (-1) an odd number of times, which yields $(-1)|\mathbf{A}|$.

EXAMPLE 6

$$\begin{vmatrix} 3 & 4 & 10 \\ 5 & 1 & 8 \\ 3 & -1 & 2 \end{vmatrix} = 6 = -\begin{vmatrix} 3 & -1 & 2 \\ 5 & 1 & 8 \\ 3 & 4 & 10 \end{vmatrix} = -(-6)$$

DP5. If any two distinct rows of $n \times n$ matrix A have identical entries, then $|\mathbf{A}| = 0$.

Proof: By DP4, if the two rows with identical entries are interchanged, then $|\mathbf{A}|$ becomes $-|\mathbf{A}|$. But the matrix is actually unchanged. Hence $|\mathbf{A}| = -|\mathbf{A}|$ or $|\mathbf{A}| = 0$.

EXAMPLE 7

(a)
$$\begin{vmatrix} 3 & 2 & 5 \\ 2 & 1 & 7 \\ 3 & 2 & 5 \end{vmatrix} = 0$$

(b)
$$\begin{vmatrix} a_1 & a_2 & a_3 & a_4 \\ b_1 & b_2 & b_3 & b_4 \\ a_1 & a_2 & a_3 & a_4 \\ d_1 & d_2 & d_3 & d_4 \end{vmatrix} = 0$$

EXAMPLE 8

$$|\mathbf{A}| = \begin{vmatrix} a_1 & a_2 & a_3 \\ ka_1 & ka_2 & ka_3 \\ c_1 & c_2 & c_3 \end{vmatrix} = a_1 \begin{vmatrix} ka_2 & ka_3 \\ c_2 & c_3 \end{vmatrix} - ka_1 \begin{vmatrix} a_2 & a_3 \\ c_2 & c_3 \end{vmatrix} + c_1 \begin{vmatrix} a_2 & a_3 \\ ka_2 & ka_3 \end{vmatrix}$$

$$= a_1(ka_2c_3 - c_2ka_3) - ka_1(a_2c_3 - c_2a_3) + c_1(a_2ka_3 - ka_2a_3)$$

$$= 0$$

In Example 8, row 2 of $\mathbf{A}$ is a scalar multiple of row 1 and $|\mathbf{A}| = 0$. Does this example illustrate a general phenomenon? That is, if one row of a matrix $\mathbf{A}$ is a scalar multiple of another row of $\mathbf{A}$, is $|\mathbf{A}| = 0$? The answer is "yes." To see this, the reader should show that this result holds for 2×2 matrices. Since we have shown it for 3×3 matrices, the inductive type of proof necessary to show the result for any $n \times n$ matrix is outlined.

We state

DP6. If one row of an $n \times n$ matrix $\mathbf{A}$ is a scalar multiple of another row of $\mathbf{A}$, then $|\mathbf{A}| = 0$.

The special case when the scalar multiple $k = 0$ is also stated as a corollary to DP2.

In Example 8 we did not consider all possible cases. That is, we did not consider what happens if the third row is a scalar multiple of the second row; and so on. Which DP rule makes such consideration unnecessary? Now, let **A** be any $n \times n$ matrix. Let **B** be identical to **A** except that the ith row of **B** is k times the mth row of **A**. Let **C** be identical to **A** and **B** except that its ith row is the sum of the ith rows of **A** and **B**. Then, by DP2,

$$|\mathbf{C}| = |\mathbf{A}| + |\mathbf{B}|$$

But, by DP6,

$$|\mathbf{B}| = 0$$

Thus

$$|\mathbf{C}| = |\mathbf{A}|$$

The matrix **C** differs from **A** only in the ith row and in the manner stated. Thus

DP7. Let **A** be any $n \times n$ matrix. Let **B** be identical to **A** except that the ith row of **B** is equal to the ith row of **A** plus k times the mth row. Then

$$|\mathbf{B}| = |\mathbf{A}|$$

EXAMPLE 9

Let

$$\mathbf{A} = \begin{bmatrix} 1 & 2 & 3 \\ 4 & 5 & 6 \\ 7 & 8 & 9 \end{bmatrix}$$

Let

$$\mathbf{B} = \begin{bmatrix} 1 & 2 & 3 \\ 0 & -3 & -6 \\ 7 & 8 & 9 \end{bmatrix}$$

where the second row of **B** is the second row of **A** plus (-4) times the first row of **A**. The reader should convince himself that

$$|\mathbf{B}| = |\mathbf{A}|$$

EXAMPLE 10

Let

$$\mathbf{A} = \begin{bmatrix} -2 & 8 & 2 & 1 \\ 0 & 1 & 0 & 0 \\ 4 & -16 & -3 & -2 \\ 0 & 0 & 0 & 1 \end{bmatrix}$$

Then $|\mathbf{A}|$ is equal to the determinant of the matrix of Exercise 13, Section 8.1. To see this, replace row 3 of **A** by row 3 plus twice row 1.

Exercises

1. Use DP1 (the homogeneity property) to simplify the evaluation of the determinant of

(a) $\begin{bmatrix} 1000 & 3 \\ 2000 & 7 \end{bmatrix}$ (b) $\begin{bmatrix} 4 & 77{,}777 \\ 1 & 22{,}222 \end{bmatrix}$ (c) $\begin{bmatrix} 4 & 180 & 49 \\ 0 & 270 & 35 \\ 0 & 360 & 21 \end{bmatrix}$

2. Use DP2 (the additive property) to add the following. (Do not evaluate.)

(a) $\begin{vmatrix} 2 & 7 & 3 \\ 7 & 2 & -8 \\ 5 & -4 & 6 \end{vmatrix} + \begin{vmatrix} 2 & 7 & 3 \\ -7 & -2 & 8 \\ 5 & -4 & 6 \end{vmatrix}$

(b) $\begin{vmatrix} a_1 & a_2 & a_3 & a_4 \\ b_1 & b_2 & b_3 & b_4 \\ x_1 & x_2 & x_3 & x_4 \\ d_1 & d_2 & d_3 & d_4 \end{vmatrix} + \begin{vmatrix} a_1 & a_2 & a_3 & a_4 \\ b_1 & b_2 & b_3 & b_4 \\ y_1 & y_2 & y_3 & y_4 \\ d_1 & d_2 & d_3 & d_4 \end{vmatrix}$

3. Verify by evaluation that DP4 is correct for

$$\begin{vmatrix} 3 & -2 & 3 \\ 2 & 0 & 1 \\ 1 & 2 & -3 \end{vmatrix}$$

when interchanging
(a) rows 1 and 2.
(b) rows 1 and 3.

4. Verify by evaluation that DP4 is correct for

$$\begin{vmatrix} 1 & 2 & 1 & 3 \\ 0 & 1 & -1 & -2 \\ 0 & 4 & 3 & -6 \\ 2 & 2 & 8 & 1 \end{vmatrix}$$

when interchanging
(a) rows 1 and 3.
(b) rows 3 and 4.

5. Using DP5 or DP6, evaluate at sight:

(a) $\begin{vmatrix} 4 & 5 \\ 8 & 10 \end{vmatrix}$ (b) $\begin{vmatrix} 1 & 6 & -4 \\ 2 & 12 & -8 \\ -1 & 6 & 7 \end{vmatrix}$ (c) $\begin{vmatrix} 2 & 1 & 4 & -5 \\ 0 & -1 & 2 & 3 \\ 20 & 10 & 40 & -50 \\ a & b & c & d \end{vmatrix}$

6. Show that

$$\begin{vmatrix} x^2 - y^2 & (x - y)^2 & 0 \\ x + y & x - y & 0 \\ x - y & 2x & 3y \end{vmatrix} = 0$$

for all x and y.

7. With the aid of DP7 show that

(a) $\begin{vmatrix} 1 & -1 & 2 \\ 5 & 2 & 0 \\ 3 & 2 & 8 \end{vmatrix} = \begin{vmatrix} 1 & -1 & 2 \\ 0 & 7 & -10 \\ 0 & 5 & 2 \end{vmatrix}$

(b) $\begin{vmatrix} 2 & -2 & 0 \\ 4 & 2 & 1 \\ 6 & 0 & 3 \end{vmatrix} = \begin{vmatrix} 2 & -2 & 0 \\ 0 & 6 & 1 \\ 0 & 6 & 3 \end{vmatrix}$

(c) $$\begin{vmatrix} 2 & 2 & 4 & 2 \\ 1 & 2 & -1 & 0 \\ 3 & 7 & 1 & 5 \\ 4 & 8 & 2 & 7 \end{vmatrix} = \begin{vmatrix} 0 & -2 & 6 & 2 \\ 1 & 2 & -1 & 0 \\ 0 & 1 & 4 & 5 \\ 0 & 0 & 6 & 7 \end{vmatrix}$$

8. For each matrix listed below and using DP7, obtain a matrix having the same determinant with zeros in the first column except for the designated row.

(a) $\begin{bmatrix} 1 & 2 & -2 \\ -2 & 3 & 0 \\ 3 & 4 & 2 \end{bmatrix}$ with the same first row.

(b) $\begin{bmatrix} 4 & 1 & 2 \\ 2 & 2 & 3 \\ 6 & 3 & 5 \end{bmatrix}$ with the same second row.

(c) $\begin{bmatrix} 3 & 1 & 4 \\ -2 & 2 & 5 \\ -1 & 3 & 6 \end{bmatrix}$ with the same third row.

(d) $\begin{bmatrix} 1 & 2 & 3 & -1 \\ 2 & 0 & -1 & 2 \\ 3 & 1 & -1 & -2 \\ 4 & 0 & 2 & 1 \end{bmatrix}$ with the same first row.

9. Using DP7, derive a matrix having the same determinant as that of

$$\begin{bmatrix} 2 & 0 & 3 \\ 4 & 1 & 5 \\ 6 & 2 & -1 \end{bmatrix}$$

whose first column contains the entries
(a) 2, 0, 0; (b) 0, 4, 0; (c) 0, 0, 6.

10. Using DP7, derive a matrix having the same determinant as that of

$$\begin{bmatrix} 1 & 2 & 3 & 4 \\ 2 & 1 & 2 & 5 \\ 3 & -1 & 0 & 4 \\ -1 & -1 & 3 & 2 \end{bmatrix}$$

whose first column contains the entries
(a) 1, 0, 0, 0; (b) 0, 2, 0, 0; (c) 0, 0, 3, 0; (d) 0, 0, 0, −1.

11. Using DP7 and DP1, find a simplification of the computations needed to evaluate

(a) $\begin{vmatrix} 3 & 4 & 5 \\ 2 & -1 & 8 \\ 5 & -2 & 7 \end{vmatrix}$ (b) $\begin{vmatrix} 2 & 1 & 3 \\ 1 & 2 & 4 \\ 1 & 0 & 6 \end{vmatrix}$

(c) $\begin{vmatrix} 3 & 0 & 3 \\ 2 & 0 & 1 \\ 1 & 1 & 1 \end{vmatrix}$ (d) $\begin{vmatrix} 2 & 0 & 1 & 0 \\ 1 & 2 & 0 & 3 \\ 2 & 1 & 0 & -1 \\ 3 & 2 & -1 & 1 \end{vmatrix}$

12. (a) Write the definition of the determinant of a 5×5 matrix, assuming that that of a 4×4 matrix has been defined.
(b) Explain how the formula for the determinant of a 2×2 matrix follows from the definition of the determinant of an $n \times n$ matrix.

13. Assuming that DP1 has been proved for 3×3 matrices, prove it true for a 4×4 matrix.

14. Prove DP3 for the exchange of rows 3 and 4 in a 4×4 matrix assuming it to be proved for 3×3 matrices.

8.3 Pivot Evaluation of Determinants

EXAMPLE 1

Let

$$\mathbf{A} = \begin{bmatrix} 3 & 4 & 1 \\ 0 & -2 & 3 \\ 0 & -8 & 5 \end{bmatrix}$$

$$|\mathbf{A}| = 3\begin{vmatrix} -2 & 3 \\ -8 & 5 \end{vmatrix} - 0\begin{vmatrix} 4 & 1 \\ -8 & 5 \end{vmatrix} + 0\begin{vmatrix} 4 & 1 \\ -2 & 3 \end{vmatrix}$$

$$= 3(14) + 0 + 0$$

$$= 42$$

The evaluation of $|\mathbf{A}|$ is quite simple because of the presence of the two zeros in column 1. Such zeros are introduced in evaluating determinants of other matrices by DP7, as shown in

EXAMPLE 2

Let

$$\mathbf{B} = \begin{bmatrix} 3 & 4 & 1 \\ 6 & 6 & 5 \\ 9 & 4 & 8 \end{bmatrix}$$

From row 2 subtract twice row 1, and from row 3 subtract three times row 1. Then by DP7 (applied twice),

$$|\mathbf{B}| = \begin{vmatrix} 3 & 4 & 1 \\ 0 & -2 & 3 \\ 0 & -8 & 5 \end{vmatrix}$$

$$= 42 \qquad \text{(see Example 1)}$$

There is a pattern that reveals itself in the general 3×3 matrix. Let

$$\mathbf{A} = \begin{bmatrix} a_{11} & a_{12} & a_{13} \\ a_{21} & a_{22} & a_{23} \\ a_{31} & a_{32} & a_{33} \end{bmatrix}$$

To obtain a zero in place of a_{21}, replace row 2 by

$$\text{row } 2 - \frac{a_{21}}{a_{11}} (\text{row } 1)$$

To obtain a zero in place of a_{31}, replace row 3 by

$$\text{row } 3 - \frac{a_{31}}{a_{11}} (\text{row } 1)$$

Then

$$\mathbf{B} = \begin{bmatrix} a_{11} & a_{12} & a_{13} \\ 0 & a_{22} - \dfrac{a_{21}a_{12}}{a_{11}} & a_{23} - \dfrac{a_{21}a_{13}}{a_{11}} \\ 0 & a_{32} - \dfrac{a_{31}a_{22}}{a_{11}} & a_{33} - \dfrac{a_{31}a_{23}}{a_{11}} \end{bmatrix}$$

By two applications of DP7, we get $|\mathbf{B}| = |\mathbf{A}|$ and therefore, using minors,

$$|\mathbf{A}| = |\mathbf{B}| = a_{11} \begin{vmatrix} a_{22} - \dfrac{a_{21}a_{12}}{a_{11}} & a_{23} - \dfrac{a_{21}a_{13}}{a_{11}} \\ a_{32} - \dfrac{a_{31}a_{22}}{a_{11}} & a_{33} - \dfrac{a_{31}a_{23}}{a_{11}} \end{vmatrix}$$

Observe that the last determinant obtained from **B** by expansion by minors is a_{11} times the condensed form of **A** when using the pivot condensation algorithm. Thus it is obtainable directly from **A** in one pivot step in **A**, pivoting on a_{11}!

EXAMPLE 3

Find the determinant of

(1)
$$\begin{bmatrix} 1^* & 2 & -1 \\ 2 & 6 & 3 \\ 5 & 7 & 1 \end{bmatrix}$$

Pivoting as indicated, condensation yields

(2)
$$\begin{bmatrix} 2 & 5 \\ -3 & 6 \end{bmatrix}$$

The determinant of (1) is

$$1 \cdot \begin{vmatrix} 2 & 5 \\ -3 & 6 \end{vmatrix} = 1 \cdot 27 = 27$$

By DP4, the determinant of (1) is equal to minus the determinant of

(3)
$$\begin{bmatrix} 2^* & 6 & 3 \\ 1 & 2 & -1 \\ 5 & 7 & 1 \end{bmatrix}$$

This is verified by pivoting as indicated in (3), yielding

(4)
$$\begin{bmatrix} -1 & -5/2 \\ -8 & -13/2 \end{bmatrix}$$

The determinant of (3) is

(5)
$$2\begin{vmatrix} -1 & -5/2 \\ -8 & -13/2 \end{vmatrix} = 2\left(\frac{-27}{2}\right) = -27$$

Thus the determinant of (1) is obtainable by pivot condensation on the 2 of column 1 provided a minus sign is used. That is,

$$\begin{vmatrix} 1 & 2 & -1 \\ 2 & 6 & 3 \\ 5 & 7 & 1 \end{vmatrix} = -2\begin{vmatrix} -1 & -5/2 \\ -8 & -13/2 \end{vmatrix}$$

Repeated use of DP4 shows that the determinant of (1) is also obtainable by pivot condensation, pivoting on the 5 of column 1, without the minus sign. That is,

$$\begin{vmatrix} 1 & 2 & -1 \\ 2 & 6 & 3 \\ 5 & 7 & 1 \end{vmatrix} = 5\begin{vmatrix} 3/5 & -6/5 \\ 16/5 & 13/5 \end{vmatrix}$$

The general theorem for $n \times n$ matrices is clear from Example 3 and the preceding analysis. Informally stated, use pivot condensation in the first column, multiplying the determinant of the reduced matrix by the pivot and possibly by (-1). Consideration of DP4 shows that the rule for multiplying by (-1) is to use the (-1) if the pivot is selected from an even-numbered row.

Stated formally, we have

DP8. Let **A** be an $n \times n$ matrix. Then

$$|\mathbf{A}| = (-1)^{i+1} a_{i1} |\mathbf{B}|$$

where **B** is the condensed form of **A** when pivoting on a_{i1}.

A look back at Section 4.6 shows that this idea is inherent in Gaussian elimination.

EXAMPLE 4

Find the determinant of

$$\begin{bmatrix} 2 & 1 & 0 & -1 \\ 1^* & 0 & 2 & -4 \\ 3 & -1 & -3 & 2 \\ 5 & 4 & 1 & -1 \end{bmatrix} \tag{1}$$

The reason for wanting to be able to pivot anywhere in column 1 is now clear. a_{21} in this example is 1, and that simplifies the pivoting. Pivoting as indicated yields, by condensation,

$$\begin{bmatrix} 1^* & -4 & 7 \\ -1 & -9 & 14 \\ 4 & -9 & 19 \end{bmatrix} \tag{2}$$

$$\begin{vmatrix} 2 & 1 & 0 & -1 \\ 1 & 0 & 2 & -4 \\ 3 & -1 & -3 & 2 \\ 5 & 4 & 1 & -1 \end{vmatrix} = (-1) \begin{vmatrix} 1 & -4 & 7 \\ -1 & -9 & 14 \\ 4 & -9 & 19 \end{vmatrix} \tag{3}$$

To find

$$\begin{vmatrix} 1 & -4 & 7 \\ -1 & -9 & 14 \\ 4 & -9 & 19 \end{vmatrix}$$

we pivot as indicated in (2), yielding

$$\begin{bmatrix} -13 & 21 \\ 7 & -9 \end{bmatrix} \tag{4}$$

$$(5) \qquad \begin{vmatrix} 1 & -4 & 7 \\ -1 & -9 & 14 \\ 4 & -9 & 19 \end{vmatrix} = 1 \begin{vmatrix} -13 & 21 \\ 7 & -9 \end{vmatrix}$$

Substituting from (5) into (3) yields

$$(6) \qquad \begin{vmatrix} 2 & 1 & 0 & -1 \\ 1 & 0 & 2 & -4 \\ 3 & -1 & -3 & 2 \\ 5 & 4 & 1 & -1 \end{vmatrix} = (-1)(1) \begin{vmatrix} -13 & 21 \\ 7 & -9 \end{vmatrix} = (-1)(-30) = 30$$

EXAMPLE 5

$$\begin{vmatrix} 2^* & 0 & -1 & 1 \\ 3 & 1 & 1 & -1 \\ 1 & 1 & 0 & 2 \\ 2 & 3 & 1 & 0 \end{vmatrix} = 2 \begin{vmatrix} 1^* & 5/2 & -5/2 \\ 1 & 1/2 & 3/2 \\ 3 & 2 & -1 \end{vmatrix} = (2)(1) \begin{vmatrix} -2 & 4 \\ -11/2 & 13/2 \end{vmatrix}$$

$$= (2)(1)(9) = 18$$

Consider the case of

$$\mathbf{A} = \begin{bmatrix} 3 & 1 & 4 \\ 5 & 2 & -3 \\ -2 & -1 & 5 \end{bmatrix}$$

in which there is no "1" in column 1, but there is a "1" in column 2. It would be convenient to have a determinant property by which column 2 can become column 1. Indeed, there is one.

DP9. If any two adjacent columns of $n \times n$ matrix $\mathbf{A}$ are interchanged, $|\mathbf{A}|$ changes its sign.

Proof:

Case 1: $n = 2$. Let

$$\mathbf{A} = \begin{bmatrix} a_{11} & a_{12} \\ a_{21} & a_{22} \end{bmatrix}$$

Interchange columns and let

$$\mathbf{B} = \begin{bmatrix} a_{12} & a_{11} \\ a_{22} & a_{21} \end{bmatrix}$$

$$|\mathbf{A}| = a_{11}a_{22} - a_{21}a_{12}$$

and

$$|\mathbf{B}| = a_{12}a_{21} - a_{22}a_{11}$$

Thus $|\mathbf{B}| = -|\mathbf{A}|$.

Case 2: $n = 3$. Let

$$\mathbf{A} = \begin{bmatrix} a_{11} & a_{12} & a_{13} \\ a_{21} & a_{22} & a_{23} \\ a_{31} & a_{32} & a_{33} \end{bmatrix}$$

Interchange columns 1 and 2 and let

$$\mathbf{B} = \begin{bmatrix} a_{12} & a_{11} & a_{13} \\ a_{22} & a_{21} & a_{23} \\ a_{32} & a_{31} & a_{33} \end{bmatrix}$$

The reader should expand **A** and **B** to see that

$$|\mathbf{B}| = -|\mathbf{A}|$$

The inductive approach to establishing this determinant property is clear.

By repeated use of DP9, column 2 of an $n \times n$ matrix can be moved to become the first column with one exchange of columns; column 3 can become first with two exchanges of adjacent columns; in general column j can become first with $j - 1$ exchanges of adjacent columns. With these $j - 1$ exchanges the determinant of the matrix is multiplied by $(-1)^{j-1}$. As a practical matter, if the column to be first is originally even, then multiply by -1 when getting the determinant. Since in pivot condensation this column is eliminated, then as a practical matter in the evaluation of determinants it is not moved at all.

When determining whether or not to use a minus sign, we now have to consider rows and columns. The checkerboard pattern below will yield the proper sign. Just start with a plus sign in the upper left of the matrix from which the pivot is selected and then alternately think of pluses and minuses.

$$\begin{matrix} + & - & + & - & \cdots \\ - & + & - & + & \cdots \\ + & - & + & - & \cdots \\ - & + & - & + & \cdots \\ \vdots & \vdots & \vdots & \vdots & \end{matrix}$$

It is clear that plus signs are on the main diagonal, so if it is convenient to select pivots on the main diagonal, then all signs are positive.

EXAMPLE 6

In evaluating $|\mathbf{A}|$ below, by the checkerboard rule there is a + in row 2, column 2, the position of the pivot. Thus

$$|\mathbf{A}| = \begin{vmatrix} 3 & 2 & 4 & 1 \\ -1 & 1^* & 2 & 2 \\ 3 & -1 & 1 & -3 \\ -2 & 0 & 3 & -4 \end{vmatrix} = (1)\begin{vmatrix} 5 & 0 & -3 \\ 2 & 3 & -1 \\ -2 & 3^* & -4 \end{vmatrix}$$

Now pivoting on 3, as indicated, the checkerboard yields a minus sign, so

$$|\mathbf{A}| = (1)(-1)(3)\begin{vmatrix} 5 & -3 \\ 4 & 3 \end{vmatrix} = (-3)(27) = -81$$

DP10. Without regard to sign, $|\mathbf{A}|$ is equal to the product of the pivots when using pivot condensation. The proper sign for each pivot is determined by the checkerboard rule, and the sign for the determinant of **A** is the product of the signs for the pivots.

There is an interesting fallout from this development. The value of $|\mathbf{A}|$, by definition, is unique. Therefore, the product of pivots, however selected, in a sequence of pivot condensations for an $n \times n$ matrix, each multiplied by the proper power of -1, is constant for any choice of pivots.

If, as a number of pivot condensations steps are carried out, only zeros remain in the fully condensed form, then the determinant of the original matrix is zero. But this is also the condition that the set of vectors in the rows of an $n \times n$ matrix is linearly dependent. Hence a set of n vectors in n components is linearly dependent if the $n \times n$ matrix formed by these vectors as rows has a zero determinant. Some of our theorems from previous chapters, restated in determinant language, are

1. $\mathbf{A}$ has an inverse if and only if $|\mathbf{A}| \neq 0$.
2. Let $T: R^n \to R^n$ be a linear mapping with matrix $\mathbf{A}$. Then T is a bijection if and only if $|\mathbf{A}| \neq 0$.
3. Let S be a system of n linear equations in n variables with coefficient matrix $\mathbf{A}$. Then S has a unique solution if and only if $|\mathbf{A}| \neq 0$.
4. The pivot rank r of an $n \times n$ matrix $\mathbf{A}$ is equal to n if and only if $|\mathbf{A}| \neq 0$.
5. A set of n vectors in R^n is a basis for R^n if and only if the matrix $\mathbf{A}$ which has these vectors as its rows has a determinant not equal to zero.

Since determinants are only defined for square matrices, none of the above comments apply in cases of $m \times n$ matrices where m and n are not equal.

There is another interesting conclusion that emerges from pivot evaluations of determinants. It is illustrated with 3×3 matrices, but it is valid for all $n \times n$ matrices.

Let

$$\mathbf{A} = \begin{bmatrix} a_{11} & a_{12} & a_{13} \\ a_{21} & a_{22} & a_{23} \\ a_{31}^* & a_{32} & a_{33} \end{bmatrix}$$

Then

$$\mathbf{A}^T = \begin{bmatrix} a_{11} & a_{21} & a_{31}^* \\ a_{12} & a_{22} & a_{32} \\ a_{13} & a_{23} & a_{33} \end{bmatrix}$$

(Recall that the transpose of $\mathbf{A}$, denoted $\mathbf{A}^T$, is the matrix derived from $\mathbf{A}$ by writing its columns as rows.) By pivot evaluations of $\mathbf{A}$ and $\mathbf{A}^T$ choosing in both cases a_{31} as pivot (arbitrarily),

$$|\mathbf{A}| = a_{31} |\mathbf{B}_{31}|$$

$$|\mathbf{A}^T| = a_{31} |\mathbf{B}_{31}^T|$$

$\mathbf{B}_{31}$ ($\mathbf{B}_{31}^T$) is obtained by pivot condensation on a_{31} (a_{31}^T).

By direct evaluation

$$|\mathbf{B}_{31}| = |\mathbf{B}_{31}^T|$$

Thus

$$|\mathbf{A}| = |\mathbf{A}^T|$$

for 3×3 matrices. The general proof is inductive. We state

DP11. For all $n \times n$ matrices $\mathbf{A}$,

$$|\mathbf{A}| = |\mathbf{A}^T|$$

Since transposing matrices changes rows to columns and columns to rows, we have that every DP proved (or asserted) in this section is true when "column" is replaced with "row" and "row" is replaced with "column."

Exercises

1. Evaluate the determinants of the following matrices by pivot condensation.

(a) $\begin{bmatrix} 1 & 0 & 3 \\ 2 & 8 & 3 \\ 3 & 6 & 2 \end{bmatrix}$ (b) $\begin{bmatrix} 3 & -2 & 4 \\ 2 & 1 & 6 \\ 0 & 2 & 5 \end{bmatrix}$

(c) $\begin{bmatrix} 1 & 1 & 0 & 1 \\ 0 & 1 & 1 & 1 \\ 2 & 1 & 2 & 1 \\ 3 & 1 & 0 & 2 \end{bmatrix}$ (d) $\begin{bmatrix} 4 & -1 & -2 & 3 \\ 4 & 2 & 4 & -2 \\ 0 & 1 & 3 & 2 \\ -2 & 5 & 6 & 7 \end{bmatrix}$

2. Evaluate

$$\begin{vmatrix} 2 & 3 & 4 \\ -1 & 0 & 5 \\ -2 & -3 & 1 \end{vmatrix}$$

by pivoting on (a) 2; (b) 5. Compare the results.

3. Let $\mathbf{A}$ be a 4×4 matrix. Let $\mathbf{B}$ be the condensed form of $\mathbf{A}$ when pivoting on a_{23}. Let $\mathbf{C}$ be the condensed form of $\mathbf{A}^T$ when pivoting on a_{23}. Show that:

(a) $\mathbf{C} = \mathbf{B}^T$.
(b) $|\mathbf{C}| = |\mathbf{B}^T|$.
(c) How does this show that $|\mathbf{A}| = |\mathbf{A}^T|$?

4. Let $\mathbf{A}$ be a 3×3 matrix. One of the six expressions for $|\mathbf{A}|$ is

$$a_{11}|\mathbf{A}_{11}| - a_{21}|\mathbf{A}_{21}| + a_{31}|\mathbf{A}_{31}|$$

Write five others.

5. One way to find $|\mathbf{A}|$, when $\mathbf{A}$ is a 4×4 matrix, is

$$a_{11}|\mathbf{A}_{11}| - a_{21}|\mathbf{A}_{21}| + a_{31}|\mathbf{A}_{31}| - a_{41}|\mathbf{A}_{41}|$$

How many similar ways are there?

6. A 3×3 matrix $\mathbf{A}$ has nonzero entries. Using pivot methods, in how many ways can $|\mathbf{A}|$ be evaluated?

7. Let

$$\mathbf{A} = \begin{bmatrix} 1 & 2 & 1 \\ 2 & 1 & 0 \\ 0 & -1 & 1 \end{bmatrix}$$

Express

$$\begin{bmatrix} |\mathbf{A}_{11}| & -|\mathbf{A}_{12}| & |\mathbf{A}_{13}| \\ -|\mathbf{A}_{21}| & |\mathbf{A}_{22}| & |\mathbf{A}_{23}| \\ |\mathbf{A}_{31}| & -|\mathbf{A}_{32}| & |\mathbf{A}_{33}| \end{bmatrix}$$

where $\mathbf{A}_{ij}$ is the minor a_{ij} in $\mathbf{A}$.

8.4 Some Historic Uses of Determinants

One of the uses of determinants is to measure the area of a triangle. Relative to a rectangular coordinate system let points $\mathbf{A}$, $\mathbf{B}$, and $\mathbf{C}$ have coordinates (a_1, a_2), (b_1, b_2), and $(0, 0)$, respectively, and let $\mathbf{ABC}$ be a triangle. (See Figure 8.1.) Let $\mathbf{B}'$ and $\mathbf{A}'$ be the respective orthogonal projections of $\mathbf{B}$ and $\mathbf{A}$ on the x_1 axis.

The area of $\Delta\ \mathbf{ABC}$ is the area of $\Delta\mathbf{CB'B}$ + area of trapezoid $\mathbf{B'A'AB}$ − area of $\Delta\ \mathbf{CA'A} = \frac{1}{2}(b_1 b_2) + \frac{1}{2}(a_2 + b_2)(a_1 - b_1) - \frac{1}{2}(a_1 a_2)$

$$= \tfrac{1}{2}(a_1 b_2 - a_2 b_1)$$

$$= \tfrac{1}{2}\begin{vmatrix} a_1 & a_2 \\ b_1 & b_2 \end{vmatrix}$$

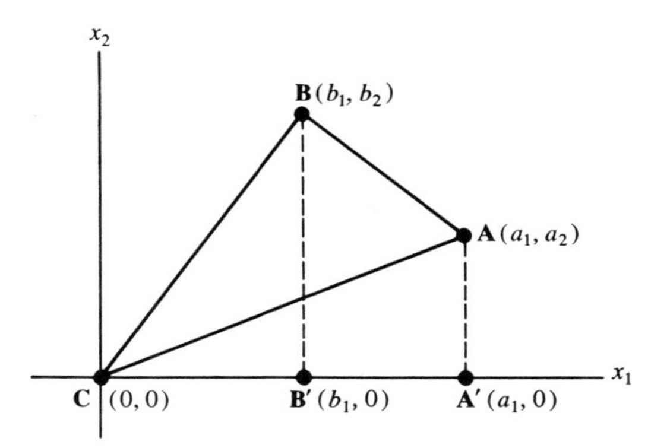

Figure 8.1

This determinant may be positive or negative. The significance of this fact is discussed in connection with the triangles in

EXAMPLE 1

(a) Let $\mathbf{A} = (3, 2)$, $\mathbf{B} = (-1, 4)$, and $\mathbf{C} = (0, 0)$.

$$\text{The area of } \Delta\ \mathbf{ABC} = \begin{vmatrix} 3 & 2 \\ -1 & 4 \end{vmatrix} = 7$$

(b) The area of Δ **BAC** (observe change of order of vertices) is obviously also 7. However, the determinant formula gives

$$\tfrac{1}{2}\begin{vmatrix} -1 & 4 \\ 3 & 2 \end{vmatrix} = -7$$

In Δ **ABC** the motion from **A** to **B** to **C** is a counterclockwise motion. In Δ **BAC** the motion from **B** to **A** to **C** is clockwise. We say that the triangles have different *orientations,* positive or counterclockwise in one case, and negative or clockwise in the other. Thus the determinant that measures the area of a triangle also tells its orientation. A plus sign indicates counterclockwise, a minus sign clockwise. The area is the absolute value of the determinant.

By an argument similar to that illustrated in Figure 8.1 we can show that if the vertices of Δ **ABC** are (a_1, a_2), (b_1, b_2), and (c_1, c_2) with no vertex at the origin, then the area and orientation of Δ **ABC** are given by the determinant

$$\frac{1}{2}\begin{vmatrix} a_1 & a_2 & 1 \\ b_1 & b_2 & 1 \\ c_1 & c_2 & 1 \end{vmatrix}$$

EXAMPLE 2

Let $\mathbf{A} = (2, 3)$, $\mathbf{B} = (-1, -4)$, and $\mathbf{C} = (-2, 2)$. Then

$$\frac{1}{2}\begin{vmatrix} 2 & 3 & 1 \\ -1 & -4 & 1 \\ -2 & 2 & 1 \end{vmatrix} = -\frac{25}{2}$$

Thus the area of Δ **ABC** is $\frac{25}{2}$ and its orientation is clockwise.

A second use of determinants is to measure the volume of a tetrahedron (triangular pyramid). Let the vertices **A**, **B**, **C**, and **D** in E^3 be (a_1, a_2, a_3), (b_1, b_2, b_3), (c_1, c_2, c_3), and $(0, 0, 0)$. (See Figure 8.2.) Geometric arguments show that the volume of the tetrahedron is the absolute value of

$$\frac{1}{6}\begin{vmatrix} a_1 & a_2 & a_3 \\ b_1 & b_2 & b_3 \\ c_1 & c_2 & c_3 \end{vmatrix}$$

If **D** is not at $(0, 0, 0)$ but at (d_1, d_2, d_3), then the volume is the absolute value of

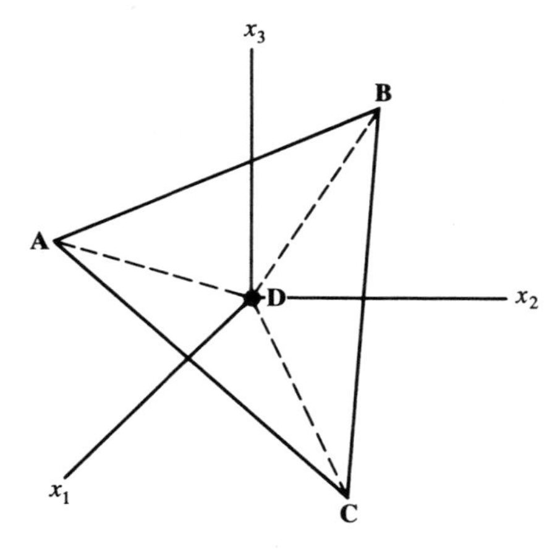

Figure 8.2

$$\frac{1}{6}\begin{vmatrix} a_1 & a_2 & a_3 & 1 \\ b_1 & b_2 & b_3 & 1 \\ c_1 & c_2 & c_3 & 1 \\ d_1 & d_2 & d_3 & 1 \end{vmatrix}$$

EXAMPLE 3

The volume of the tetrahedron with vertices at (1, 0, 0), (0, 1, 0), (0, 0, 1), and (1, 1, 1) is the absolute value of

$$\frac{1}{6}\begin{vmatrix} 1 & 0 & 0 & 1 \\ 0 & 1 & 0 & 1 \\ 0 & 0 & 1 & 1 \\ 1 & 1 & 1 & 1 \end{vmatrix} = -\frac{1}{3}$$

As with triangles, orientation (positive or negative) is applicable to tetrahedrons.

The next four applications illustrate how determinants describe certain geometric aspects of mappings.

APPLICATION 1

Let $S: E^2 \to E^2$ with $(x_1, x_2) \mapsto (x_1 + cx_2, x_2)$. This mapping is called a *shearing* with matrix

$$\begin{bmatrix} 1 & 0 \\ c & 1 \end{bmatrix}$$

It maps Δ **0AB** (see Figure 8.3) into Δ **0A′B**.

The area of Δ **0AB** is equal to that of Δ **0A′B**. This is true for any triangle and its image under the shearing. This is indicated in the fact that

$$\begin{vmatrix} 1 & 0 \\ c & 1 \end{vmatrix} = 1$$

In general the area of a triangle is preserved under a linear mapping if the determinant of its matrix is 1 or -1.

APPLICATION 2

Consider the projection $P: E^3 \to E^3$ with $(x_1, x_2, x_3) \mapsto (x_1, x_2, 0)$. Under this projection the image of every point in E^3 is the foot of the perpendicular from the point to the x_1x_2 plane. (See Figure 8.4.)

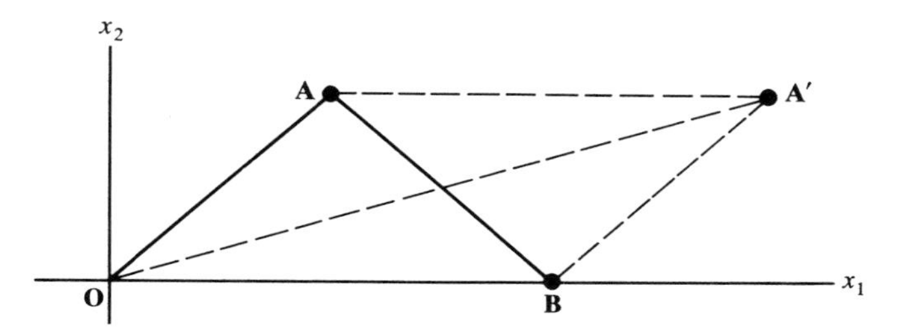

Figure 8.3

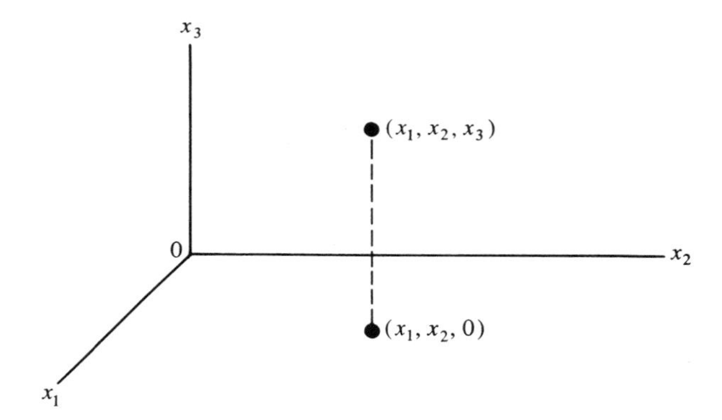

Figure 8.4

The matrix of P is

$$\mathbf{A} = \begin{bmatrix} 1 & 0 & 0 \\ 0 & 1 & 0 \\ 0 & 0 & 0 \end{bmatrix}$$

The projection is not invertible since $|\mathbf{A}| = 0$.

In contrast to the above projection, consider the *reflection* $R : E^3 \to E^3$ with $(x_1, x_2, x_3) \mapsto (x_1, x_2, -x_3)$. (See Figure 8.5.)

The matrix of R is

$$\mathbf{A} = \begin{bmatrix} 1 & 0 & 0 \\ 0 & 1 & 0 \\ 0 & 0 & -1 \end{bmatrix}$$

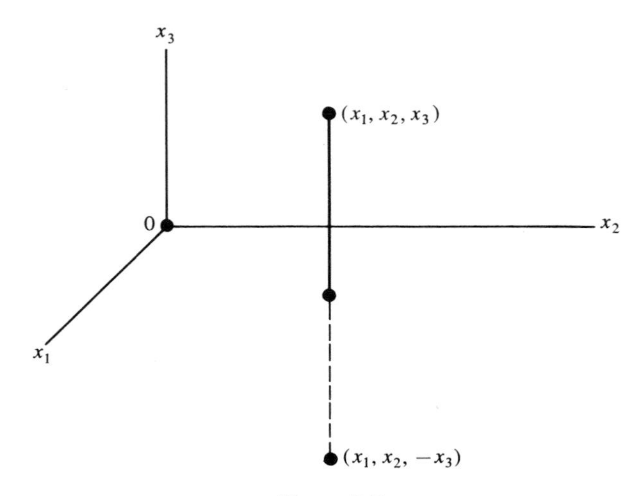

Figure 8.5

From $|\mathbf{A}| = -1$ we conclude that R is invertible, preserves areas and volumes, and reverses orientation.

APPLICATION 3

Consider the *dilation* (magnification) $D: E^3 \to E^3$ with $(x_1, x_2, x_3) \mapsto (cx_1, cx_2, cx_3)$. Under this dilation, segment lengths are multiplied by c, areas by c^2, and volumes by c^3. Its matrix is

$$\mathbf{A} = \begin{bmatrix} c & 0 & 0 \\ 0 & c & 0 \\ 0 & 0 & c \end{bmatrix}$$

and $|\mathbf{A}| = c^3$. If $c > 0$, orientation is preserved. If $c < 0$, orientation is reversed. If $c = 0$ all points in E^3 go into $(0, 0, 0)$.

APPLICATION 4 CRAMER'S RULE

Given the system S of n linear equations in n variables,

$$\begin{array}{ccccc|c} x_1 & x_2 & \cdots & x_n & -1 & \\ \hline a_{11} & a_{12} & \cdots & a_{1n} & b_1 & = 0 \\ a_{21} & a_{22} & \cdots & a_{2n} & b_2 & = 0 \\ \vdots & \vdots & & \vdots & \vdots & \vdots \\ a_{n1} & a_{n2} & \cdots & a_{nn} & b_n & = 0 \end{array}$$

Let **A** be the coefficient matrix of this system and let ${}_j\mathbf{A}$ (the subscript j precedes **A**) be the matrix obtained from **A** by replacing its jth column by the b's. If $|\mathbf{A}| \neq 0$, then the system has exactly one solution, where each x_j for $j = 1, 2, \ldots, n$ is given by

$$x_j = \frac{|{}_j\mathbf{A}|}{|\mathbf{A}|}$$

EXAMPLE 4

Using Cramer's rule, solve

$$\begin{cases} x + 4y + 2z = 19 \\ 2x + y + 2z = 19 \\ 2x + 3y + z = 18 \end{cases}$$

x_1	x_2	x_3	-1	
1	4	2	19	= 0
2	1	2	19	= 0
2	3	1	18	= 0

Using the notation of Cramer's rule,

$$\mathbf{A} = \begin{bmatrix} 1 & 4 & 2 \\ 2 & 1 & 2 \\ 2 & 3 & 1 \end{bmatrix}, \qquad |\mathbf{A}| = 11$$

$${}_1\mathbf{A} = \begin{bmatrix} 19 & 4 & 2 \\ 19 & 1 & 2 \\ 18 & 3 & 2 \end{bmatrix}, \qquad |{}_1\mathbf{A}| = 51$$

$${}_2\mathbf{A} = \begin{bmatrix} 1 & 19 & 2 \\ 2 & 19 & 2 \\ 2 & 18 & 1 \end{bmatrix}, \qquad |{}_2\mathbf{A}| = 17$$

$${}_3\mathbf{A} = \begin{bmatrix} 1 & 4 & 19 \\ 2 & 1 & 19 \\ 2 & 3 & 18 \end{bmatrix}, \qquad |{}_3\mathbf{A}| = 45$$

Thus $x_1 = 51/11$, $x_2 = 17/11$, and $x_3 = 45/11$.

Exercises

1. Find the area of the triangle whose vertices are:
(a) (0, 0), (3, 2), (1, 8).
(b) (0, 0), (−2, 1), (−3, −2).
(c) (2, 0), (4, 2), (−1, 1).
(d) (2, 0), (−1, 1), (4, 2).
(e) (−2, 3), (−3, 5), (2, 5).

2. Find the area of the quadrilateral whose vertices are (0, 0), (3, 1), (5, 4), and (1, 8). (*Hint:* Find the areas of two triangles.)

3. Find the volume of the tetrahedron whose vertices are:
(a) (0, 0, 0), (3, 0, 0), (2, 1, 5), (1, −2, 3).
(b) (2, 1, 2), (3, −1, 3), (4, 5, 2), (−1, −2, −3).

4. Using determinants, determine whether or not each of the following mappings is a bijection.
(a) $R^2 \to R^2, (x_1, x_2) \mapsto (x_1 + x_2, x_1 - x_2)$.
(b) $R^3 \to R^3, (x_1, x_2, x_3) \mapsto (2x_1, 3x_2, 0)$.
(c) $R^3 \to R^3, (x_1, x_2, x_3) \mapsto (x_1 + x_2 - x_3, 2x_1 - x_2 + x_3, 3x_2)$.
(d) $R^4 \to R^4, (x_1, x_2, x_3, x_4) \mapsto (x_4, x_3, x_2, x_1)$.

5. Determine whether or not each of the following preserves (i) area, (ii) orientation.
(a) $E^2 \to E^2, (x_1, x_2) \mapsto (2x_1, x_1 + x_2)$.
(b) $E^2 \to E^2, (x_1, x_2) \mapsto (x_1, 3x_1 + x_2)$.
(c) $E^2 \to E^2, (x_1, x_2) \mapsto (x_1, 3x_1 - x_2)$.

6. Using Cramer's rule, find the unique solution, if any, of:

(a)

x	y	z	-1	
1	3	−1	0	= 0
1	1	1	0	= 0
0	1	−1	1	= 0

(b)
$$\begin{cases} 3x + y - 2z = 0 \\ 3x - 4y + z = 2 \\ x - 2y + z = 1 \end{cases}$$

(c)

x_1	x_2	x_3	x_4	-1	
4	1	1	1	2	= 0
1	−1	2	−3	0	= 0
2	1	3	5	0	= 0
1	1	−1	−1	4	= 0

(d) $$\begin{cases} x_1 + x_2 + x_3 - x_4 = 8 \\ x_1 + x_2 + x_3 + x_4 = 0 \\ 2x_1 + x_2 - 4x_3 - x_4 = 2 \\ -x_1 - 2x_2 + 3x_3 - 5x_4 = 0 \end{cases}$$

8.5 More Properties of Determinants

In Section 3.6 we used pivot matrices to decompose a matrix **A.** We now go through a similar type of decomposition using different pivot matrices. Since our interest in this chapter is with determinants, we restrict ourselves to square matrices.

EXAMPLE 1

We recall from Chapter 3 that a matrix of the type illustrated as (1) is called a pivot matrix:

(1) $$\begin{bmatrix} 1 & 0 & 2 \\ 0 & 1 & 3 \\ 0 & 0 & -2 \end{bmatrix}$$

Using pivoting we see that

(2) $$\begin{vmatrix} 1 & 0 & 2 \\ 0 & 1 & 3 \\ 0 & 0 & -2 \end{vmatrix} = -2$$

The determinant -2 is immediate as soon as we convince ourselves of the validity of

DP12. Let **P** be a pivot matrix and let p be the element in the nonunit column which lies on the main diagonal. Then $|\mathbf{P}| = p$.

In example 1 the third column is the nonunit column and -2 is on the main diagonal. That DP12 is true follows from the following:

(a) Pivoting on the 1 of a unit column in a matrix does not alter the matrix. The truth of this statement becomes obvious the moment you test the statement.

(b) Using the pivot procedure for finding determinants, on a pivot matrix, always pivoting in a unit column will condense the matrix to the required p.

By definition of a pivot matrix $p \neq 0$, and so $|\mathbf{P}| \neq 0$.

EXAMPLE 2

$$\begin{vmatrix} 1 & 0 & 0 & 3 & 0 & 0 \\ 0 & 1 & 0 & 17 & 0 & 0 \\ 0 & 0 & 1 & -98 & 0 & 0 \\ 0 & 0 & 0 & 4 & 0 & 0 \\ 0 & 0 & 0 & 17 & 1 & 0 \\ 0 & 0 & 0 & 623 & 0 & 1 \end{vmatrix} = 4$$

EXAMPLE 3

Let

$$\mathbf{A} = \begin{bmatrix} 1 & 2^* & 3 \\ 4 & 5 & 6 \\ -1 & 0 & 3 \end{bmatrix}$$

Then

$$\begin{bmatrix} 1 & 2 & 3 \\ 4 & 5 & 6 \\ -1 & 0 & 3 \end{bmatrix} = \begin{bmatrix} 2 & 0 & 0 \\ 5 & 1 & 0 \\ 0 & 0 & 1 \end{bmatrix} \begin{bmatrix} 1/2 & 1 & 3/2 \\ 3/2 & 0 & -3/2 \\ -1 & 0 & 3 \end{bmatrix} \tag{1}$$

as you should verify by direct multiplication.

This decomposition of $\mathbf{A}$ into a product of two other matrices is accomplished as follows:

(a) Select a pivot, in this case 2.

(b) Replace a unit column of the identity matrix, in this case $\mathbf{I}_3$, with the column of the pivot. Select the column to be replaced so that the pivot element falls on the main diagonal. In this example replace the first column of $\mathbf{I}_3$ by the column

$$\begin{bmatrix} 2 \\ 5 \\ 0 \end{bmatrix}$$

yielding the pivot matrix

$$\mathbf{P}_1 = \begin{bmatrix} 2 & 0 & 0 \\ 5 & 1 & 0 \\ 0 & 0 & 1 \end{bmatrix}$$

(c) Now perform a pivot step in **A**. In the resulting matrix **B** replace the column of the pivot with the unit column removed from the identity matrix. In this example pivoting on 2 yields

$$\mathbf{B} = \begin{bmatrix} 1/2 & 1/2 & 3/2 \\ 3/2 & -5/2 & -3/2 \\ -1 & 0 & 3 \end{bmatrix}$$

Replacing the column

$$\begin{bmatrix} 1/2 \\ -5/2 \\ 0 \end{bmatrix}$$

by the unit column

$$\begin{bmatrix} 1 \\ 0 \\ 0 \end{bmatrix}$$

yields

$$\mathbf{B}_1 = \begin{bmatrix} 1/2 & 1 & 3/2 \\ 3/2 & 0 & -3/2 \\ -1 & 0 & 3 \end{bmatrix}$$

(d) The decomposition of **A** is given by

(2) $$\mathbf{A} = \mathbf{P}_1\mathbf{B}_1$$

In Example 3 the matrix equation (2) is given in (1). Using Example 3 as a guide we indicate a proof that (2) is always true. Consider the tableau.

(3)

r	1	2*	3
s	4	5	6
t	−1	0	3
	$= x_1$	$= x_2$	$= x_3$

This tableau has **A** as its matrix. The decomposition of **A** given in (2) is the equivalent of representing tableau (3) by the composite tableaus of (4).

(4)

r	1	2	3
s	4	5	6
t	−1	0	3
	$= x_1$	$= x_2$	$= x_3$

=

r	2	0	0
s	5	1	0
t	0	0	1
	$= x_2$	$= s$	$= t$

∘

x_2	$1/2$	1	$3/2$
s	$3/2$	0	$-3/2$
t	−1	0	3
	$= x_1$	$= x_2$	$= x_3$

In the last tableau on the right, x_1 and x_3 columns are those that result from pivoting on 2. Since composition of tableaus results in multiplication of matrices (see Chapter 3), we have the desired result.

We need not restrict ourselves to square matrices. Using an identity matrix with as many rows as **A**, allows for the identical analysis as given above.

EXAMPLE 4

By decomposing $\mathbf{B}_1$ of Example 3 we can further decompose **A**.

$$\mathbf{B}_1 = \begin{bmatrix} \frac{1}{2} & 1 & \frac{3}{2} \\ \frac{3}{2}^* & 0 & -\frac{3}{2} \\ -1 & 0 & 3 \end{bmatrix}$$

Pivoting as indicated and using the method of Example 3 we get that

$$\mathbf{B}_1 = \mathbf{P}_2\mathbf{B}_2$$

where

$$\mathbf{P}_2 = \begin{bmatrix} 1 & \frac{1}{2} & 0 \\ 0 & \frac{3}{2} & 0 \\ 0 & -1 & 1 \end{bmatrix}$$

and

$$\mathbf{B}_2 = \begin{bmatrix} 0 & 1 & 2 \\ 1 & 0 & -1 \\ 0 & 0 & 2 \end{bmatrix}$$

Thus

$$\begin{bmatrix} \frac{1}{2} & 1 & \frac{3}{2} \\ \frac{3}{2} & 0 & -\frac{3}{2} \\ -1 & 0 & 3 \end{bmatrix} = \begin{bmatrix} 1 & \frac{1}{2} & 0 \\ 0 & \frac{3}{2} & 0 \\ 0 & -1 & 1 \end{bmatrix} \begin{bmatrix} 0 & 1 & 2 \\ 1 & 0 & -1 \\ 0 & 0 & 2 \end{bmatrix} \tag{5}$$

Substituting in (1) of Example 3 gives

$$\begin{bmatrix} 1 & 2 & 3 \\ 4 & 5 & 6 \\ -1 & 0 & 3 \end{bmatrix} \tag{6}$$

$$= \begin{bmatrix} 2 & 0 & 0 \\ 5 & 1 & 0 \\ 0 & 0 & 1 \end{bmatrix} \begin{bmatrix} 1 & \frac{1}{2} & 0 \\ 0 & \frac{3}{2} & 0 \\ 0 & -1 & 1 \end{bmatrix} \begin{bmatrix} 0 & 1 & 2 \\ 1 & 0 & -1 \\ 0 & 0 & 2 \end{bmatrix}$$

In symbolic terms

$$\mathbf{A} = \mathbf{P}_1\mathbf{P}_2\mathbf{B}_2 \tag{7}$$

Using the same decomposition method,

$$\mathbf{B}_2 = \mathbf{P}_3\mathbf{B}_3$$

where

$$\mathbf{P}_3 = \begin{bmatrix} 1 & 0 & 2 \\ 0 & 1 & -1 \\ 0 & 0 & 2 \end{bmatrix} \quad \text{(On what are we pivoting?)}$$

and

$$\mathbf{B}_3 = \begin{bmatrix} 0 & 1 & 0 \\ 1 & 0 & 0 \\ 0 & 0 & 1 \end{bmatrix}$$

Thus the final decomposition is

$$\mathbf{A} = \mathbf{P}_1\mathbf{P}_2\mathbf{P}_3\mathbf{B}_3 \tag{8}$$

or in the example

$$\begin{bmatrix} 1 & 2 & 3 \\ 4 & 5 & 6 \\ -1 & 0 & 3 \end{bmatrix} = \begin{bmatrix} 2 & 0 & 0 \\ 5 & 1 & 0 \\ 0 & 0 & 1 \end{bmatrix} \begin{bmatrix} 1 & \frac{1}{2} & 0 \\ 0 & \frac{3}{2} & 0 \\ 0 & -1 & 1 \end{bmatrix} \begin{bmatrix} 1 & 0 & 2 \\ 0 & 1 & -1 \\ 0 & 0 & 2 \end{bmatrix} \begin{bmatrix} 0 & 1 & 0 \\ 1 & 0 & 0 \\ 0 & 0 & 1 \end{bmatrix} \tag{9}$$

$$|\mathbf{A}| = |\mathbf{P}_1|\ |\mathbf{P}_2|\ |\mathbf{P}_3|\ |\mathbf{B}_3| = 2 \cdot \tfrac{3}{2} \cdot 2 \cdot (-1) = -6$$

To see this, consider the following:

(a) Since the **P**'s are pivot matrices, their determinants are given by DP12: in this case 2, $\frac{3}{2}$, and 2, respectively.
(b) $\mathbf{B}_3$ is a permutation matrix. As such its determinant is either $+1$ or -1, as we establish momentarily.

(c) By DP10 the determinant of $\mathbf{A}$ is the product of the pivots when using pivot condensation, multiplied by the appropriate sign as determined by the checkerboard rule. The pivots using pivot condensation are precisely the determinants of $\mathbf{P}_1$, $\mathbf{P}_2$, and $\mathbf{P}_3$ as given. We show that the sign of $\mathbf{B}_3$ is the correct sign.

To verify that $|\mathbf{B}_3| = -1$, we broaden our investigation to that of evaluating the determinant of any permutation matrix. From the definition of a permutation matrix (see Section 3.5) we conclude that a permutation matrix is an identity matrix whose rows and/or columns are interchanged (permuted). Thus each column is a unit column, and therefore pivoting does not alter the matrix. Thus the determinant is the product of the pivots (all 1's) with signs determined by the checkerboard rule. That this sign is the correct one for evaluating the determinant of $\mathbf{A}$ follows since each 1 in the permutation matrix is in precisely the same location as the pivots are in pivot condensation.

Examples 3 and 4 illustrate:

DP13. Let $\mathbf{A}$ be an $n \times n$ matrix. If $|\mathbf{A}| \neq 0$, then

$$|\mathbf{A}| = |\mathbf{P}_1|\ |\mathbf{P}_2| \cdots |\mathbf{P}_n|\ |\mathbf{B}_n|$$

where $\mathbf{P}_1, \mathbf{P}_2, \ldots, \mathbf{P}_n$ are the pivot matrices of the decomposition and $\mathbf{B}_n$ is the associated permutation matrix. If $|\mathbf{A}| = 0$, then in the decomposition process, prior to the nth step, a matrix with a row of zeros appears.

EXAMPLE 5

Evaluate

$$\begin{vmatrix} 2 & 3 & 4 \\ 1 & -1 & 2 \\ 3 & 2 & 6 \end{vmatrix}$$

$$\begin{vmatrix} 2 & 3 & 4 \\ 1^* & -1 & 2 \\ 3 & 2 & 6 \end{vmatrix} = \begin{vmatrix} 1 & 2 & 0 \\ 0 & 1 & 0 \\ 0 & 3 & 1 \end{vmatrix} \begin{vmatrix} 0 & 5^* & 0 \\ 1 & -1 & 2 \\ 0 & 5 & 0 \end{vmatrix}$$

$$= \begin{vmatrix} 1 & 2 & 0 \\ 0 & 1 & 0 \\ 0 & 3 & 1 \end{vmatrix} \begin{vmatrix} 5 & 0 & 0 \\ -1 & 1 & 0 \\ 5 & 0 & 1 \end{vmatrix} \begin{vmatrix} 0 & 1 & 0 \\ 1 & 0 & 2 \\ 0 & 0 & 0 \end{vmatrix}$$

Thus

$$\begin{vmatrix} 2 & 3 & 4 \\ 1 & -1 & 2 \\ 3 & 2 & 6 \end{vmatrix} = 0$$

EXAMPLE 6

$$\begin{bmatrix} 0 & 0 & 1 \\ 1 & 0 & 0 \\ 0 & 1 & 0 \end{bmatrix} \begin{bmatrix} 1 & 2 & 3 \\ 4 & 5 & 6 \\ -1 & 0 & 3 \end{bmatrix} = \begin{bmatrix} -1 & 0 & 3 \\ 1 & 2 & 3 \\ 4 & 5 & 6 \end{bmatrix}$$

Example 6 illustrates a property of permutation matrices: if a matrix $\mathbf{A}$ is multiplied on the left by a permutation matrix, then rows of $\mathbf{A}$ are interchanged. The number of interchanges depends on the permutation matrix and is determined as follows. Look at the permutation matrix and see how it differs from the identity matrix. Thus in Example 6 the first row of the permutation matrix is the third row of $\mathbf{I}_3$ and so differs by two rows. It thus induces two interchanges, moving the third row, $[-1 \quad 0 \quad 3]$, to the first row. The second row of the permutation matrix is the first row of $\mathbf{I}_3$ and so differs by one row. It thus induces one interchange moving the first row, $[1 \quad 2 \quad 3]$, to the second row. The third row of the permutation matrix is the second row of $\mathbf{I}_3$ and thus induces one interchange. By repeated application of DP4 we have that the four interchanges in Example 6 leave the determinant unchanged and thus

$$\begin{vmatrix} 1 & 2 & 3 \\ 4 & 5 & 6 \\ -1 & 0 & 3 \end{vmatrix} = -6 = \begin{vmatrix} -1 & 0 & 3 \\ 1 & 2 & 3 \\ 4 & 5 & 6 \end{vmatrix}$$

The general conclusion is

DP14. Let $\mathbf{A}$ be any square matrix of order n and π a permutation matrix of order n. Then $|\pi\mathbf{A}| = (-1)^m |\mathbf{A}|$, where m is the number of interchanges of adjacent rows induced on $\mathbf{A}$ by π.

Let

$$\mathbf{A} = \begin{bmatrix} a & 0 & 0 \\ b & 1 & 0 \\ c & 0 & 1 \end{bmatrix}$$

be a pivot matrix. Then

$$|\mathbf{A}| = a$$

Let

$$\mathbf{B} = \begin{bmatrix} d & e & f \\ g & h & k \\ m & n & p \end{bmatrix}$$

$$\mathbf{AB} = \begin{bmatrix} ad & ae & af \\ bd + g & be + h & bf + k \\ cd + m & ce + n & cf + p \end{bmatrix}$$

Then, by DP1,

$$|\mathbf{AB}| = a \begin{vmatrix} d & e & f \\ bd + g & be + h & bf + k \\ cd + m & ce + n & cf + p \end{vmatrix}$$

But by a double application of DP7,

$$\begin{vmatrix} d & e & f \\ bd + g & be + h & bf + k \\ cd + m & ce + n & cf + p \end{vmatrix} = |\mathbf{B}|$$

$$|\mathbf{AB}| = a\,|\mathbf{B}| = |\mathbf{A}|\;|\mathbf{B}|$$

We have outlined a proof of

DP15. If $\mathbf{A}$ is any $n \times n$ pivot matrix and $\mathbf{B}$ is any $n \times n$ matrix, then

$$|\mathbf{AB}| = |\mathbf{A}|\;|\mathbf{B}|$$

EXAMPLE 7

Let

$$\mathbf{A} = \begin{bmatrix} 1 & 0 & 2 \\ 0 & 1 & -1 \\ 0 & 0 & 3 \end{bmatrix}$$

$$\mathbf{B} = \begin{bmatrix} 2 & 1 & 1 \\ 3 & 2 & -1 \\ -1 & 3 & -2 \end{bmatrix}$$

Since **A** is a pivot matrix

$$|\mathbf{A}| = 3$$
$$|\mathbf{B}| = 16$$

Now

$$\mathbf{AB} = \begin{bmatrix} 0 & 7 & -3 \\ 4 & -1 & 1 \\ -3 & 9 & -6 \end{bmatrix}$$

$$|\mathbf{AB}| = 3\begin{vmatrix} 0 & 7 & -3 \\ 4 & -1 & 1 \\ -1 & 3 & -2 \end{vmatrix} = 3 \cdot 16 = 48$$

What if **A** and **B** are any $n \times n$ matrices? Can we generalize DP15? The answer is "yes." We state, give an example, and prove

DP16. Let **A** and **B** be any $n \times n$ matrices. Then

$$|\mathbf{AB}| = |\mathbf{A}|\ |\mathbf{B}|$$

EXAMPLE 8

$$|\mathbf{A}| = \begin{vmatrix} 1 & 2 & 3 \\ 4 & 5 & 6 \\ -1 & 0 & 3 \end{vmatrix} = -6$$

$$|\mathbf{B}| = \begin{vmatrix} 2 & 1 & 1 \\ 3 & 2 & -1 \\ -1 & 3 & -2 \end{vmatrix} = 16$$

$$\begin{bmatrix} 1 & 2 & 3 \\ 4 & 5 & 6 \\ -1 & 0 & 3 \end{bmatrix}\begin{bmatrix} 2 & 1 & 1 \\ 3 & 2 & -1 \\ -1 & 3 & -2 \end{bmatrix} = \begin{bmatrix} 5 & 14 & -7 \\ 17 & 32 & -13 \\ -5 & 8 & -7 \end{bmatrix} = \mathbf{AB}$$

$$|\mathbf{AB}| = \begin{vmatrix} 5 & 14 & -7 \\ 17 & 32 & -13 \\ -5 & 8 & -7 \end{vmatrix} = 5(-22)(96/110) = -96 = |\mathbf{A}|\ |\mathbf{B}|$$

To prove DP16, first consider the case if $\mathbf{A}$ is not invertible. We know that $|\mathbf{A}| = 0$ if and only if $\mathbf{A}$ is not invertible. We also know from Theorem 3.3 that if $\mathbf{A}$ or $\mathbf{B}$ is not invertible, then neither is $\mathbf{AB}$. Thus

$$|\mathbf{AB}| = 0 = |\mathbf{A}|\ |\mathbf{B}|$$

If $\mathbf{A}$ and $\mathbf{B}$ are both invertible, then

$$\mathbf{AB} = \mathbf{P}_1\mathbf{P}_2 \cdots \mathbf{P}_n\mathbf{B}_n\mathbf{B}$$

where $\mathbf{P}_1\mathbf{P}_2\cdots\mathbf{P}_n\mathbf{B}_n$ is the decomposition of $\mathbf{A}$, the $\mathbf{P}$'s the pivot matrices, and $\mathbf{B}_n$ the permutation matrix. By DP14,

$$|\mathbf{B}_n\mathbf{B}| = |\mathbf{B}| \qquad \text{or} \qquad |\mathbf{B}_n\mathbf{B}| = -|\mathbf{B}|$$

By repeated application of DP15,

$$|\mathbf{P}_n(\mathbf{B}_n\mathbf{B})| = p_n\,|\mathbf{B}_n|\ |\mathbf{B}|$$

where

$$|\mathbf{P}_n| = p_n$$

$$|\mathbf{P}_{n-1}(\mathbf{P}_n\mathbf{B}_n\mathbf{B})| = p_{n-1}p_n\,|\mathbf{B}_n|\ |\mathbf{B}|$$

$$\vdots$$

$$|\mathbf{AB}| = |\mathbf{P}_1(\mathbf{P}_2\cdots\mathbf{P}_n\mathbf{B}_n\mathbf{B})| = p_1p_2\cdots p_n\,|\mathbf{B}_n|\ |\mathbf{B}| = |\mathbf{A}|\ |\mathbf{B}|$$

by DP13.

EXAMPLE 9

The inverse of

$$\mathbf{P} = \begin{bmatrix} 2^* & 0 & 0 \\ 5 & 1 & 0 \\ 1 & 0 & 1 \end{bmatrix} \quad \text{is} \quad \mathbf{P}^{-1} = \begin{bmatrix} 1/2 & 0 & 0 \\ -5/2 & 1 & 0 \\ -1/2 & 0 & 1 \end{bmatrix}$$

In general from Chapter 3 we know that if $\mathbf{P}$ is a pivot matrix, its inverse is also a pivot matrix, obtained by pivoting on the diagonal element in the nonunit column. Thus it is clear that

DP17. If $\mathbf{P}$ is a pivot matrix, then

$$|\mathbf{P}^{-1}| = |\mathbf{P}|^{-1}$$

or

$$|\mathbf{P}|\ |\mathbf{P}^{-1}| = 1$$

It is easy to show that if π is a permutation matrix, then $\pi^{-1} = \pi$. Using this fact and DP17 we show the generalization of DP17.

DP18. If $\mathbf{A}$ is any invertible $n \times n$ matrix.

$$|\mathbf{A}^{-1}| = |\mathbf{A}|^{-1}$$

To see this, let

$$\mathbf{A} = \mathbf{P}_1\mathbf{P}_2\cdots\mathbf{P}_n\mathbf{B}_n$$

Then

$$\mathbf{A}^{-1} = \mathbf{B}_n^{-1}\mathbf{P}_n^{-1}\cdots\mathbf{P}_2^{-1}\mathbf{P}_1^{-1}$$

as is verifiable by direct multiplication. Thus

$$|\mathbf{A}^{-1}| = |\mathbf{B}_n|\ |\mathbf{P}_n|^{-1}\cdots|\mathbf{P}_2|^{-1}|\mathbf{P}_1|^{-1} = |\mathbf{A}|^{-1}$$

Exercises

1. Express each of the following matrices as the product of pivot matrices and a permutation matrix.

(a) $\begin{bmatrix} 3 & 1 \\ 4 & 2 \end{bmatrix}$

(b) $\begin{bmatrix} 1 & 0 & 3 \\ 2 & 4 & 7 \\ 3 & 1 & 6 \end{bmatrix}$

(c) $\begin{bmatrix} 1 & 2 & 3 \\ 4 & 5 & 6 \end{bmatrix}$

(d) $\begin{bmatrix} 1 & 2 \\ 3 & 4 \\ 5 & 6 \end{bmatrix}$

(e) $\begin{bmatrix} 1 & 2 & 0 & 2 \\ 3 & 1 & 4 & 6 \\ 0 & 1 & 2 & -1 \end{bmatrix}$

(f) $\begin{bmatrix} 2 & -1 & 4 & 3 \\ 1 & 2 & 0 & 1 \\ 0 & 3 & 6 & 5 \\ -2 & -2 & 1 & 3 \end{bmatrix}$

2. Find the inverse of each of the following pivot matrices.

(a) $\mathbf{A} = \begin{bmatrix} 2 & 0 & 0 \\ 3 & 1 & 0 \\ 5 & 0 & 1 \end{bmatrix}$

(b) $\mathbf{B} = \begin{bmatrix} 1 & 2 & 0 \\ 0 & 3 & 0 \\ 0 & 4 & 1 \end{bmatrix}$

(c) $\mathbf{C} = \begin{bmatrix} 1 & 0 & -1 \\ 0 & 1 & 3 \\ 0 & 0 & -2 \end{bmatrix}$

(d) $\mathbf{D} = \begin{bmatrix} 1 & 0 & -8 & 0 \\ 0 & 1 & 2 & 0 \\ 0 & 0 & 4 & 0 \\ 0 & 0 & 12 & 1 \end{bmatrix}$

3. Using the data in Exercise 2, verify
(a) $|\mathbf{AB}| = |\mathbf{A}|\ |\mathbf{B}|$.
(b) $|\mathbf{A}^{-1}| = |\mathbf{A}|^{-1}$.
(c) $[\mathbf{AB}]^{-1} = \mathbf{B}^{-1}\mathbf{A}^{-1}$.
(d) $[\mathbf{ABC}]^{-1} = \mathbf{C}^{-1}\mathbf{B}^{-1}\mathbf{A}^{-1}$.

4. Prove that the inverse of a permutation matrix is itself.

5. Using the result of Exercise 1(b), find the inverse of

$$\mathbf{A} = \begin{bmatrix} 1 & 0 & 3 \\ 2 & 4 & 7 \\ 3 & 1 & 6 \end{bmatrix}$$

6. Verify DP17($|\mathbf{A}^{-1}| = |\mathbf{A}|^{-1}$) with

$$\begin{bmatrix} 1 & 0 & 0 & -1 \\ 1 & 1 & 0 & 1 \\ 1 & 0 & 1 & 1 \\ 0 & 1 & 1 & 0 \end{bmatrix}$$

7. Find $|\mathbf{AB}|$ mentally if

$$\mathbf{A} = \begin{bmatrix} 1 & 0 & 2 & 0 & 0 \\ 0 & 1 & 3 & 0 & 0 \\ 0 & 0 & 5 & 0 & 0 \\ 0 & 0 & -1 & 1 & 0 \\ 0 & 0 & 4 & 0 & 1 \end{bmatrix}$$

and

$$\mathbf{B} = \begin{bmatrix} 4 & 0 & 0 & 0 & 0 \\ 5 & 1 & 0 & 0 & 0 \\ 6 & 0 & 1 & 0 & 0 \\ 7 & 0 & 0 & 1 & 0 \\ 8 & 0 & 0 & 0 & 1 \end{bmatrix}$$

8. Discuss the statement: If the determinants of two square matrices of the same order are equal, then the matrices are equal.

8.6 Some Applications of Determinants

EXAMPLE 1

An important communication problem is to minimize the cost of transmitting data between locations. A data loop is a loop of stations:

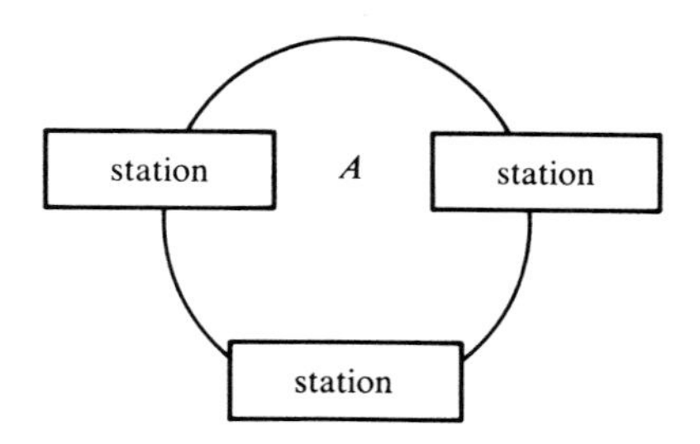

Two loops are *connected* if they have a station in common:

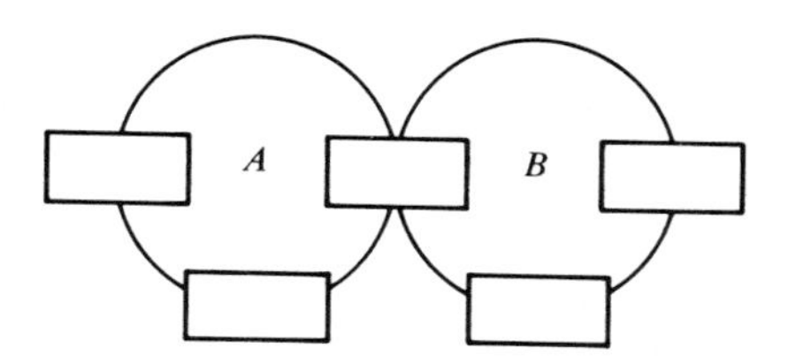

When many loops are connected together it is necessary to find the shortest route between loops. A convenient mathematical model to study this problem is a graph whose vertices represent the loops. In the graph two vertices are connected by an edge if the loops they represent are connected by a station. The following diagram shows a system of connected loops and the corresponding graph:

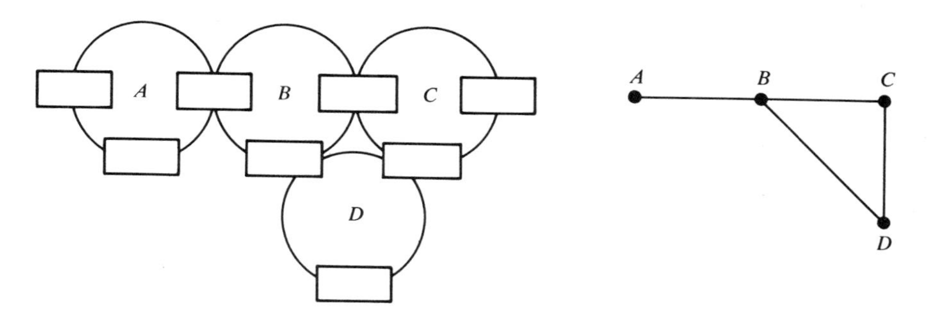

This graph theoretic approach requires an analysis of *tree* graphs, that is, graphs that are connected but themselves have no circuits. See Figure 8.6.

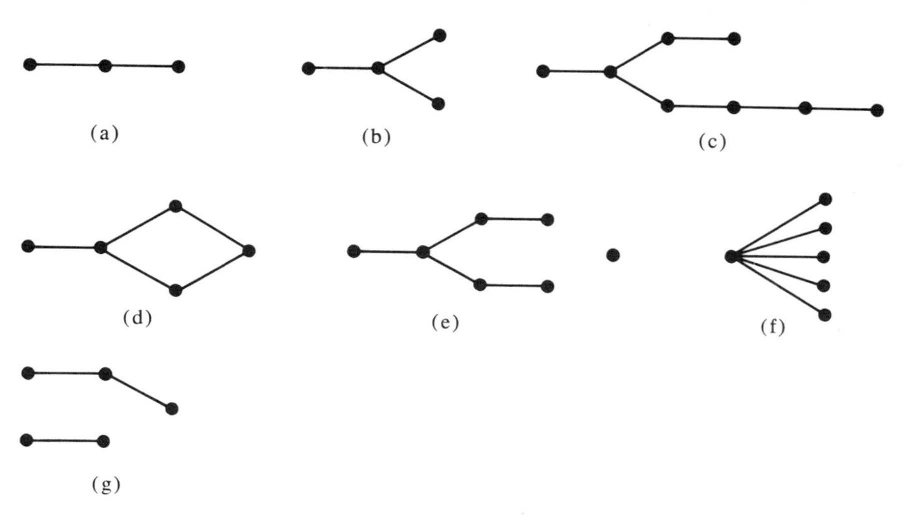

Figure 8.6

In Figure 8.6, (a), (b), (c), and (f) are trees and (d), (e), and (g) are not trees; (d) is not a tree because it has a circuit; (e) is not a tree because it is not connected (it has a vertex with no edges); (g) is not a tree because it is not connected. The two tree graphs in four vertices are shown in Figure 8.7.

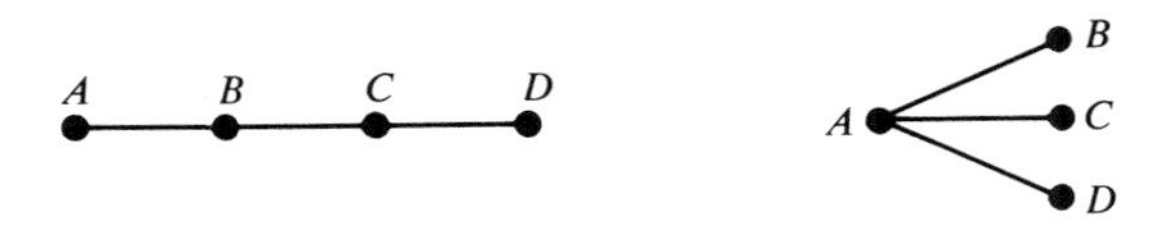

Figure 8.7

Associated with each of (a) and (b) is a matrix called the *distance* matrix of the tree. The distance matrix of (a) is

$$\mathbf{M}_1 = \begin{array}{c} \\ A \\ B \\ C \\ D \end{array} \begin{array}{c} \begin{array}{cccc} A & B & C & D \end{array} \\ \begin{bmatrix} 0 & 1 & 2 & 3 \\ 1 & 0 & 1 & 2 \\ 2 & 1 & 0 & 1 \\ 3 & 2 & 1 & 0 \end{bmatrix} \end{array}$$

The entries of $\mathbf{M}_1$ are obtained by counting the least number of edges traversed, traveling between the various vertices. Thus the two 3's in $\mathbf{M}_1$ mean that the least number of edges between A and D (or D and A) is 3. The distance matrix of (b) is

$$\mathbf{M}_2 = \begin{array}{c} \\ A \\ B \\ C \\ D \end{array} \begin{array}{c} \begin{array}{cccc} A & B & C & D \end{array} \\ \begin{bmatrix} 0 & 1 & 1 & 1 \\ 1 & 0 & 2 & 2 \\ 1 & 2 & 0 & 2 \\ 1 & 2 & 2 & 0 \end{bmatrix} \end{array}$$

$|\mathbf{M}_1| = -12 = |\mathbf{M}_2|$, as you can verify by pivoting.

The equality $|\mathbf{M}_1| = |\mathbf{M}_2|$ is tantalizing and suggests that we investigate the determinants of the distance matrices of trees with five vertices. There are three of these, shown with their corresponding distance matrices, in Figure 8.8.

$$\mathbf{M}_1 = \begin{array}{c c} & \begin{array}{ccccc} A & B & C & D & E \end{array} \\ \begin{array}{c} A \\ B \\ C \\ D \\ E \end{array} & \begin{bmatrix} 0 & 1 & 2 & 3 & 4 \\ 1 & 0 & 1 & 2 & 3 \\ 2 & 1 & 0 & 1 & 2 \\ 3 & 2 & 1 & 0 & 1 \\ 4 & 3 & 2 & 1 & 0 \end{bmatrix} \end{array}$$

$$\mathbf{M}_2 = \begin{array}{c c} & \begin{array}{ccccc} A & B & C & D & E \end{array} \\ \begin{array}{c} A \\ B \\ C \\ D \\ E \end{array} & \begin{bmatrix} 0 & 1 & 2 & 3 & 3 \\ 1 & 0 & 1 & 2 & 2 \\ 2 & 1 & 0 & 1 & 1 \\ 3 & 2 & 1 & 0 & 2 \\ 3 & 2 & 1 & 2 & 0 \end{bmatrix} \end{array}$$

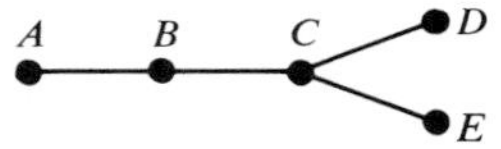

$$\mathbf{M}_3 = \begin{array}{c c} & \begin{array}{ccccc} A & B & C & D & E \end{array} \\ \begin{array}{c} A \\ B \\ C \\ D \\ E \end{array} & \begin{bmatrix} 0 & 1 & 1 & 1 & 1 \\ 1 & 0 & 2 & 2 & 2 \\ 1 & 2 & 0 & 2 & 2 \\ 1 & 2 & 2 & 0 & 2 \\ 1 & 2 & 2 & 2 & 0 \end{bmatrix} \end{array}$$

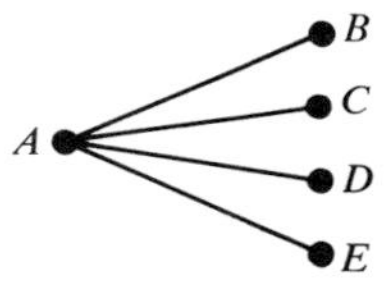

Figure 8.8

The reader should verify that $|\mathbf{M}_1| = |\mathbf{M}_2| = |\mathbf{M}_3| = 32$.

It is known that if n is the number of vertices of a tree, then the determinant of the distance matrix is

$$(-1)^{n-1}(n - 1)2^{(n-2)}$$

In the cases we have considered $n = 4$ and $n = 5$. For these values the formula gives -12 and 32.

A result just as startling is in Example 2.

EXAMPLE 2

The *cofactors* of any matrix $\mathbf{A}$ are the determinants of the minors of $\mathbf{A}$

multiplied by the plus or minus sign associated with the minors. Thus the cofactor of a in

$$\begin{bmatrix} a & b & c \\ d & e & f \\ g & h & i \end{bmatrix}$$

is

$$\begin{vmatrix} e & f \\ h & i \end{vmatrix} = ei - fh$$

while the cofactor of h is

$$-\begin{vmatrix} a & c \\ d & f \end{vmatrix} = -(af - cd) = cd - af$$

A *spanning tree* T of a graph G is a tree subgraph of G which contains each vertex of G. A graph G and its eight spanning trees are shown in Figure 8.9.

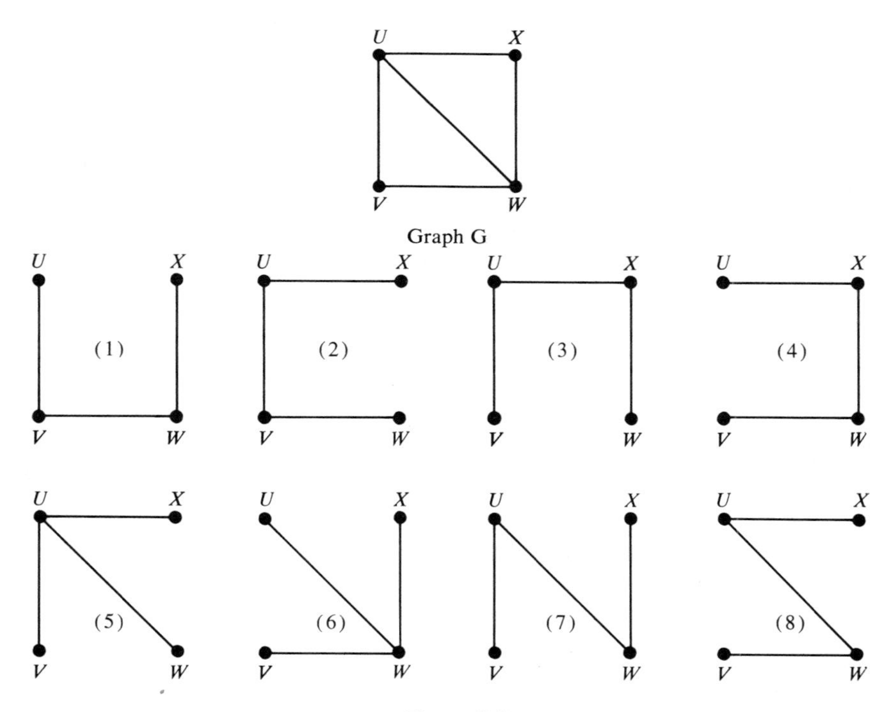

Figure 8.9

The adjacency matrix (see Section 3.1) of G is

$$\mathbf{A} = \begin{array}{c} \\ U \\ V \\ W \\ X \end{array} \begin{array}{c} \begin{array}{cccc} U & V & W & X \end{array} \\ \begin{bmatrix} 0 & 1 & 1 & 1 \\ 1 & 0 & 1 & 0 \\ 1 & 1 & 0 & 1 \\ 1 & 0 & 1 & 0 \end{bmatrix} \end{array}$$

The *valence* of U is 3. (The valence of vertex U is the number of edges connecting U to other vertices.) The valence of V is 2, that of W is 3, and that of X is 2. Adding these valences to the diagonal elements of $-\mathbf{A}$ gives

$$\mathbf{M} = \begin{bmatrix} 3 & -1 & -1 & -1 \\ -1 & 2 & -1 & 0 \\ -1 & -1 & 3 & -1 \\ -1 & 0 & -1 & 2 \end{bmatrix}$$

The startling result is that the value of any of the 16 cofactors of $\mathbf{M}$ is 8, and that 8 is the number of spanning trees of G. This result is true for any graph G. As an illustration, consider the cofactor of the 3 in row 3, column 3 of $\mathbf{M}$. It is

$$\begin{vmatrix} 3 & -1 & -1 \\ -1 & 2 & 0 \\ -1 & 0 & 2 \end{vmatrix}$$

Evaluating by pivoting gives

$$\begin{vmatrix} 3 & -1 & -1 \\ -1 & 2 & 0 \\ -1^* & 0 & 2 \end{vmatrix} = (-1) \begin{vmatrix} -1 & 5 \\ 2 & -2 \end{vmatrix} = (-1)(-8) = 8$$

EXAMPLE 3

$$|\mathbf{A}| = \begin{vmatrix} 3 & -2 \\ 3 & -4 \end{vmatrix} = -6$$

Question: Is there a number e which when subtracted from each diagonal element of **A** will result in a determinant of 0? That is, are there values of e for which

$$\begin{vmatrix} 3-e & -2 \\ 3 & -4-e \end{vmatrix} = 0?$$

If so, by the definition of determinant,

$$(3-e)(-4-e) - (3)(-2) = 0 \tag{1}$$

or

$$e^2 + e - 6 = 0 \tag{2}$$

or

$$(e+3)(e-2) = 0 \tag{3}$$

Thus for either $e = -3$ or $e = 2$,

$$\begin{vmatrix} 3-e & -2 \\ 3 & -4-e \end{vmatrix} = 0$$

We verify. If $e = -3$,

$$\begin{vmatrix} 6 & -2 \\ 3 & -1 \end{vmatrix} = 0 \tag{4}$$

If $e = 2$,

$$\begin{vmatrix} 1 & -2 \\ 3 & -6 \end{vmatrix} = 0 \tag{5}$$

In this example the numbers $e = -3$ and $e = 2$ are called *eigenvalues* of **A**. It is clear that

$$\begin{bmatrix} 3-e & -2 \\ 3 & -4-e \end{bmatrix} = \begin{bmatrix} 3 & -2 \\ 3 & -4 \end{bmatrix} - \begin{bmatrix} e & 0 \\ 0 & e \end{bmatrix}$$

$$= \begin{bmatrix} 3 & -2 \\ 3 & -4 \end{bmatrix} - e\begin{bmatrix} 1 & 0 \\ 0 & 1 \end{bmatrix} = \mathbf{A} - e\mathbf{I}$$

Thus the expression

$$\det\begin{bmatrix} 3 - e & -2 \\ 3 & -4 - e \end{bmatrix} = 0$$

is replaceable by

$$\det(\mathbf{A} - e\mathbf{I}) = 0 \tag{6}$$

The equation (6) for any $n \times n$ matrix **A** is called the *characteristic equation* of **A**. The numbers e that make this equation true are called the *eigenvalues* of **A**. In the example -3 and 2 are the eigenvalues of **A**.

For any eigenvalue e a nonzero vector **X** (thought of as a one-column n-row matrix) such that

$$(\mathbf{A} - e\mathbf{I})\mathbf{X} = \mathbf{0} \tag{7}$$

is called an *eigenvector of* **A** *associated with e*. For our example we have from (4) and (5) two tableau forms for (7):

(8)

x_1	x_2	
6	-2	$= 0$
3	-1	$= 0$

and

(9)

x_1	x_2	
1	-2	$= 0$
3	-6	$= 0$

respectively. $\mathbf{X} = k(1, 3) = (k, 3k)$ is a solution of (8) for any real number k. Therefore, for any k, $(k, 3k)$ is an eigenvector of

$$\mathbf{A} = \begin{bmatrix} 3 & -2 \\ 3 & -4 \end{bmatrix}$$

associated with the eigenvalue -3. From (9), for any real number t, $(2t, t)$ is an eigenvector of **A** associated with the eigenvalue 2.

Eigenvalues and eigenvectors are useful in many fields, such as graph theory and mechanics. In the context of linear mappings, eigenvalues and eigenvectors arise in the following manner. Let $T: R^n \to R^n$ be a linear mapping. What is the condition on the mapping under which a vector line α gets mapped into itself (not necessarily each point on the line going to the same point)? The condition is simply that if $\mathbf{V}$ generates the line, then $T(\mathbf{V}) = e\mathbf{V}$ for some real number e. This number e is an eigenvalue of the matrix of T and $\mathbf{V}$ is an associated eigenvector (as is any multiple of $\mathbf{V}$).

Exercises

1. Show that there is essentially only one tree on three vertices and find the determinant of its distance matrix. Verify your result with the formula $(-1)^{n-1}(n-1)2^{n-2}$, with $n = 3$.

2. (a) Carry out the instructions in Exercise 1 for a tree on two vertices.
(b) Is the formula valid for a "tree" on one vertex?

3. Show that the distance matrix for the tree

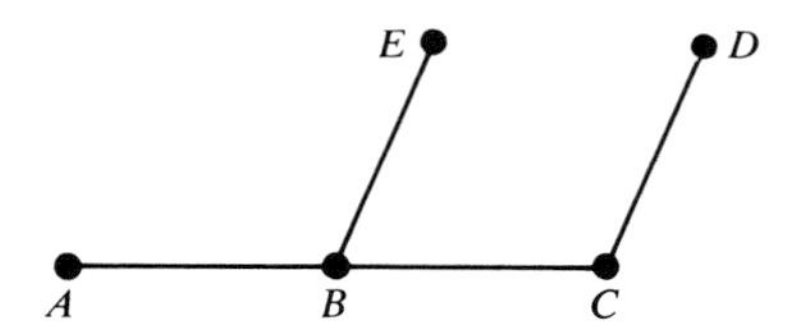

has the same determinant as the distance matrix for the tree

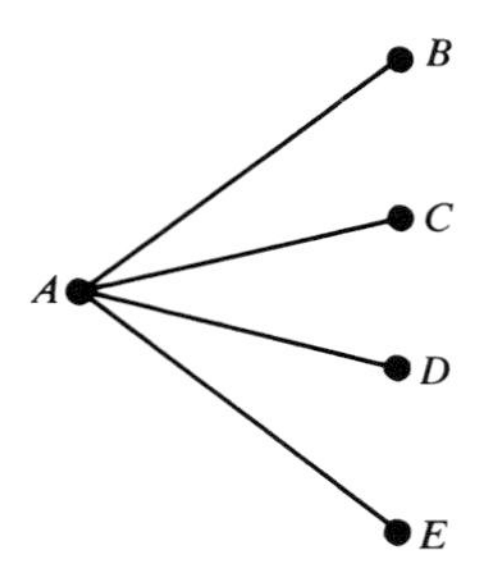

4. Start with the tree on five vertices

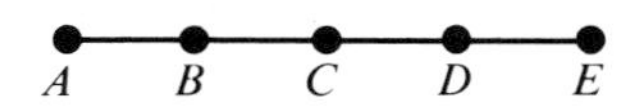

and show that the addition of a sixth vertex to form a tree can be done in essentially two ways only, and calculate the determinant of their distance matrices. Verify your results with the formula $(-1)^{n-1}(n-1)2^{n-2}$, with $n = 6$.

5. Let H be the graph

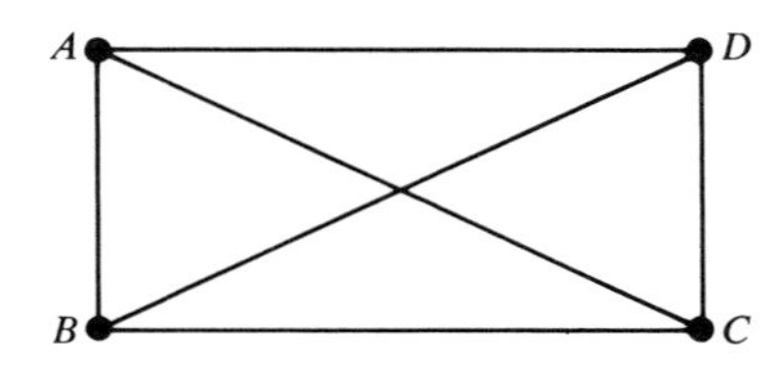

(a) Form $\mathbf{N}$ the adjacency matrix of H.
(b) List the spanning trees of H.
(c) What is the valence of each vertex?
(d) Add the valences to the diagonal entries of $-\mathbf{N}$ and find the cofactors of two distinct entries. How can you check the results?

6. Find the eigenvalues and eigenvectors, if any, of:

(a) $\begin{bmatrix} 1 & 2 \\ 5 & 4 \end{bmatrix}$ (b) $\begin{bmatrix} 1 & 2 \\ 4 & 8 \end{bmatrix}$ (c) $\begin{bmatrix} 1 & 4 \\ -2 & 3 \end{bmatrix}$

(d) $\begin{bmatrix} 1 & -1 \\ 1 & -1 \end{bmatrix}$ (e) $\begin{bmatrix} 2 & -2 \\ 2 & -2 \end{bmatrix}$

7. Show that the identity matrix $\mathbf{I}$ of order n has only one eigenvalue, namely 1. Find the eigenvectors of $\mathbf{I}$.

8. Prove that the sum of the eigenvalues of

$$\begin{bmatrix} a & b \\ c & d \end{bmatrix}$$

is $a + d$, and that their product is

$$\begin{vmatrix} a & b \\ c & d \end{vmatrix}$$

9. Given a 2×2 matrix $\mathbf{A}$, show that $\mathbf{A}$ satisfies its characteristic equation. That is, if $ae^2 + be + c = 0$ is the characteristic equation, then $a\mathbf{A}^2 +$

$b\mathbf{A} + c\mathbf{I} = 0$. (This is the Cayley–Hamilton theorem, which is valid for all $n \times n$ matrices.)

10. Find the eigenvalues and eigenvectors of

$$\begin{bmatrix} 2 & 2 & 0 \\ 1 & 2 & 1 \\ 1 & 2 & 1 \end{bmatrix}$$

8.7 Chapter Review

1. The determinant of an $n \times n$ matrix $\mathbf{A}$ is defined inductively.
2. Various theorems about determinants are proved or the outline of a proof is given. These are
   (a) *DP1* (homogeneity). Let $\mathbf{A}$ be an $n \times n$ matrix. If $\mathbf{B}$ is obtained from $\mathbf{A}$ by multiplying each entry in any row by k, then $|\mathbf{B}| = k|\mathbf{A}|$.
   (b) *DP2* (the additive property). Let $\mathbf{A}_1$ and $\mathbf{A}_2$ be identical matrices except for the kth row. Then $|\mathbf{A}_3| = |\mathbf{A}_1| + |\mathbf{A}_2|$, where $\mathbf{A}_3$ is identical to $\mathbf{A}_1$ and $\mathbf{A}_2$ except that its kth row is the sum of the kth rows of $\mathbf{A}_1$ and $\mathbf{A}_2$.
   (c) *DP3*. If any two adjacent rows in an $n \times n$ matrix $\mathbf{A}$ are interchanged to give a matrix $\mathbf{B}$, then $|\mathbf{B}| = -|\mathbf{A}|$.
   (d) *DP4*. If *any* two rows of $\mathbf{A}$ are interchanged yielding a matrix $\mathbf{B}$, then $|\mathbf{B}| = -|\mathbf{A}|$.
   (e) *DP5*. If any two rows of $n \times n$ matrix $\mathbf{A}$ are identical, then $|\mathbf{A}| = 0$.
   (f) *DP6*. If one row of an $n \times n$ matrix $\mathbf{A}$ is a scalar multiple of another row of $\mathbf{A}$, then $|\mathbf{A}| = 0$.
   (g) *DP7*. Let $\mathbf{A}$ be any $n \times n$ matrix. Let $\mathbf{B}$ be identical to $\mathbf{A}$ except that the ith row of $\mathbf{B}$ is equal to the ith row of $\mathbf{A}$ plus k times the mth row. Then $|\mathbf{B}| = |\mathbf{A}|$.
   (h) *DP8*. Let $\mathbf{A}$ be an $n \times n$ matrix. Then $|\mathbf{A}| = (-1)^{i+1}a_{i1}|\mathbf{B}|$, where $\mathbf{B}$ is the condensed form of $\mathbf{A}$ when pivoting on a_{i1}.
   (i) *DP9*. If any two adjacent columns of $n \times n$ matrix $\mathbf{A}$ are interchanged, $|\mathbf{A}|$ changes its sign.
   (j) *DP10*. Without regard to sign, $|\mathbf{A}|$ is equal to the product of the pivots when using pivot condensation. The proper sign for each pivot is determined by the checkerboard rule and the sign for the determinant of $\mathbf{A}$ is the product of the signs for the pivots.

(k) *DP11.* For all $n \times n$ matrices $\mathbf{A}$, $|\mathbf{A}| = |\mathbf{A}^T|$.
(l) *DP12.* Let $\mathbf{P}$ be a pivot matrix and let p be the element in the nonunit column which lies on the main diagonal. Then $|\mathbf{P}| = p$.
(m) *DP13.* Let $\mathbf{A}$ be an $n \times n$ matrix. If $|\mathbf{A}| \neq 0$, then $|\mathbf{A}| = \mathbf{P}_1\|\mathbf{P}_2| \cdots |\mathbf{P}_n\|\mathbf{B}_n|$.
(n) *DP14.* Let $\mathbf{A}$ be any square matrix of order n and π a permutation matrix of order n. Then $|\pi\mathbf{A}| = (-1)^m|\mathbf{A}|$, where m is the number of adjacent row interchanges induced on $\mathbf{A}$ by π.
(o) *DP15.* If $\mathbf{A}$ is any $n \times n$ pivot matrix and $\mathbf{B}$ is any $n \times n$ matrix, then $|\mathbf{AB}| = |\mathbf{A}\|\mathbf{B}|$.
(p) *DP16.* Let $\mathbf{A}$ and $\mathbf{B}$ be any $n \times n$ matrices. Then $|\mathbf{AB}| = |\mathbf{A}\|\mathbf{B}|$.
(q) *DP17.* If $\mathbf{P}$ is a pivot matrix, then $|\mathbf{P}^{-1}| = 1/|\mathbf{P}|$ or $|\mathbf{P}\|\mathbf{P}^{-1}| = 1$.
(r) *DP18.* If $\mathbf{A}$ is any invertible $n \times n$ matrix, $|\mathbf{A}^{-1}| = 1/|\mathbf{A}|$.

3. Applications of determinants to geometry, graph theory and communications, area, volumes, systems of linear equations, mappings, and eigenvalues are presented.

Review Exercises

1. Evaluate:

(a) $\begin{vmatrix} 3 & 1 \\ 2 & 5 \end{vmatrix}$ (b) $\begin{vmatrix} 2 & 1 & 2 \\ 2 & 0 & -1 \\ 0 & 3 & -1 \end{vmatrix}$ (c) $\begin{vmatrix} 1 & 2 & -1 \\ 5 & 0 & -1 \\ 3 & 4 & -2 \end{vmatrix}$

(d) $\begin{vmatrix} 13 & 12 & 50 \\ 6 & 1 & 2 \\ 13 & 12 & 50 \end{vmatrix}$ (e) $\begin{vmatrix} 4 & -1 & 1 \\ 2 & 0 & 0 \\ 8 & 4 & 3 \end{vmatrix}$ (f) $\begin{vmatrix} 0 & 1 & 1 & 3 \\ 1 & 0 & -3 & 1 \\ 2 & -2 & 0 & -3 \\ 3 & 1 & 2 & 5 \end{vmatrix}$

(g) $\begin{vmatrix} 1 & 0 & 0 & 0 \\ 6 & -3 & 0 & 0 \\ 1 & 4 & 5 & 0 \\ 6 & 6 & 7 & -8 \end{vmatrix}$ (h) $\begin{vmatrix} 1 & 4 & 9 & 2 \\ 0 & 3 & 5 & 6 \\ 0 & 0 & 2 & 1 \\ 0 & 0 & 0 & 5 \end{vmatrix}$

(i) $\begin{vmatrix} 3 & 1 & 5 & 6 \\ 0 & 2 & 1 & 3 \\ 0 & 4 & 1 & 2 \\ -1 & 1 & 2 & 0 \end{vmatrix}$

2. Let **A** be a 3×3 matrix and c any number. Prove that $|c\mathbf{A}| = c^3|\mathbf{A}|$. Generalize this statement for an $n \times n$ matrix.

3. Prove that

$$\begin{vmatrix} 1 & 3 & 1 & 1 \\ 1 & -2 & 4 & 1 \\ 0 & -4 & 1 & 2 \\ 4 & 1 & -3 & 7 \end{vmatrix} = -115$$

4. Prove that the determinant of a diagonal $n \times n$ matrix is $a_{11}a_{22}a_{33} \cdots a_{nn}$.

5. Prove that the determinant of a triangular $n \times n$ matrix is the product of its diagonal entries.

6. Prove that

$$\begin{vmatrix} 1 & a_1 & a_1^2 & a_1^3 \\ 1 & a_2 & a_2^2 & a_2^3 \\ 1 & a_3 & a_3^2 & a_3^3 \\ 1 & a_4 & a_4^2 & a_4^3 \end{vmatrix}$$

$$= (a_4 - a_3)(a_4 - a_2)(a_4 - a_1)(a_3 - a_2)(a_3 - a_1)(a_2 - a_1)$$

7. (a) Find the area of the triangle whose coordinates are (3, 2), (−1, 4), (6, −8).
(b) Using determinants, show that (−3, −1), (1, 1), (7, 4) are collinear (in one line).

8. Find the volume of the tetrahedron whose vertices are at (0, 0, 0), (1, 0, 0), (0, 2, 0), and (0, 0, 3).

9. Using determinants, determine whether the following sets of vectors are linearly dependent or independent.
(a) {(2, 3, 5), (−2, 1, 4), (6, 1, −3)}.
(b) {(1, 1, 0, 1), (0, 1, 0, 1), (1, 0, −1, 2), (0, 0, 0, 1)}.

10. Using Cramer's rule, find the unique solution, if any, of:

(a) $\begin{cases} 6x + 4y = 2 \\ 2x + 4 = 0. \end{cases}$

(b) $\begin{cases} 2x + 2y - z = 1 \\ x + y - 3z = -7 \\ x + 3y = 4. \end{cases}$

(c) $\begin{cases} x_1 + x_2 = 3 \\ x_2 + x_3 = 6 \\ x_1 + x_3 = 1. \end{cases}$

(d) $\begin{cases} 2x_1 + 2x_2 + 3x_3 = 6 \\ 4x_1 + x_2 = 0 \\ 2x_1 + x_2 + x_3 = 1. \end{cases}$

11. Let

$$\mathbf{A} = \begin{bmatrix} 2 & -2 & 4 \\ 1 & 2 & 2 \\ 1 & -3 & 6 \end{bmatrix}$$

(a) Express **A** as the product of three pivot matrices only. (*Hint:* Pivot on diagonal entries only.)
(b) Using the result in (a), find $\mathbf{A}^{-1}$.
(c) Verify that $|\mathbf{A}^{-1}| = \dfrac{1}{|\mathbf{A}|}$.

12. Let

$$\mathbf{B} = \begin{bmatrix} 1 & 4 & 0 \\ 3 & 5 & 6 \\ 0 & 7 & 8 \end{bmatrix}$$

and take **A** as in Exercise 11. Verify
(a) $|\mathbf{AB}| = |\mathbf{A}||\mathbf{B}|$.
(b) $(\mathbf{AB})^{-1} = \mathbf{B}^{-1}\mathbf{A}^{-1}$.

Selected Bibliography

ADAMS, WILLIAM J., ALLAN GEWIRTZ, and LOUIS V. QUINTAS, *Elements of Linear Programming*. New York: Van Nostrand Reinhold Co., 1969. A modest textbook for the beginner.

BIRKHOFF, GARRETT, and SAUNDERS MACLANE, *A Survey of Modern Algebra,* 3rd ed. New York: The Macmillan Company, 1965. This classic work, for mathematically mature students, develops linear algebra within modern abstract algebra.

DANTZIG, GEORGE B., *Linear Programming and Extensions.* Princeton, N.J.: Princeton University Press, 1963. The treatise by the inventor of the subject, with introductory chapters on concepts, origins, and formulation of linear programs.

DAVIS, PHILIP J., *The Mathematics of Matrices.* Waltham, Mass.: Ginn-Blaisdell, 1965. A pleasant introduction for the beginner, informal yet informative.

HOFFMAN, KENNETH, and RAY KUNZE, *Linear Algebra,* 2nd ed. Englewood Cliffs, N.J.: Prentice-Hall, Inc., 1971. A complete textbook in linear algebra for mathematically mature students.

HOHN, FRANZ E., *Elementary Matrix Algebra,* 3rd ed. New York: The Macmillan Company, 1973. Designed for students of diverse interests, at an intermediate level, with an extensive bibliography of linear algebra and applications.

KEMENY, JOHN G., ARTHUR SCHLEIFER, JR., J. LAURIE SNELL, and GERALD L. THOMPSON, *Finite Mathematics with Business Applications,* 2nd ed. Englewood Cliffs, N.J.: Prentice-Hall, Inc., 1972. The overlap between our book and this FM book shows how well CLA teams with FM.

NERING, EVAR D., *Elementary Linear Algebra.* Philadelphia: W. B. Saunders Co., 1974. This new text, for a first course, makes central use of pivot reduction.

NOBLE, BEN, *Applied Linear Algebra.* Englewood Cliffs, N.J.: Prentice-Hall, Inc., 1969. This is a comprehensive development of linear algebra and its wealth of applications, with careful attention to numerical computation.

SINGLETON, ROBERT R., and WILLIAM F. TYNDALL, *Games and Programs: Mathematics for Modeling.* San Francisco: W. F. Freeman and Co., 1974. A new introduction to mathematical models for management and social sciences.

STIEFEL, EDUARD L., *An Introduction to Numerical Mathematics* (translated by Werner C. and Cornelie J. Rheinboldt). New York: Academic Press, Inc., 1963. Pivot exchange provides the algorithmic standpoint for the first third of this introduction to numerical computation.

WILLIAMSON, RICHARD E., RICHARD H. CROWELL, and HALE F. TROTTER, *Calculus of Vector Functions,* 3rd ed. Englewood Cliffs, N.J.: Prentice-Hall, Inc., 1972. The first third of this textbook develops linear algebra for the purposes of multivariable calculus.

Index

B

C

D

E

F

G

K

L

M

R

V